Next-Generation Heat Pump Technology

Kiyoshi Saito, a leading heat pump researcher, provides a comprehensive guide to this technology and its related trends, issues, and policies. Using a solutions-driven approach, this book provides a roadmap for building carbon-neutral and recycling-oriented societies with the help of heat pump technology.

Situated in the context of advanced initiatives taking place in Japan, the book addresses the expansion of heat pump technology in recent years and the extension of its application into fields such as industrial and environmental technology. In doing so, it discusses not only technical but also digital aspects to explore various challenges related to the functioning of heat pumps, such as energy saving and refrigerant leakage. Saito underscores the importance of holistic and practical solutions that take into account individual factors (such as increased equipment size, increased refrigerant consumption, and increased product prices) to secure resources and optimize the value chain with the larger goal of building a circular economy and a decarbonized society. By offering valuable recommendations that involve technological solutions as well as policy initiatives, this book emphasizes the importance of heat pump technology in efficiently converting electricity and power into cooling and heating, and its indispensability in achieving carbon neutrality for building a more sustainable future.

This book is an essential guide for graduate and postgraduate students as well as consumers and corporate engineers in the field. It is also an invaluable resource for governments and civil-society organizations working toward the realization of a circular economy.

"Heat generated by burning fossil fuels produces greenhouse gases, which are the primary cause of global warming. Heat pump technology uses electricity to generate heat that is several times greater than the amount of electricity consumed. When powered by renewable energy, it does not impose a burden on the global environment.

"This book clearly outlines the future society enabled by heat pump technology, as well as the roadmap and policies necessary for its realization.

"We encourage everyone concerned about the global environment to read this book."
Dr. Hiroshi Komiyama, *The Chairman of the Mitsubishi Research Institute/The Chairman of the Platinum Society Network/The 28th President of the University of Tokyo.*

Next-Generation Heat Pump Technology

Cold and Thermal Utilization Technology to Achieve Carbon Neutrality

Edited by
Kiyoshi Saito
With contributions from Waseda University Consortium for the Research Strategy of Next-Generation Heat Pump Technology

Waseda
Heat Pump

CRC Press
Taylor & Francis Group
Boca Raton London New York

CRC Press is an imprint of the
Taylor & Francis Group, an **informa** business

Designed cover image: Kiyoshi Saito

First published in English 2027
by CRC Press
2385 NW Executive Center Drive, Suite 320, Boca Raton FL 33431

and by CRC Press
4 Park Square, Milton Park, Abingdon, Oxon, OX14 4RN

CRC Press is an imprint of Taylor & Francis Group, LLC

Published in Japanese as *Zukai de wakaru jisedai hītoponpu gijutsu: Kābon nyūtoraru o jitsugen suru hiya onnetsu riyō gijutsu* by Gijutsu-Hyohronsha 2024

A draft translation of the book was produced by DeepL Pro. This was subsequently reviewed and edited by the author.

ISBN: 9781041128526 (hbk)
ISBN: 9781041128571 (pbk)
ISBN: 9781003666950 (ebk)

DOI: 10.1201/9781003666950

Typeset in Times
by KnowledgeWorks Global Ltd.

Contents

About the editor

Kiyoshi Saito is a Professor at Waseda University. He is also Dean of the Sustainable Energy & Environmental Society Open Innovation Research Organization and President of the Consortium for Research Strategy of Next-Generation Heat Pump Technology. Since 2023, he has also been President of the Japanese Society of Refrigerating and Air Conditioning Engineers.

WORKING GROUP MEMBERS

Kenji Ueda
General Manager, Large Refrigeration Systems Engineering Department, Mitsubishi Heavy Industries Thermal Systems, Ltd.

Hiroaki Okamoto
Senior Researcher, Jutaku Kankyo Keikaku Laboratory Co., Ltd.

Osami Kataoka
Deputy General Manager, Planning Department, Air Conditioning Manufacturing Division, Daikin Industries, Ltd.

Tetsuro Kishimoto
President, Environmental Energy Network 21 (NPO)

Takeshi Kitagawa
General Manager, CSR and Global Environment Center, Daikin Industries, Ltd.

Satoshi Sakashita
Chief Engineer / Senior Engineer, Technology Strategy, Housing System Equipment Development Center, Air Quality and Air Conditioning Company, Panasonic Corporation

Masanobu Sasaki
Deputy General Manager, Sales Division, TEPCO Energy Partner, Inc.

Hideaki Suzuki
Chief Engineer, Toshiba Carrier Corporation

Shigeharu Taira
Section Manager, CSR and Global Environment Center (concurrently Section Manager, Planning Department, Air Conditioning Manufacturing Division), Daikin Industries, Ltd.

Kuniyuki Nishimura
Senior Research Fellow, Next-Generation Heat Pump Technology Strategy Consortium, Waseda University

Satoru Fujimoto
Director, CSR and Global Environment Center, Daikin Industries, Ltd.

Kazuhiro Furushō
General Manager, Technology Strategy Office, Technology and Innovation Center, Daikin Industries, Ltd.

Ichiro Hongo
Former Senior Chief Engineer, Toshiba Carrier Corporation

Akito Machida
Executive Officer, Technical Planning Division, Mayekawa Mfg. Co., Ltd.

Masaru Matsui
Director, Technology Policy Strategy Office, Innovation Center, Panasonic Corporation

Kenji Matsuda
Senior Research Fellow, Next-Generation Heat Pump Technology Strategy Consortium, Waseda University
Former Deputy General Manager, Air Conditioning Engineering Department, Mitsubishi Heavy Industries Thermal Systems, Ltd.

Kazuhiko Mihara
General Manager, Air Conditioner & Cold Chain Development Center, Appliances Company, Panasonic Corporation

Koji Yamashita
Former Chief Engineer, Air-Conditioning and Refrigeration Systems Division, Mitsubishi Electric Corporation

Masato Yosomiya
Senior Chief Engineer, Refrigeration Systems Works, Mitsubishi Electric Corporation

Shinichi Tanabe
Professor, Department of Architecture, Waseda University

Chiharu Tokoro
Professor, Resources and Environmental Engineering, Waseda University

Kei Nagashima
Professor, Faculty of Human Sciences, Waseda University

SECRETARIAT

Mitsubishi Research Institute, Inc.

Note: Company names and job titles are presented exactly as written in the original Japanese publication.

Introduction

The world is undergoing a period of major change as digital transformation (DX) and green transformation (GX) are being promoted to realize a sustainable, carbon neutral, recycling-oriented society.

Although 50% of the final energy consumption in the world is as heat, its effective utilization has not progressed. Heat pump technology is expected to be capable of ushering in the next era because it can efficiently convert electricity and power into cold and heat. If its use spreads, it is estimated that it will help to reduce CO_2 emissions by more than 100 million tonnes per year in Japan alone by 2050. With the achievement of carbon neutrality becoming the greatest challenge for science and technology, heat pumps will continue to be one of the most important technologies.

Although heat pump technology originally started as a means of refrigeration, in recent years its applications have expanded to air conditioning, hot water supply, and industrial technology. It is becoming recognized as an air conditioning technology that can prevent heatstroke and the spread of infectious diseases while using air heat as a renewable energy source. As a core technology of the cold chain, it can realize safe and secure food distribution. In industrial processes, it enables the re-use of energy by raising the temperature of low-temperature waste heat. In this way, thermal energy technology is developing such that it can serve as the foundation for a future society in terms of living spaces, food distribution, and industry, going beyond environmental technology. The market is expected to continue to expand significantly, and it is one of the technologies that can drive the global economy.

On the other hand, the widespread use of heat pump technology will lead to increased electricity consumption, so significant performance improvements are required for heat pump technology. Even refrigerants that do not deplete the ozone layer have been found to be greenhouse gases, and there is a need to switch to refrigerants with low global warming potential (GWP). However, improving the performance of equipment to reduce energy consumption leads to larger equipment, which in turn increases the amount of refrigerant used and requires the use of relatively expensive metals, which could lead to significant price increases. Many low GWP refrigerants reduce equipment performance, so forcing a refrigerant conversion without sufficient refrigerant recovery and recycling could actually lead to global warming. Due to these complex trade-offs, forcing individual measures as in the past could actually delay environmental measures and the spread of high-performance equipment.

For this reason, it is essential to make effective use of DX and other technologies to further improve equipment performance and reduce the GWP of refrigerants, as well as to optimize the overall value chain, taking into account the circular economy, including the recovery of refrigerants and equipment. One of the fundamental reasons why Japan has lagged behind in many technologies is due to a lack of such comprehensive efforts, and it is necessary to work with people from diverse fields, transcending traditional academic disciplines, government ministries, and boundaries between manufacturers and users, to co-create a future society enabled by heat pump technology, including the value chain, and to solve the complex issues surrounding heat pump technology through backcasting.

Therefore, Waseda University, as a comprehensive university with a wide range of academic disciplines, has brought together its collective wisdom and established the Next-Generation Heat Pump Technology Strategy Research Consortium, an open forum with nearly 20 private companies to discuss next-generation heat pump technology and strategies for realizing a future society. The consortium has clarified this vision for the 2030s in which heat pump technology plays a central role, as well as the technologies necessary to build such a society, and has been examining the roadmap and policies required to achieve it. We have now compiled the results of our work into a book for wide distribution.

We believe that it is our mission to leave behind a bright future for our society. Against this backdrop, heat pump technology will continue to drive science and technology as a means of protecting not only the environment but also life itself. We hope that this book will be of some help in building a sustainable society that utilizes heat pump technology.

2025

Waseda University, Saito Kiyoshi

Table of Basic Terms

For understanding next-generation heat pump technology

Basic Terms	Explanation
A2L refrigerant	A2L, a mildly flammable refrigerant, is a combination of safety classifications A (low toxicity) and 2L (lower flammability) established by the American Society of Heating, Refrigerating and Air-Conditioning Engineers (ASHRAE). It includes hydrofluorocarbon (HFC) refrigerants, hydrofluoroolefins (HFOs) and mixtures thereof, which are characterized by their low flammability, low toxicity, and low global warming potential (GWP).
APF (annual performance factor)	Indicates the annual performance under various operating conditions that change throughout the year.
Carbon dioxide capture, utilization, and storage	Technology for capturing CO_2 emitted from industry and other sources and fixing or utilizing it. In addition to CO_2 capture and storage (CCS), it is used in the production of fuels, chemicals, and other valuable materials, such as synthetic gas, methane, methanol, algae cultivation, carbon materials, and concrete hardening through reactions with hydrogen.
CO_2 refrigerant	One of the natural refrigerants used in water heaters and other appliances. When used for cooling, CO_2 has a critical temperature of approximately 31°C, and when the outside temperature exceeds this, its performance deteriorates because latent heat cannot be utilized.
HFO refrigerants	Hydrofluoroolefin (HFO) is also known as a green refrigerant because it is low in toxicity and has little impact on the environment and human health.
Nationally determined contribution	Nationally determined contribution refers to the greenhouse gas (GHG) emission reduction targets that each country must create, report, and maintain based on the Paris Agreement.
SBT (science-based targets)	GHG emission reduction targets set by companies for a target year five to 10 years after the base year, with a reduction of 4.2% or more per year as a guideline, in line with the levels required by the Paris Agreement.
SDGs (Sustainable Development Goals)	The 17 Sustainable Development Goals were adopted at the United Nations General Assembly in 2015 to be achieved by 2030, addressing universal challenges facing the international community. Goal 7 states that access to affordable, reliable, sustainable, and modern energy for all is essential for the achievement of all other goals, while Goal 13 calls for urgent action to combat climate change and its impacts.
ZEB (zero energy building)/ZEH (zero energy house)	Buildings and houses that consume the same amount of energy as they generate from renewable energy sources, resulting in net zero energy consumption.
Inverter technology	Technology that allows the rotation speed of a motor to be freely changed and the output of air conditioner to be freely controlled by turning power semiconductors on and off. It offers significantly higher energy-saving performance than models without inverters.

(Continued)

Basic Terms	Explanation
Basic energy plan	A plan formulated by the government based on the Basic Energy Policy Act in order to indicate the basic direction of energy policy, taking into account changes in the domestic and international energy situation. The 6th Basic Energy Plan was formulated in 2021. Based on the S+3E perspective of safety, security of supply, economic efficiency, and environmental compatibility, it outlines the path for energy policy toward achieving carbon neutrality by 2050.
Energy carriers	Chemical substances suitable for the transportation and storage of energy. In general, it refers to substances that are difficult to transport or store in gaseous form, such as hydrogen, which is converted into organic hydrides, ammonia, or methanol to make it easier to liquefy. When using renewable energy, the future supply chain will be determined by considering not only transportation and storage costs, but also safety, transportation distance, and usage systems when producing large amounts of electricity or hydrogen.
GHGs (greenhouse gases)	Gases in the atmosphere that absorb infrared rays emitted from the earth's surface and have the effect of warming the earth. The Kyoto Protocol, adopted at the Third Conference of the Parties to the United Nations Framework Convention on Climate Change (COP3) in 1997, lists carbon dioxide (CO_2), methane (CH_4), nitrous oxide (N_2O), hydrofluorocarbons (HFCs), perfluorocarbons (PFCs), and sulphur hexafluoride (SF_6) were designated as GHGs to be reduced. Nitrogen trifluoride (NF_3) was added later.
Carbon neutral	A state in which the amount of CO_2 emitted is equal to the amount of CO_2 absorbed. The Japanese government's goal of carbon neutrality is not limited to CO_2, but includes all GHGs, including methane, nitrous oxide, and fluorinated gases. Since completely eliminating emissions is practically impossible, the goal is to achieve net zero by offsetting the unavoidable emissions through absorption or removal, thereby realizing "neutrality."
Carbon pricing	Carbon pricing refers to economic measures that promote CO_2 reduction by assigning a price to carbon. This includes a "carbon tax," which is a tax levied on emissions proportional to the use of fuels or electricity. A representative example is "emissions trading," where companies buy and sell emissions allowances. When using "J-Credits," which trade domestic emission reductions, or "bilateral credits (JCM)," which share emission reductions between countries in cooperation with developing nations, this is collectively referred to as "credit trading." Additionally, border adjustment measures, which seek to impose economic burdens equivalent to the difference in CO_2 prices on imports of products made in countries with lower CO_2 prices, have been proposed.
Carbon recycling	In order to reduce CO_2 emissions into the atmosphere, CO_2 is positioned as a carbon resource, collected, and reused as various carbon compounds. It is expected to be used in chemicals, fuels, minerals, and other products. Although there are various technologies for converting CO_2 into useful substances, many of them utilize reactions with hydrogen, and the key to this is technology for producing hydrogen from renewable energy at low cost.

(Continued)

Basic Terms	Explanation
Kigali Amendment	The Montreal Protocol, adopted in 1987, has already phased out chlorofluorocarbons (CFCs) used as refrigerants in air conditioning, and it has been decided that hydrochlorofluorocarbons (HCFCs), which are substitutes for CFCs, will have been phased out in developed countries by 2020 and in developing countries by 2030. Furthermore, HFCs, which have been used as substitutes for HCFCs, have recently become a concern due to their impact on global warming, and in October 2016, it was decided to establish obligations to reduce their production and consumption within the framework of the Montreal Protocol. This decision is called the Kigali Amendment and came into effect in January 2019.
Green Growth Strategy	In October 2020, Prime Minister Suga declared that Japan would become carbon neutral by 2050. Based on this declaration, the Ministry of Economy, Trade and Industry, in collaboration with other relevant ministries and agencies, formulated the Green Growth Strategy for Carbon Neutrality by 2050 in December of the same year. This industrial policy aims to create a virtuous cycle between the economy and the environment" by transforming industrial structures and socio-economic systems through proactive measures that shift away from conventional thinking, thereby driving significant growth.
Grid interconnection	The ability to connect power generation facilities to the power grid (transmission and distribution lines of general power transmission and distribution companies) for power transmission. Rules for grid interconnection include guidelines on technical requirements for grid interconnection related to power quality (Agency for Natural Resources and Energy).
Border Carbon Tax	A carbon levy imposed by countries implementing climate change measures on imports from countries with insufficient measures, applied at the border. Additionally, some countries may provide refunds equivalent to carbon costs on exports from their own countries. The aim is to mitigate adverse effects on international competition and prevent carbon leakage, where domestic product reduction and increased imports lead to higher CO_2 emissions than before.
Renewable energy	Resources that are replenished from nature at a rate faster than they are used. In Japan, seven types of renewable energy are listed in laws and government ordinances: solar, wind, hydro, geothermal, solar thermal, heat from the atmosphere and other heat existing in nature, as well as biomass (organic matter derived from animals and plants). The introduction of renewable energy is an essential measure for achieving carbon neutrality and reducing CO_2 emissions to zero overall.
Sustainable finance	Integrating sustainability factors into the financing of economic activities in order to achieve a sustainable society. This concept includes a wide range of financing methods, such as investment, debt, and loans, and means incorporating responsible finance principles into management decisions and strategies. Specific trends are accelerating on a global scale through guidelines such as the Sustainable Development Goals and the Paris Agreement, as well as guidance, evaluation criteria, and information disclosure methods issued by various countries and organizations.
Natural refrigerant	Refrigerants that are environmentally friendly and mainly consist of ammonia, which is found in nature.

(Continued)

Basic Terms	Explanation
Supply and demand adjustment market	The adjustment capacity required by general power transmission and distribution companies to control frequency, and balance supply and demand in their supply areas has been procured through public tenders since October 2016. However, in order to secure inexpensive and stable adjustment capacity across areas, a new supply and demand adjustment market was established in April 2021. Adjustment capacity is categorized into five product categories based on response time and duration, with trading starting with the third-tier adjustment capacity (2) with the slowest response time, and gradually expanding to product categories with faster response times.
Hydrogen carrier	A general term for states or substances in which hydrogen has been converted to facilitate transportation and storage through changes in state or chemical composition in order to increase its density per unit weight or volume. The appropriate hydrogen carrier varies depending on the energy consumption during conversion, the type of reaction (endothermic or exothermic), the purity of hydrogen after conversion, the demand equipment, and the scale. Typical hydrogen carriers include compressed hydrogen, liquefied hydrogen, organic hydrides (such as MCH), ammonia, synthetic methane, and hydrogen storage materials.
Decarbonization	Typical approaches include transitioning the entire economy, or specific sectors such as industry, residential services, and transportation, or individual companies, to systems that do not emit CO_2 in their activities, and increasing the proportion of electricity and hydrogen produced using CO_2-free methods. A typical approach is to shift the entire economy, specific sectors such as industry, civil life, and transportation, or companies to systems that do not emit CO_2 in their activities, and to increase the proportion of electricity and hydrogen produced by methods that do not generate CO_2.
Energy storage	In a broad sense, this refers to the temporary storage of energy by some physical or chemical means. From a carbon neutral perspective, it often refers to the conversion of electrical energy into other forms of energy for storage. Typical methods include storage batteries, pumped storage, hydrogen, heat, and compressed/liquefied air. The stored energy is used as electricity or heat as needed.
Thermal storage	Thermal storage exists in various forms, ranging from fully commercialized products to those still in the research stage. Commercial examples include hot blast stoves (1,000°C) installed next to steel furnaces and two-tank nitrate systems (around 600°C). In recent years, a highly safe thermal storage system using crushed stone and air has been attracting attention.
DC (direct current) transmission	This is mainly used for submarine power cables of 50 km or more. AC–DC conversion equipment is required to convert alternating current (AC) to DC at both the transmission and reception ends, which increases equipment costs. However, DC transmission is used under the above conditions because it has the characteristic of less loss than AC transmission.
Chiller	A general term for equipment that circulates heat transfer media such as water to maintain a constant temperature in various facilities and spaces. The term "chiller" conjures up images of cooling, but some chillers also provide heating.

(Continued)

Basic Terms	Explanation
Electricity market	In Japan, the market for trading electricity includes the wholesale electricity market (spot market) for trading electricity (kWh value), the capacity market for trading the ability to supply electricity (kW value), the supply and demand adjustment market for trading supply and demand adjustment capacity (ΔkW value), and the non-fossil value trading market for trading the environmental value of power sources that do not emit CO_2, such as renewable energy.
Electricity system reform	The Great East Japan Earthquake in 2011 brought to light issues with the power system, such as supply shortages due to the shutdown of large-scale centralized power sources and uniform demand restrictions, such as planned power outages. In response, the government promoted electricity system reform in three stages by April 2020, based on three pillars: (1) expansion of wide-area system operation, (2) full liberalization of retail and power generation, and (3) further ensuring the neutrality of the power transmission and distribution sector through legal unbundling, with the aim of ensuring a stable supply, minimizing electricity rates, and expanding consumer choice and business opportunities for operators. Broadly defined electricity system reforms, such as the design of a trading market and the next-generation power network, are continuing.
Negative emissions	The opposite of GHG emissions. In other words, the artificial removal of GHGs from the atmosphere. Also known as greenhouse gas removal (GGR). When referring specifically to CO_2, it is also known as carbon dioxide removal (CDR) or carbon removal.
Paris Agreement	A new international framework on climate change for the period after 2020, adopted at the 21st Conference of the Parties (COP21) to the United Nations Framework Convention on Climate Change held in Paris, France, in 2015. It establishes a global long-term goal of limiting global temperature rise to 2°C, the submission and updating of national emission reduction targets every 5 years, the implementation of adaptation plans and actions, financial support from developed countries and voluntary contributions from developing countries, reporting and review of implementation progress by all countries using common and flexible methods, and the use of market mechanisms, among other provisions.
United Nations Framework Convention on Climate Change	Adopted in 1992, this treaty aims to promote global efforts to combat global warming and stabilize the concentration of GHGs in the atmosphere. Based on this treaty, the COP has been held every year since 1995. At the Third Conference of the Parties (COP3) held in Kyoto in 1997, the Kyoto Protocol was agreed upon, which clearly stipulated reduction targets for developed countries, marking a step forward in reducing GHG emissions.
Product Lifetime Climatic Impact	Life Cycle Climate Performance (LCCP) is an index that comprehensively evaluates the impact on global warming, taking into account not only the GWP but also the impact of the refrigerant life cycle and the impact of GHGs generated by the equipment's power consumption.
TEWI (Total Equivalent Warming Impact)	An index that comprehensively evaluates the direct global warming impact of CO_2 diffused into the atmosphere due to the use of refrigerants in equipment and leakage during disposal, as well as the indirect global warming impact of CO_2 generated by energy consumption during equipment operation.

Abbreviation Table

Abbreviation	Full form
APF	Annual performance factor
ASHRAE	American Society of Heating, Refrigerating and Air-Conditioning Engineers
CFC	Chlorofluorocarbon
CN	Carbon neutral
COP	Coefficient of performance
DPP	Digital product passport
DR	Demand response
EMS	Energy management systems
EU	European Union
EV	Electric vehicle
FC	Fluorocarbons
FCV	Fuel cell vehicle
FIP	Feed-in premium
FIT	Feed-in tariff
GHG	Greenhouse gas
GWP	Global warming potential
HCFC	Hydrochlorofluorocarbon
HEMS	Home Energy Management System
HEV	Hybrid electric vehicle
HFC	Hydrofluorocarbon
HFO	Hydrofluoroolefin
HP	Heat pump
IEA	International Energy Agency
IPCC	Intergovernmental Panel on Climate Change
IRENA	International Renewable Energy Agency
ISO	International Organization for Standardization
JST	Japan Science and Technology Agency
LCA	Life Cycle Assessment
LCCP	Life Cycle Climate Performance
LIB	Lithium-ion battery
LNG	Liquefied natural gas
METI	Ministry of Economy, Trade, and Industry
NDC	Nationally determined contribution
NEDO	New Energy and Industrial Technology Development Organization
PFAS	Perfluoroalkyl and polyfluoroalkyl substances

(Continued)

Abbreviation	Full form
PHV	Plug-in hybrid electric vehicle
PV	Photovoltaic
RE100	Renewable energy 100%
RTOC	Refrigeration Technical Option Committee
SBT	Science-based targets
SEA	Strategic environmental assessment
TEWI	Total Equivalent Warming Impact
UNEP	United Nations Environment Programme
UNFCCC	United Nations Framework Convention on Climate Change
VPP	Virtual power plant
VRF	Variable refrigerant flow
WMO	World Meteorological Organization
ZEB	Net zero energy building
ZEH	Net zero energy housing

Unit Conversion Table

	SI Unit	Engineering Unit	British Unit	Customary Unit
Length	1 m	1 m	39.37 inches	–
	25.40 mm	25.40 mm	1 inches	–
Area	10,000 m^2	–	–	1ha
Mass	1 kg	1 kg	2.205 pounds (lb)	–
	1,000 kg	1t	2,205 pounds (lb)	–
	0.4536 kg	0.4536 kg	1 pound (lb)	–
Pressure	1 kPa	0.01020 kg/cm^2	0.1450 psi	0.009869 atm
	1 MPa	10.20 kg/cm^2	145.0 psi	9.869 atm
Energy/heat	1J	0.0002389 kcal	0.0009488 Btu	0.0002778 Wh
	1kJ	0.2389 kcal	0.9488 Btu	0.2778 Wh
	1Mj	238.9 kcal	948.8 Btu	0.2778 kWh
	1GL	2.389×105 kcal	0.9488 MMBtu	277.8 kWh
	1PJ	2.389×108 kcal	948.8 MMBtu	2.778×105 kWh
	1EJ	2.389×1011 kcal	948,800 MMBtu	2.778×108 kWh
	1.054 kJ	0.252 kcal	1 Btu	0.2928 Wh
	1.054 GJ	2.520×105 kcal	1 MMBtu	292.8 kWh
Heat flow	–	860 kcal/h	–	1 kW
	–	8.60×105 kcal/h	–	1 MW
	–	8.60×108 kcal/h	–	1 GW
Temperature	273.15 K	0°C	–	–

Unit Prefixes Table

10^n	Prefixes	Symbol
10^{18}	exa	E
10^{15}	peta	P
10^{12}	tera	T
10^{9}	giga	G
10^{6}	mega	M
10^{3}	kilo	k
10^{2}	hecto	h
10^{1}	deca	da
10^{0}	–	–
10^{-1}	deci	d
10^{-2}	centi	c
10^{-3}	milli	m
10^{-6}	micro	μ
10^{-9}	nano	n

1 The Necessity of Heat Pump Technology for a Decarbonized Society

OVERVIEW

Heat pump technology started as refrigeration technology, but in recent years it has been adaptable to a wide variety of temperature ranges and its applications have expanded to air conditioning, water heating, and industrial technology.

As a result, a large number of heat pump equipment has emerged, ranging from low temperature to high temperature heat pump technology. This section first provides an overview of heat pump technology as a whole, including the operating principle of heat pumps, the types of equipment, and even the materials and types of refrigerant operating fluid used (Figure 1.1).

FIGURE 1.1 What is heat pump technology?

DOI: 10.1201/9781003666950-1

1.1 WHAT IS A HEAT PUMP?

1.1.1 Characteristics of Heat Pumps

Heat pump is a generic term that describes equipment and systems that use a small amount of drive energy (mainly electricity), using some kind of heat medium or semiconductor, to take heat from the low temperature side (also known as heat source) and transfer and dissipate heat efficiently to the high temperature side.

As shown in Figure 1.2, the low temperature side of a heat pump can take heat from an object and cool it, so it can be used as a refrigerator, freezer, or cooling unit (or as the cooling operation of an air conditioning unit that can switch between cooling and heating). In contrast, the high temperature side can be used for heating (or as the heating operation of air conditioning) and water heaters. Heat pumps originally started out as refrigeration equipment, but are often used as a name for equipment that uses the high temperature side, to distinguish it from equipment that uses the low temperature side.

1.1.2 What Is Next-Generation Heat Pump Technology?

1.1.2.1 Differences between Heat Pumps and Heat Pump Technology

As discussed in Section 1.1, heat pumps have been utilized differently as equipment, depending on whether they use the high temperature or low temperature side. On the other hand, a single equipment, such as an air conditioning unit, has in recent

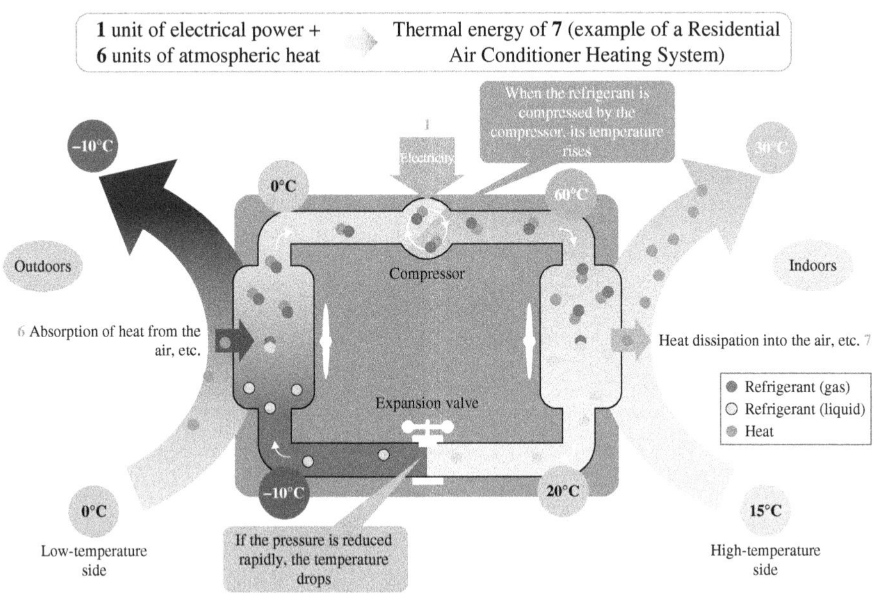

FIGURE 1.2 How a heat pump works. (From Heat Pump & Thermal Storage Technology Center of Japan "Let's learn about heat pumps and thermal storage systems—About heat pumps," "What is a heat pump" [1].)

years also been used for both cooling and heating. As a result, there has often been confusion as to whether the term "heat pump" refers to a heating or cooling equipment, or whether it is a generic term for all temperature control technologies.

Therefore, in this publication, "heat pump technology" is used as a generic term for technologies that include cold and heat generation.

1.1.2.2 What is Next-Generation Heat Pump Technology?

The form that heat pump technology should aim to become in the 2030s in order to achieve carbon neutrality, which is also the main point of discussion in this publication, will be referred to as "next-generation heat pump technology."

The term is used here to describe the ideal state of heat pump technology in the next era, which includes not only equipment but also technologies strongly related to heat pump technology, such as digital transformation (DX), supply chains including resource circulation, and social systems.

1.2 OPERATING PRINCIPLES AND CYCLES OF HEAT PUMP TECHNOLOGY

1.2.1 Operating Principles of Heat Pump Technology

1.2.1.1 Principles of Heat

To understand heat pump technology, it is necessary to understand two scientific events. The most important is heat. As we know, the term heat is strongly related to the phenomena of hot and cold. The definition of heat is that it is a form of energy that is transferred from the high temperature side to the low temperature side.

For example, if a hot object is brought into contact with a low temperature object, as shown in Figure 1.3, heat is transferred from the hot object to the low temperature one, which means that the hot side is cooled and the low temperature side is heated. Conversely, heat transfer from the low temperature side object to the high temperature side object does not normally occur. It cannot happen, for example, that cold water left in the summer will dissipate heat to the hot object outside and cool it down further.

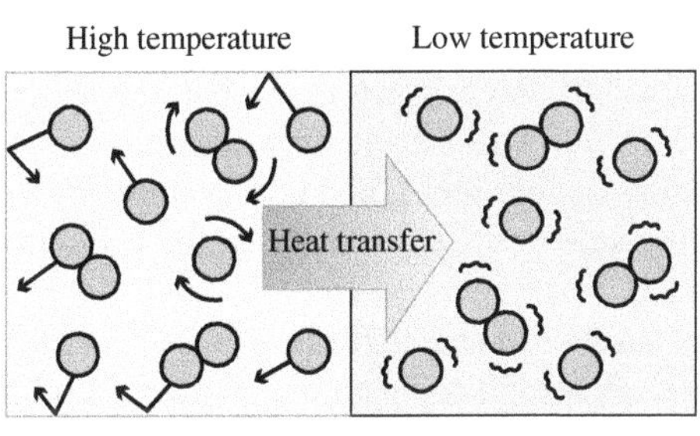

FIGURE 1.3 Heat transfer principles.

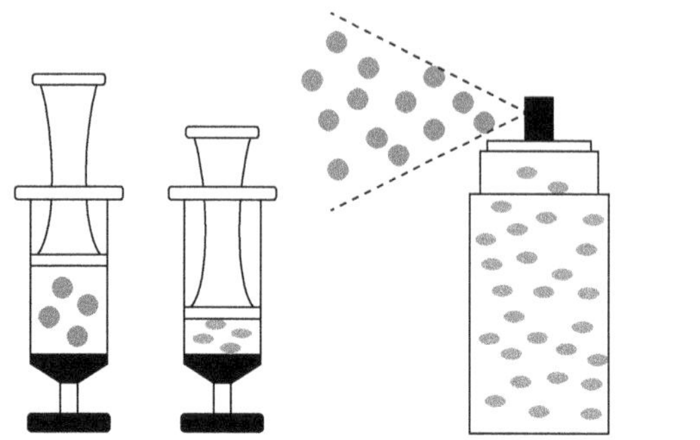

FIGURE 1.4 Temperature changes due to gas compression and expansion.

1.2.1.2 Temperature Changes Due to the Compression and Expansion of Gases

Another important scientific phenomenon is the temperature change caused by the compression and expansion of gases, as shown in Figure 1.4. When gas is compressed to a high pressure in a piston, its temperature will also increase. On the other hand, if the piston is pulled back and expanded to a lower pressure, it will cool down.

Heat pump technology can be seen as a generic term for technologies that utilize this property of gases to transfer heat from the low temperature side to the high temperature side (which would not normally occur) using a small amount of energy, via compression to raise the temperature and expansion to lower the temperature.

1.2.2 Air Conditioning as a Representative of Heat Pump Technology

1.2.2.1 Heat Transfer Principles of Air Conditioner

Air conditioner is the most familiar equipment in which heat pump technology is employed.

The principle of heat pump technology is therefore explained using air conditioner in heating mode as an example.

As shown in Figure 1.5, an air conditioning system consists of an outdoor unit and an indoor unit, with refrigerant circulating between the two units as the working fluid. In the outdoor unit, the refrigerant temperature is lower than the air temperature outside. Here, the outdoor air and the refrigerant come into indirect contact through the walls such as copper pipes, and heat is transferred from the outdoor air on the high temperature side to the refrigerant on the low temperature side. This is done in a heat exchanger.

The refrigerant that has gained heat in the heat exchanger moves to the compressor, where it is compressed to a high temperature and pressure, which is higher than the temperature in the room. A small amount of electricity is required to drive this compressor.

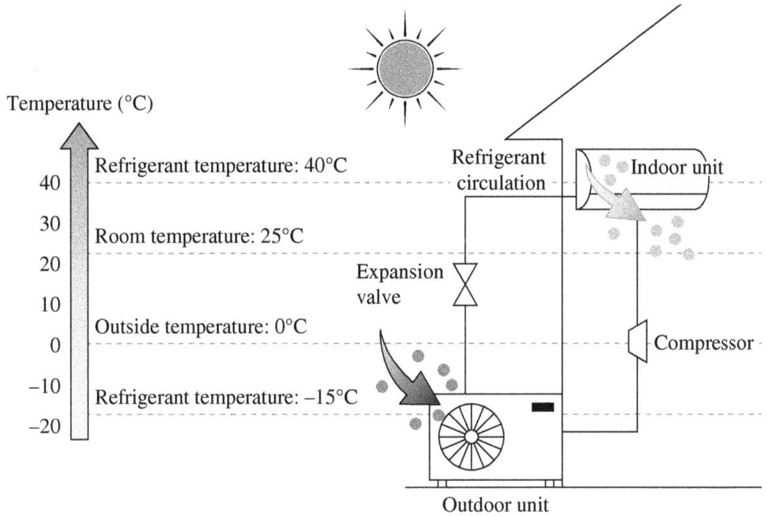

FIGURE 1.5 How air conditioner works.

This is the main source of drive energy for air conditioner.

Refrigerant flows out of the compressor and is sent to the indoor unit. As the indoor unit mainly consists of a heat exchanger, the indoor air on the low temperature side and the refrigerant on the high temperature side come into indirect contact, heat is transferred from the refrigerant to the indoor air and the indoor air is warmed. The refrigerant flows out of the heat exchanger while maintaining high pressure, passes through the expansion valve, expands and drops to a low temperature and low pressure. It then flows into the outdoor unit again and receives heat from the air outside.

Overall, heat pump technology is a reverse phenomenon that does not normally occur, as it transfers heat from the outdoor air on the low temperature side as a heat source to the indoor air on the high temperature side. This is made possible by the small amount of electricity used by the compressor.

1.2.2.2 Heat Pumps Based on the Principle of Water Pumps

Heat pump technology is easier to understand when compared to water pumps: water normally flows from a higher to a lower position, as shown in Figure 1.6. Water pumps use electricity to pump water from a lower to a higher position. Heat pumps are so called because they are based on exactly the same principle.

In water pumps, the smaller the difference in height, the less electricity is required. The same applies to heat pumps: the smaller the temperature difference between the low and high temperatures, the less power is required to drive the pump.

For example, in air conditioning, it is recommended that the set temperature should be lower in winter and higher in summer. This is because this reduces the temperature difference with the outside air, which reduces the power required for the compressor, thereby reducing power consumption.

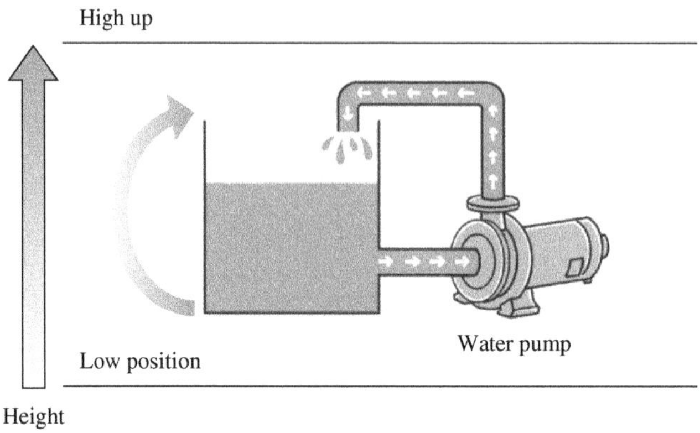

FIGURE 1.6 How water pumps pump water.

1.2.3 The Cycle of Heat Pump Technology

1.2.3.1 Types of Heat Pump Technology

Heat pump technology can be divided into two main categories: vapor compression type[1] (hereinafter referred to as compression type) and thermal-driven absorption and adsorption type[2]. In addition, there is also magnetic refrigeration and other types of refrigeration, although their applications are limited.

Heat pump technology may attract renewed attention in the future, as it can be driven by renewable energy or untapped energy.

As the current heat pump technology is expected to mainly use the compression type, which can be driven efficiently by electric power, this book will focus on the compression type.

1.2.3.2 Compression Cycle

The aforementioned cycle of the compression formula will now be explained in more detail. As shown in Figure 1.7, the cycle consists of a compressor, a condenser as a heat exchanger, an expansion valve, and an evaporator as a heat exchanger as the main elements, and the cycle is established by refrigerant circulating through them.

During heating, the heat exchanger on the outdoor unit side functions as an evaporator and the heat exchanger on the unit indoors functions as a condenser. In heat exchangers, heat is transferred through indirect contact between air, water, and refrigerant through the walls such as copper tubes. In addition, the refrigerant flows through a phase change[3] between gas and liquid, called evaporation or condensation, and repeatedly evaporates and liquefies.

In the evaporator, the low temperature, low pressure refrigerant liquid removes heat from the outside through air or water and evaporates. At this time, the latent heat of vaporization[4] efficiently removes heat from the air and water, and cools them.

As shown in Figure 1.8, the refrigerant vapor (1) evaporates in the evaporator and is sent to the compressor, where it is compressed by the power applied from

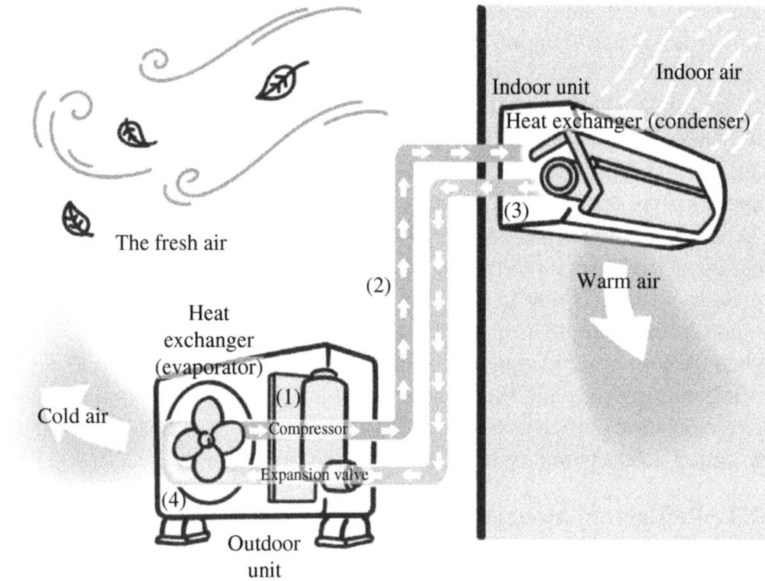

FIGURE 1.7 Compression cycle heat pump and its components. (From Hitachi Global Life Solutions Inc., "I want to know how air conditioner works." [2].)

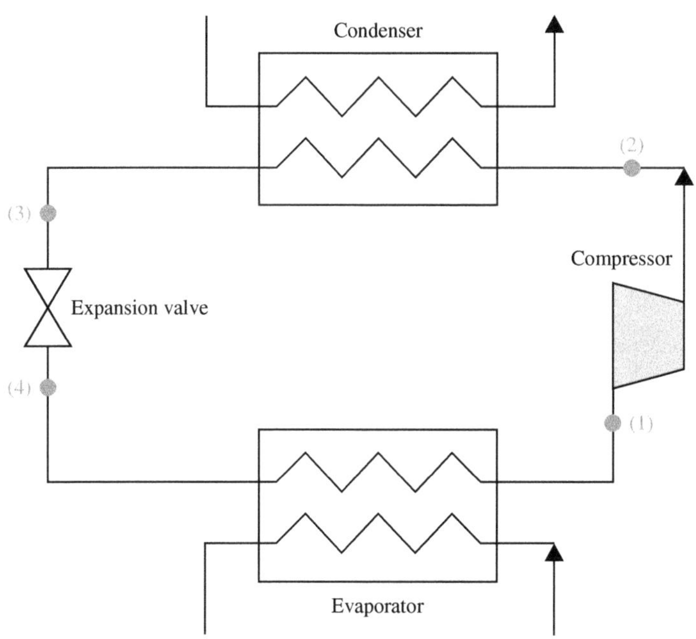

FIGURE 1.8 Flow of a compression heat pump cycle.

outside to become superheated vapor[5] (2) with high temperature and pressure. The power (electricity) added to the compressor is the driving source for the compression cycle.

The superheated refrigerant vapor in the compressor condenses from gas to liquid in the condenser, which uses the heat of condensation when the refrigerant condenses, and efficiently heats the fluid, water, or air with the condensation heat to become condensate (3). The heating capacity of the heat pump[6] is the amount of heat generated by the refrigerant heating the fluid. The refrigerant that has become condensate at high pressure expands in the expansion valve to a low temperature and low pressure (4) and is sent to the evaporator again. These processes are carried out continuously in a compression cycle.

When the unit is set to cool, and fluid is cooled by the evaporator, it is referred to as a refrigeration cycle, and the ability to cool heat at a given volume in a given time is the refrigeration capacity. As the fluid to be heated is warmed in the condenser, it is also called a heat pump cycle if heating is the main action.

1.2.3.3 Refrigerant Movement in a Compression Cycle

In air conditioner, a four-way valve changes the direction of refrigerant flow, with the indoor unit heat exchanger as the evaporator and the outdoor unit heat exchanger as the condenser during cooling and vice versa during heating. This allows the functions of the evaporator and condenser to be switched without changing the indoor and outdoor units.

The *P-h* diagram in Figure 1.9 shows the operating conditions of the refrigerant in this compression cycle.

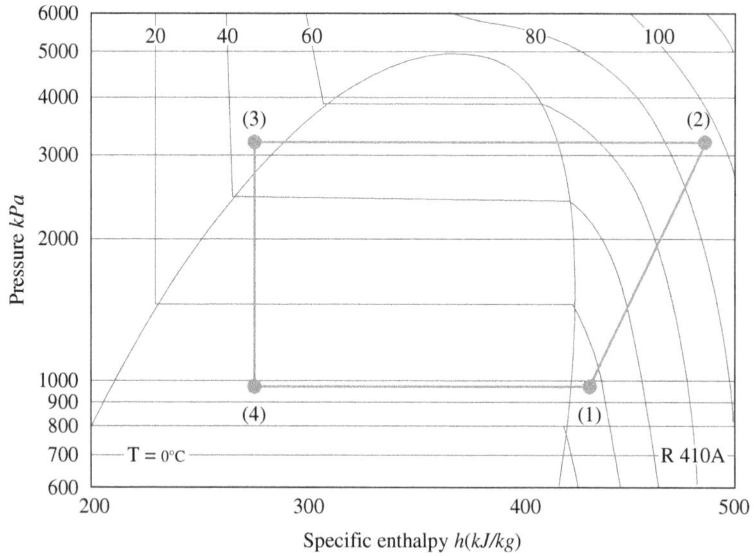

FIGURE 1.9 *P-h* diagram. *Note:* For clarity, only temperature is shown as a parameter in this diagram.

The numbers shown in Figure 1.8 correspond to those shown in Figures 1.7 and 1.9.

The horizontal axis of Figure 1.9 shows the specific enthalpy *h [kJ/kg]*, which is the heat content per kg of refrigerant; the vertical axis shows the pressure *P [kPa]* of the refrigerant and the parameters such as temperature, specific volume, and specific entropy are also shown.

The curve in the center represents the saturation boundary, the saturated vapor line on the right, with the apex being the critical point. The inner part of the curve represents the wet vapor, which is the area where gas and liquid are mixed. The left side of the saturated liquid line represents the supercooled liquid, which is the liquid, and the right side of the saturated vapor line represents the region of superheated vapor of the gas. The specific enthalpy represents the energy state of the refrigerant.

Therefore, the specific enthalpy difference between the inlet and outlet of each element is the amount of energy per unit refrigerant flow rate transferred by the refrigerant of that element.

The aforementioned compression cycle on a *P-h* diagram can be divided into the following four processes.

Compression process	Electric power is applied to the gaseous refrigerant in the compressor to compress it, resulting in an increase in temperature and pressure. The specific enthalpy increases accordingly ((1)→(2)). The power required to compress the refrigerant, w [kJ/kg], is expressed using the specific enthalpy of the refrigerant at the point corresponding to the subscript number as follows. $$w = h_2 - h_1 \ldots \quad (1)$$
Condensation process	The refrigerant condenses into a liquid in the condenser. It exchanges heat with air and water (external fluids) and these external fluids are heated, the refrigerant is cooled, and its enthalpy decreases ((2)→(3)). The heat exchanged qc [kJ/kg] is then given by the following formula: $$qc = h_2 - h_3 \ldots \quad (2)$$
Expansion process	The pressure and temperature of the refrigerant decrease at the expansion valve. Since there is no external power supply or heat transfer, the enthalpy is constant ((3→(4)) $$0 = h_3 - h_4 \ldots \quad (3)$$
Evaporation process	Heat is absorbed from the external fluid when the refrigerant evaporates in the evaporator. The specific enthalpy increases ((4)→(1)) because the air and water (external fluid) are cooled without much temperature change and the refrigerant side is heated. The heat exchanged qe [kJ/kg] is then given by the following formula: $$qe = h_1 - h_4 \ldots \quad (4)$$

1.2.4 Performance of Compression Cycles Measured by Performance Evaluation Indicators

1.2.4.1 Standards for Equipment Performance

The performance of a compression cycle under certain operating conditions is expressed as a COP (coefficient of performance). It is defined as the ratio of the heating capacity q_c in the condenser (q_e in the evaporator in the case of cooling) to the power w applied to the compressor in the case of heating as follows[7].

$$COPh = qc/w\ldots \quad (5)$$

Since the COP is well above 1, heating can be achieved with several times the energy input.

This is the great thing about heat pump technology and it can be achieved by making good use of air heat and untapped energy as heat sources.

In addition, while COP represents the performance under certain operating conditions, the annual performance factor (APF) indicates how well the system performs on an annual basis for a variety of varying annual operating conditions, and is expressed as follows.

$$APF = \frac{\text{annual cooling capacity} + \text{annual heating capacity}}{\text{annual power consumption of the equipment}}\ldots \quad (6)$$

This allows manufacturers to develop highly efficient equipment that is more suited to actual annual operation, and allows users to know equipment performance that is closer to actual operation. In this way, equipment performance has moved from one specific performance to representing equipment performance throughout the year, and standards for equipment performance have become more diverse due to differences in operating conditions and other factors.

Good equipment performance means that less power is required to achieve the same capacity. This indirectly reduces the CO_2 emissions[8] required during electricity generation.

In addition, the use of renewable energies and other energy sources that do not emit CO_2 for electricity generation is expected to increase rapidly in the future, further reducing CO_2 emissions.

This is why the shift from combustion appliances, which generate heat by combustion, to heat pumps, which can generate heat by electricity, is becoming increasingly important. This is the so-called electrification of heat generation technology.

1.2.4.2 Relationship between Heat Pump Technology and the Top Runner Program

The movement to improve the performance of heat pump technology has been accelerated by the introduction of the Top Runner program for the Act on the Rational Use of Energy and Conversion to Non-Fossil Energy Sources (Energy Conservation Law), which was amended in 1999.

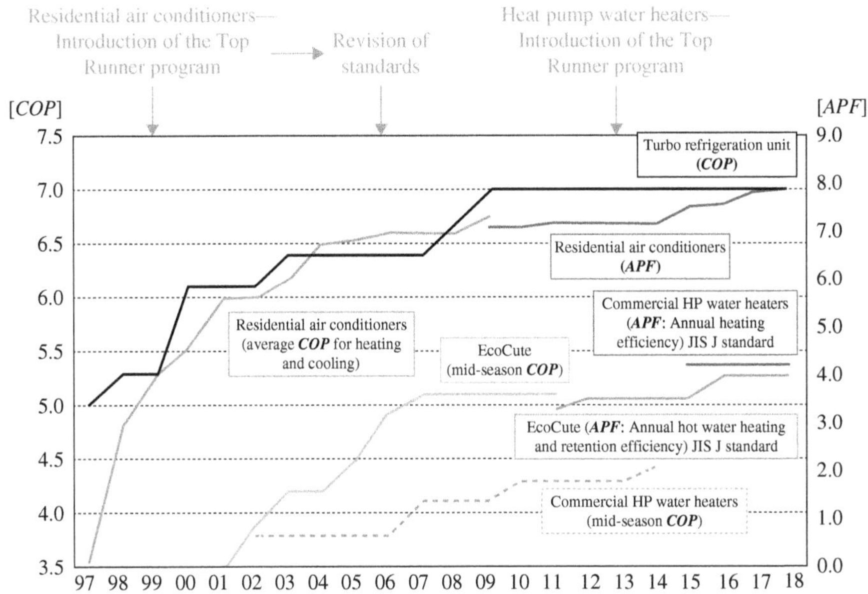

FIGURE 1.10 Efficiency trends of heat pump technology. (From Heat Pump & Thermal Storage Technology Center of Japan, "Highly efficient heat pump technology—Residential air conditioner efficiency APF (Annual Performance Factor): Is 7.0 or more!" [3].)

The Top Runner program is a system that encourages the achievement of higher technically feasible targets than those set based on the most energy efficient equipment currently on the market (top runner). Target items are selected from equipment with the highest energy consumption. Residential air conditioners have been included in the system since its inception, but now commercial air conditioners are also included, as are heat pump water heaters and showcases.

Energy efficiency and conservation standards under the Top Runner program used to be assessed by the COP under cooling and heating conditions, but in the May 2009 revision of the energy conservation law standards, the APF, which is closer to actual use, was introduced. It is used for residential air conditioners, commercial air conditioners, and residential and commercial heat pump water heaters.

Energy efficiency standards have been reviewed and strengthened step by step, and the performance of heat pump technology has improved accordingly: as shown in Figure 1.10, some heat pumps now have an APF of nearly 8.

1.3 COMPONENTS AND EQUIPMENT RELATED TO HEAT PUMP TECHNOLOGY

As heat pump technology is applied from low to high temperature ranges, there are a great many types and names for the equipment. Therefore, the components and equipment related to heat pump technology are classified here.

1.3.1 Heat Pump Components

It has been noted above that heat pump technology comprises a compressor, condenser, expansion valve, and evaporator as the main elements. This section provides a brief introduction to the functions involved in the components.

1.3.1.1 Heat Exchangers: Evaporators and Condensers

Heat exchangers are components that exchange heat between a refrigerant and a heated or cooled fluid, such as air or water, and are the most important components in heat utilization equipment.

In heat pump technology, fin and tube types are often used when the fluid to be heated or cooled is air, and shell and tube types are often used when the fluid to be cooled is a liquid such as water. The types of heat exchanger applied in heat pump technology are numerous, see, for example, the *Handbook of Air Conditioning and Refrigeration*.

In air conditioner, fin and tube type heat exchangers are generally used, which consist of copper heat transfer tubes mechanically expanded and press-fitted with plate fins made of aluminum. As they are also mass-produced equipment, the performance of these heat exchangers has been significantly improved in recent years.

Performance improvements in heat exchangers are achieved by increasing the heat transfer area of the tube walls and improving the heat transfer coefficient and thermal conductivity.

In fin and tube type heat exchangers, the fins on the air side have been changed from plate fins to slit fins (Figure 1.11), the heat transfer tubes on the refrigerant side

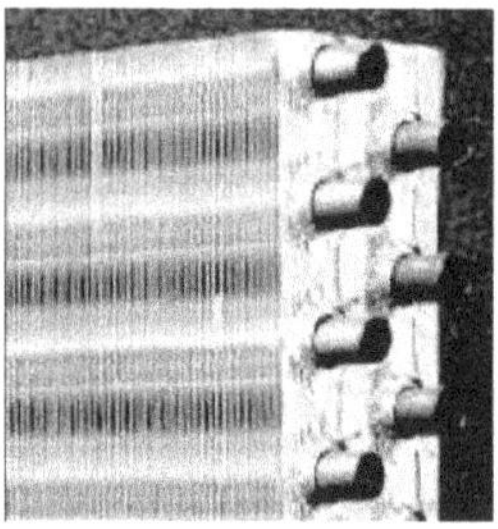

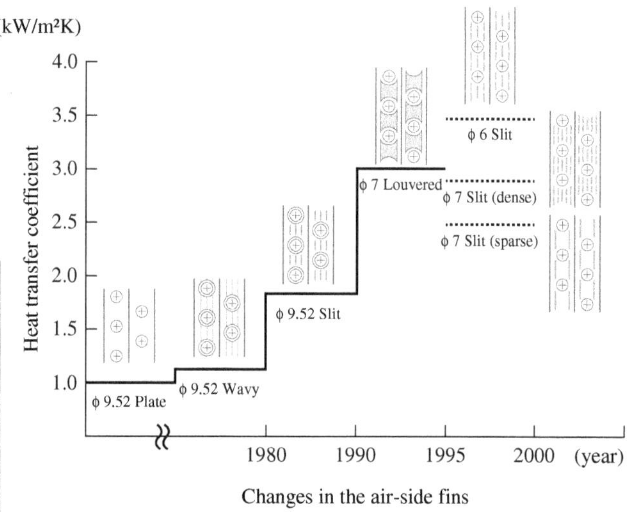

FIGURE 1.11 Fin and tube type heat exchanger. (From Yutaka Shibata/Japanese Society of Mechanical Engineers (JSME), "Research and development in improving the performance of heat exchangers for air conditioning equipment" [4, 5].)

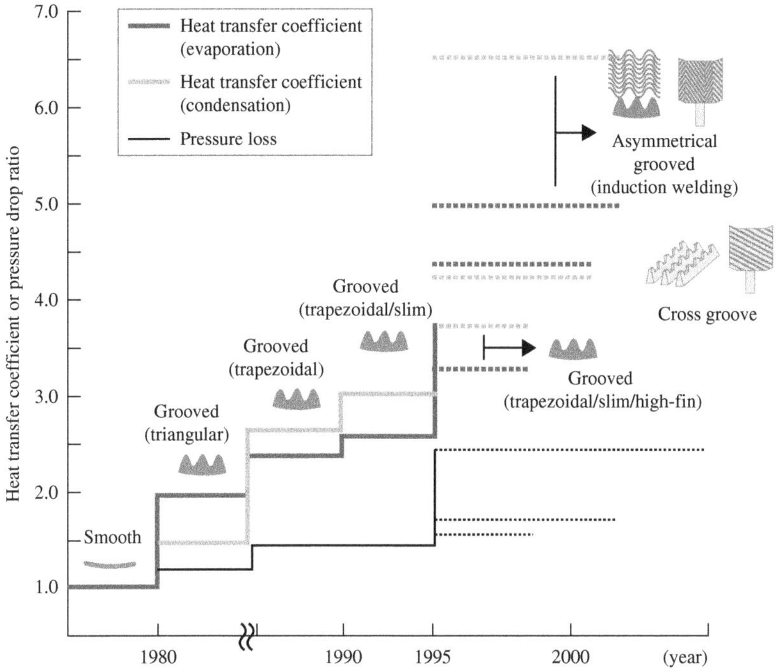

FIGURE 1.12 Transition in heat transfer tubes on the refrigerant side in heat exchangers. (From Yutaka Shibata/Japanese Society of Mechanical Engineers (JSME), "Research and development in improving the performance of heat exchangers for air conditioning equipment" [6].)

have been changed from smooth tubes to internally grooved tubes (Figure 1.12), and the tube diameter has been reduced from Φ9.52 to Φ7 or smaller, resulting in higher performance and more compact size.

The shell and tube type is also used in chillers (Figure 1.13).

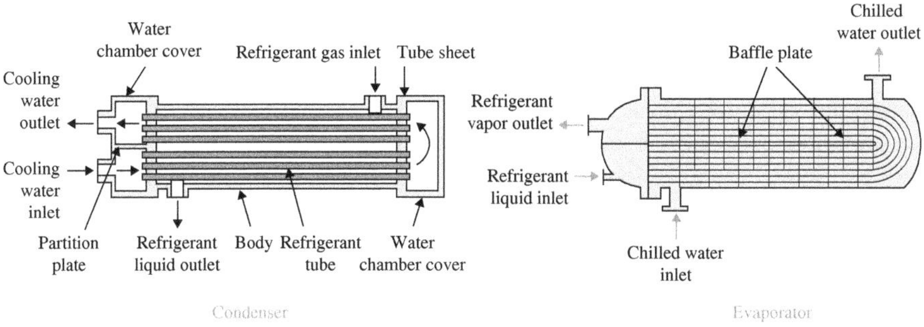

FIGURE 1.13 Horizontal shell and tube type heat exchangers. (Based on Japanese Society of Refrigerating and Air Conditioning Engineers, Advanced Standard Text Refrigeration and Air Conditioning Technology, Refrigeration Edition (4th Edition) (Left) [7], (Right) [8].)

Shell and tube types have not undergone the same major technological innovation as fin and tube types, but their performance has been steadily improving and is already sufficiently high.

1.3.1.2 Compressors

As shown in Figure 1.14, a compressor is a component that physically compresses low temperature, low pressure refrigerant gas to a high temperature, high pressure state and can be broadly classified into positive displacement and centrifugal types. There are several compression methods in the positive displacement compression type, with different, corresponding ranges of refrigeration capacity and structure.

By construction, the hermetic type is mainly used in small machines, as the electric motor and compressor are welded together to form an integrated type and cannot be repaired. The semi-hermetic type is used for medium to large machines, and contains the electric motor and compressor in a sealed container, but they are not welded together and can be repaired.

1.3.1.3 Expansion Valves

As shown in Figure 1.15, expansion valves expand the condensed liquid refrigerant by passing it through a narrow section to reduce the pressure, and also have the role of regulating the refrigerant flow rate.

Many valves that open and close automatically according to temperature have been used, but in recent years electronic expansion valves that can be opened and closed by electronic control have been used.

Thus, equipment based on heat pump technology is characterized by its relatively simple construction and the fact that it consists of a small number of components. This, coupled with the fact that they are energy saving, means that they have been the core of heat utilization technology for a long time.

1.3.2 Classification of Equipments

As heat pump technology is applied to a wide variety of equipment, various equipment names are used. In particular, these are divided according to the temperature range in which they are used. Table 1.1 shows the classification of heat pump technology-based appliances according to temperature range. As can be seen, they can be used in the ultra-low temperature range (–100°C) to the high temperature range (200°C).

The applications are characterized by their use in diverse heat utilization fields, ranging from freezing, refrigeration, air conditioning, hot water supply, and steam generation. They can also be used in a wide range of sectors, from residential to commercial, industrial to transportation.

Category			Shape	Closed mechanism	Main Applications	Drive power range [kW]	Features
Positive displacement	Reciprocating type	Piston-crank type	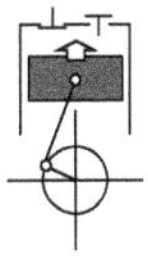	Open	Refrigeration, heat pump	0.4–120	Easy to use, wide variety of models, inexpensive, not suitable for large capacities
				Semi-hermetic	Refrigeration, air conditioning, heat pumps	0.75–45	
				Hermetic	Freezer, air conditioner	0.1–15	
		Piston-inclined plane type	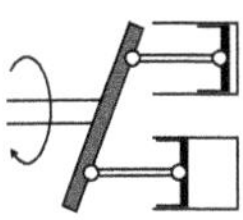	Open	Car air conditioner	0.75–5	Dedicated for car air conditioner Capacity controllable
	Rotary type	Rolling piston type	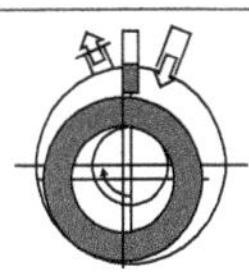	Hermetic	Car air conditioning, small refrigeration units, showcases, water heaters	0.1–5.5	Small capacity, high speed
		Rotary vane type	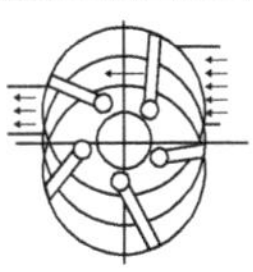	Open	Car air conditioner	0.75–2.2	Compact for its capacity
	Scroll type			Open	Car air conditioner	0.75–2.2	
				Semi-hermetic	EV car air conditioner		
				Hermetic	Air conditioning, refrigeration, hot water supply	0.75–30	Small capacity, high speed
	Screw type	Twin rotor	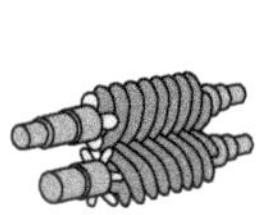	Open	Refrigeration, air conditioning, heat pumps	20–1,800	Compared to centrifugal compressors, they are better suited for high-pressure applications and are therefore widely used in heat pumps and refrigeration systems. Smaller models are increasingly being designed as hermetically sealed units
				Semi-hermetic	Refrigeration, air conditioning, heat pumps	30–300	
		Single rotor	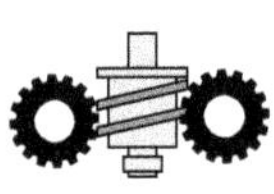	Open	Refrigeration, air conditioning, heat pumps	100–1,100	
				Semi-hermetic	Refrigeration, air conditioning, heat pumps	22–90	
Centrifugal			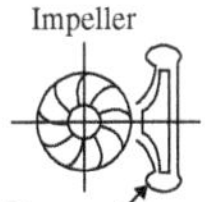	Open	Refrigeration, air conditioning	90–10,000	Suitable for large capacities. Not suitable for high pressure ratios
				Hermetic			

FIGURE 1.14 Types of compressors. (From Japanese Society of Refrigerating and Air Conditioning Engineers, Advanced Standard Text Refrigeration and Air Conditioning Technology, Refrigeration Edition (4th Edition) [9].)

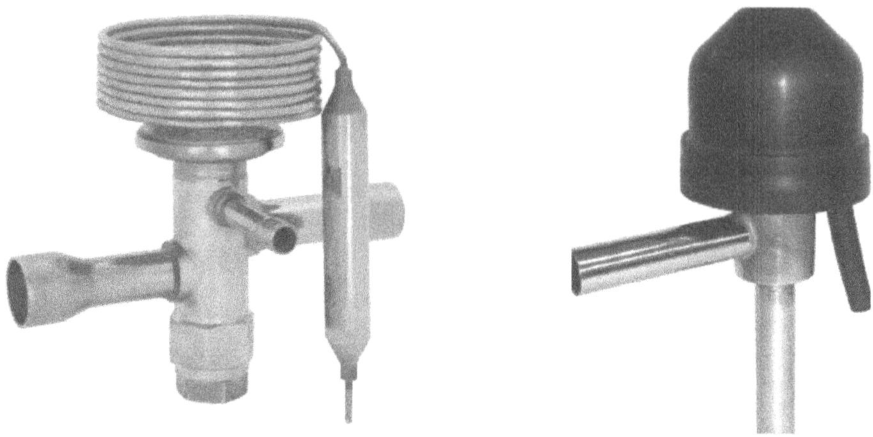

FIGURE 1.15 Temperature expansion valve and electronic expansion valve. ((Left) Based on Saginomiya Seisakusho, Inc, Expansion Valve, "Temperature expansion valve/Medium type temperature expansion valve SCX" [10]. (Right) Based on Saginomiya Seisakusho, Inc, Expansion Valve, "Electronic expansion valve VKV" [11].)

TABLE 1.1
Applications by Temperature Range and Sector of Use

Temperature Range	Use of the Product	Equipment	Sector	Outline
Cryogenic (near absolute temperature), ultra-low temperature (−50°C and below), low temperature (−45°C to +10°C)	Freezing and refrigeration	Built-in refrigeration and freezing equipment	Residential and commercial	Freezing, refrigeration, drying, and warming for storage
	Freezing and refrigeration	Transport refrigeration and freezing units	Transportation	Freezing and refrigeration in cargo holds of vehicles, etc.
	Freezing and refrigeration	Condensing units	Commercial and industrial	Refrigeration in showcases, stockyards, etc. in convenience stores and supermarkets
Medium temperature (+10°C to +30°C)	Air conditioning	Residential air conditioners	Residential	Air conditioning in living rooms, bedrooms, etc.
	Air conditioning	Commercial packaged air conditioners	Commercial	Air conditioning for small and medium-sized businesses
	Air conditioning	Multi-air conditioner for buildings	Commercial	Air conditioning of relatively large commercial spaces
	Air conditioning	Vehicle air conditioners	Transportation	Air conditioning for vehicles

(Continued)

TABLE 1.1 *(Continued)*
Applications by Temperature Range and Sector of Use

Temperature Range	Use of the Product	Equipment	Sector	Outline
	Air conditioning and cooling	Chillers (scroll and screw)	Commercial/industrial	Commercial and factory air conditioning Cooling of equipment, etc.
	Air conditioning	Centrifugal chiller	Industrial	Air conditioning of factories and large facilities
High temperature (+30°C to +90°C)	Hot water supply and air conditioning	Exhaust heat recovery Hot water heat pumps	Industrial	Heat utilization of waste heat from factories to produce hot and cold water.
	Hot water supply and air conditioning	Heat pump water heaters	Residential and commercial	Heat utilization of air to hot water
	Drying/heating	Heat pump type hot air generator	Industrial	Drying and heating in factories
High temperature (+90°C to +200°C)	Steam	Steam generation heat pumps	Industrial	Steam supply in factories

An overview of each type of equipment is given below.

1.3.2.1 Built-in Refrigeration and Freezing Equipment

As shown in Figure 1.16, this refers to refrigeration and freezing equipment in which the constituent compressor, condenser, expansion valve, and evaporator are housed in (built into) a single structure. Examples include built-in showcases commonly found in supermarkets and convenience stores, and chillers in prefabricated warehouses.

FIGURE 1.16 Built-in refrigeration and freezing equipment. (From Panasonic Corporation, "Cooling unit top slim type" [12].)

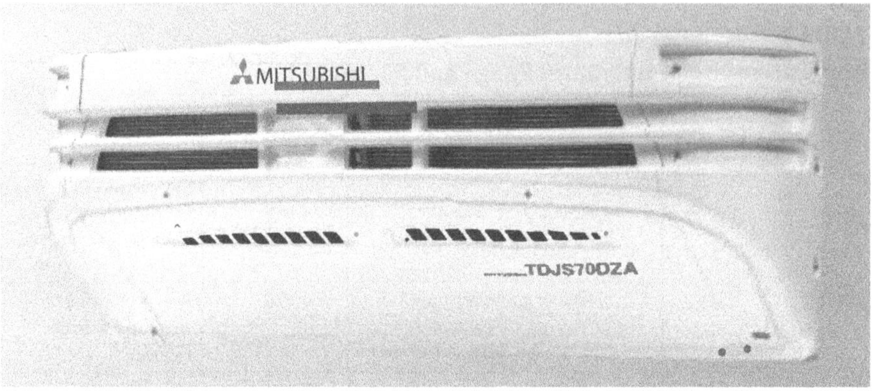

FIGURE 1.17 Transport refrigeration and freezing units. (From Mitsubishi Heavy Industries Thermal Systems Ltd, "Transportation refrigeration units" [13].)

Commercial refrigerators and freezers also fall into this category.

1.3.2.2 Transport Refrigeration and Freezing Units

As shown in Figure 1.17, refrigeration and freezing units are units that are installed in the cargo hold of trucks and other land, rail, sea, and air transportation to maintain a constant temperature.

1.3.2.3 Condensing Unit

Figure 1.18 shows refrigeration and freezing equipment consisting of the compressor and condenser, also commonly referred to as heat source units. Condensing units are not used alone but are connected to showcases commonly found in supermarkets and convenience stores[9], and are commercialized as new refrigeration and freezing equipment in combination with evaporators (including expansion valves).

FIGURE 1.18 Condensing unit. (From Panasonic Corporation, "Condensing unit/remote condensing units" [14].)

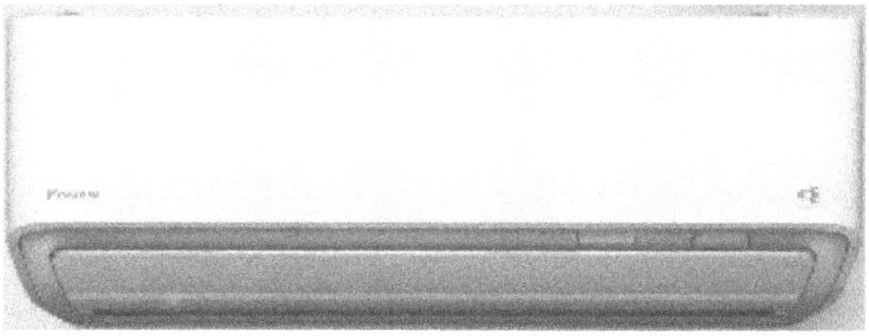

FIGURE 1.19 Residential air conditioning. (From Daikin Industries Ltd, "Residential air conditioners" [15].)

1.3.2.4 Residential Air Conditioners

Figure 1.19 shows a room air conditioner, which are generally used in homes and are also referred to as residential air conditioners. Systems in which separate outdoor units and indoor units are connected by piping are the most common type, and wall-hung air conditioners are commonly found in Japan.

1.3.2.5 Commercial Packaged Air Conditioners

Commercial air conditioners (Figure 1.20) refers to small and medium-sized air conditioners used in shops, offices, and factories. Various forms of commercial air

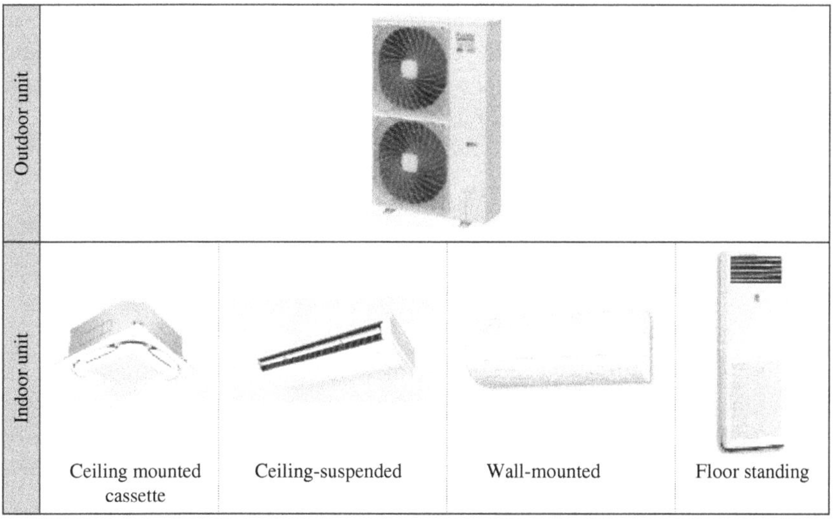

FIGURE 1.20 Commercial packaged air conditioners. (Outdoor and indoor ceiling-mounted and ceiling-suspended units based on Daikin Industries Ltd, "Commercial air conditioners (shop and office air conditioners) 'SkyAir' series" [16]. Indoor wall-mounted unit based on Daikin Industries Ltd, "Product features of wall-mounted type commercial air conditioners" [17]. Indoor floor-standing unit based on Daikin Industries Ltd, "Floor-standing type product features" [18].)

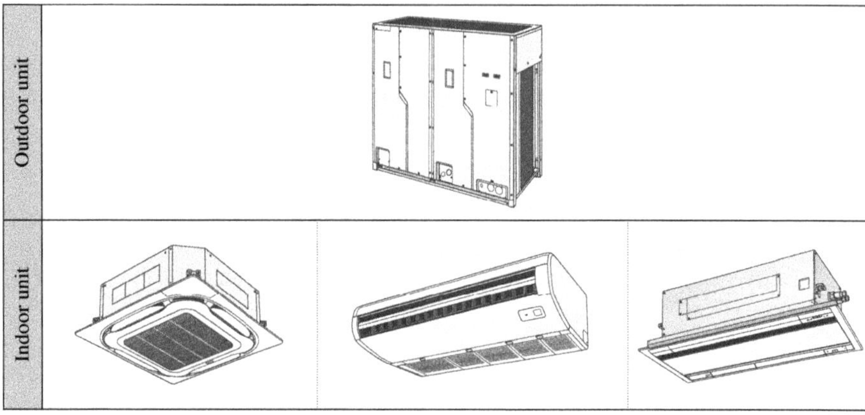

FIGURE 1.21 Multi-air conditioner for buildings. (Outdoor unit based on Daikin Industries Ltd, "VRVX series high COP type" [19]. Indoor units based on Daikin Industries Ltd, "Commercial multi-air conditioner for buildings" [20].)

conditioners are available, such as ceiling-mounted, ceiling-suspended, wall-mounted, and floor-mounted, to suit the installation location.

1.3.2.6 Multi-Air Conditioner for Buildings

As shown in Figure 1.21, these are air conditioners for buildings and other large facilities. They are larger air conditioners than commercial packaged air conditioners and are called multi-air conditioners because several (up to several dozen) indoor units are connected to one outdoor unit.

1.3.2.7 Vehicle Air Conditioners

Vehicle air conditioners provide air conditioning for the interior of standard passenger cars, buses, trucks, and other vehicles (see Figure 1.22).

1.3.2.8 Scroll and Screw Chiller

Scroll and screw chiller is a generic term for equipment that circulates a heat medium, such as water, to maintain a constant temperature in various equipment and spaces. While the term "chiller" suggests cooling, some chillers also perform heating. Depending on the form of compressor, they are also referred to as scroll chillers and screw chillers (Figure 1.23).

1.3.2.9 Centrifugal Chiller

A centrifugal chiller is a type of chiller equipment that uses a centrifugal compressor as the compressor. It is mainly used for cooling large equipment and facilities (see Figure 1.24).

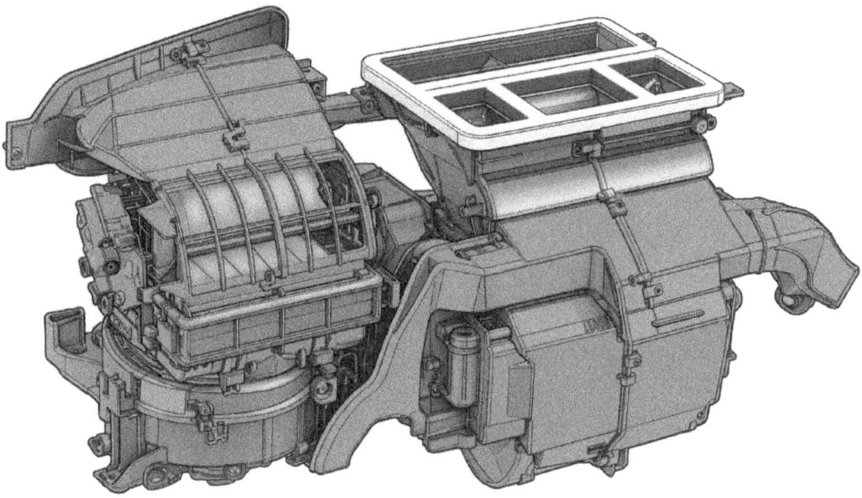

FIGURE 1.22 Vehicle air conditioner. (From Mitsubishi Heavy Industries Thermal Systems Ltd, "Car air conditioner" [21].)

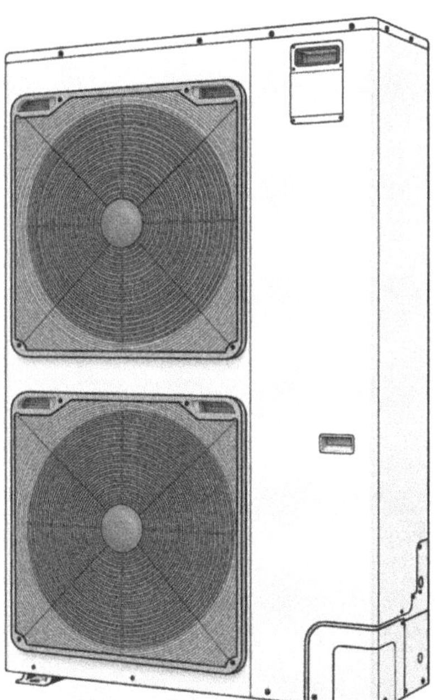

FIGURE 1.23 Chiller. (From Toshiba Carrier Corporation, General Catalogue of Heat Source Systems [22].)

FIGURE 1.24 Centrifugal chiller. (From Mitsubishi Heavy Industries Thermal Systems Ltd, "Centrifugal chillers" [23].)

1.3.2.10 Heat Pump Water Heaters

A water heating system uses a heat pump to boil water (see Figure 1.25). Of these, those using natural refrigerant as the refrigerant are called Eco-Cute in Japan.

FIGURE 1.25 Heat pump water heater. (From Mitsubishi Electric Corporation, "Mitsubishi Eco-Cute" [24].)

FIGURE 1.26 Industrial heat pumps. (From Maekawa Manufacturing Co. Ltd, "Heat pumps" [25].)

1.3.2.11 Industrial Heat Pumps: Heat Pump Type Hot Air Generator, Steam Generation Heat Pump, Etc.

Heat pumps are mainly used for air conditioning, cooling, drying, heating, sterilizing, and cleaning processes in manufacturing industries, such as machining and food processing, with a temperature range of −60°C to as high as 200°C (see Figure 1.26). For example, there are various types of heat pumps, such as waste heat recovery hot water heat pumps that produce hot water from waste heat from factories, heat pump hot air generators that produce hot air between 60°C and 90°C, and steam generation heat pumps as an alternative to boilers.

1.3.3 Classification of Equipment in Terms of Structure and Refrigerant Supply Method

Heat pump equipment is also classified according to its structure and configuration and the method of refrigerant distribution. Some of the most typical examples are described in this section.

First, as shown in Figure 1.27, an integrated type is a product in which the elemental components of the heat pump equipment, such as compressor, condenser, expansion valve, and evaporator, are incorporated into a single housing. The integrated type is used for localized cooling, such as in refrigerators, spot coolers, and showcases for freezing and refrigerating food.

The opposite of this integrated, built-in type is the separate installation, which is used not only in residential, shop, and office air conditioners, but also in large showcases for freezing and refrigerating foodstuffs.

The most basic combination of separate types is a 1:1 combination of an outdoor unit (condenser for cooling) and an indoor unit (evaporator for cooling).

Built-in type
(integrated type)

Separate type
(separately installed type)

FIGURE 1.27 Integrated type and separate type. ((Left) Based on Panasonic Corporation, "Cooling unit top slim type" [12]. (Right) Based on Daikin Industries Ltd, "Commercial air conditioners (shop and office air conditioners) 'SkyAir' series" [16].)

Residential air conditioners, for example, use this configuration and are referred to as single units.

It is possible to connect multiple indoor units to one outdoor unit, which is called multi-air conditioner, and combinations of up to 100 or more indoor units and multiple outdoor units are also possible.

Some showcases in convenience stores and supermarkets have several different types of showcases connected to a single condensing unit, operating at different set temperatures (Figure 1.28).

1.3.3.1 Inter-Expansion Type

Inter-expansion units are a type of heat pump in which a compressor, condenser, expansion valve, and evaporator are installed in an outdoor unit, refrigerant circulates in the outdoor unit, and a medium (air or brine: antifreeze) different from the refrigerant in the outdoor unit exchanges heat with the refrigerant and goes to the indoor unit, where it exchanges heat at a required temperature and circulates to exchange heat with the refrigerant in the outdoor unit again. This is equipment where the refrigeration cycle is completed in the outdoor unit (heat source unit) and is often found in relatively large refrigeration units, such as chillers, including centrifugal types.

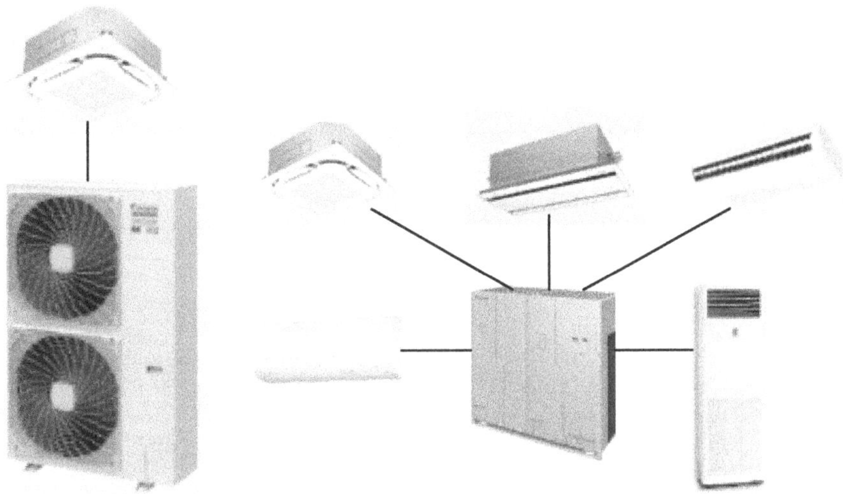

FIGURE 1.28 Single and multiple separate installations. (Based on Daikin Industries Ltd, (left) "Commercial air conditioners (shop and office air conditioners) 'SkyAir' series" [26]; (right) "VRVX series high COP type" [19]; "Commercial multi-air conditioners (multi air conditioner for buildings) indoor unit" [27]; "Product features of wall-mounted type commercial air conditioner" [17]; "Floor standing type product features" [18].)

Large centrifugal chillers, screw chillers, and centrifugal chillers are collected in one place, such as a machine room, and use a medium other than refrigerant to cool individual spaces (e.g. different floors) in a building or factory collectively. This is also called the central method.

It is suitable for keeping large spaces under constant conditions, but individual adjustment is difficult.

1.3.3.2 Direct Expansion Type

In contrast, the type in which the refrigerant itself goes from the outdoor unit to the indoor unit and circulates back to the outdoor unit again, as in separate type residential air conditioners to multi air conditioners, is called the direct expansion type. A system in which individual spaces are cooled separately by individual refrigeration equipment is an individual decentralized system and uses a direct expansion type. Although the system can be adjusted to meet individual demand, separate pipes are used for each appliance, which means longer pipes and more refrigerant, so installation space and measures to prevent refrigerant leakage are necessary.

The classification of the central method and individually distributed systems is often used for air conditioning, but there are also central and individually distributed systems for heat retention, cooling, and heating (Figure 1.29).

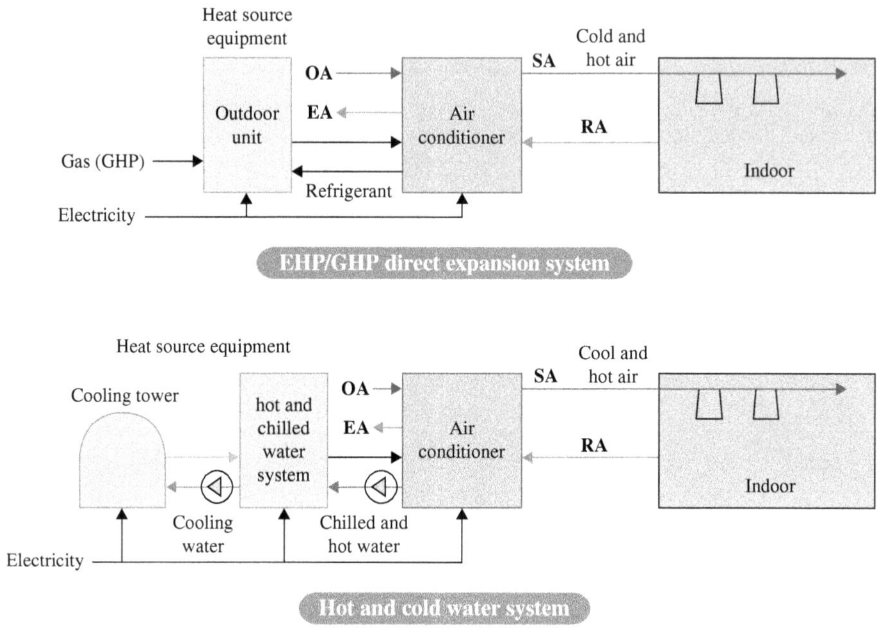

FIGURE 1.29 Refrigerant circulation system. (Top) Direct expansion type. (Bottom) Inter-expansion type. In the lower part, refrigerant circulates inside the heat source equipment and exchanges heat with chilled and hot water. (From Mitsubishi Heavy Industries Thermal Systems Ltd, "EHP/GHP Direct Expansion Air Han LX/GAP Series" [28].)

1.3.3.3 Methods Adopted for Next-Generation Heat Pump Technology

Direct expansion types are widely used in Japan because, although they are more difficult to control and the refrigerant and technical hurdles are higher, they can omit the process of heat exchange between different media, which offers overwhelmingly higher performance. Japan's adoption of this technically difficult direct expansion type has resulted in the world's highest performing equipment.

The difference in format due to construction is a very important point toward refrigerant conversion. This will be discussed below, but many of the next-generation refrigerants expected to be in use in the future will have negative aspects, including flammability and toxicity. In integrated and inter-expansion types, the potential risks associated with using flammable or toxic refrigerants are lowered because the refrigerant can be sealed on the side of the equipment installed outside the room, for example.

However, the direct expansion type requires these refrigerants to be brought directly indoors, so the equipment requires strict safety measures, so the technical level of the contractors connecting the outdoor unit to the indoor unit must also be high. Needless to say, it is also necessary to develop new connection methods that enable these units to be connected easily and safely without high technical skills.

1.3.4 Classification of Equipment by Heat Source

When heat is extracted from a heat pump, the type of heat source used in the low temperature side is important. In air conditioning, atmospheric air is used as a heat source, with heat taken from the atmosphere, which is almost inexhaustibly supplemented by the sun's energy, and a heat pump is thus defined as a renewable energy source in Europe, where its introduction is being promoted. While these efforts have contributed significantly to the proliferation of heat pumps, Japan tends to lag behind in its efforts to promote them, although regulations have been put in place to deal with the negative aspects of heat pumps. Expectations are high for proactive measures to promote heat pumps.

Water is used as a heat source when a large amount of heat exchange is required. This is because water has a higher heat transfer coefficient and heat exchange efficiency than air. Using water as a heat source also allows geothermal heat and waste heat from factories to be used, making it possible to use a variety of heat sources available at the heat pump location. It should be noted that the selection of heat sources is also a major issue in the introduction of industrial heat pumps, but if renewable energy and waste heat are used effectively, it is possible to build an extremely efficient heat utilization system.

Figure 1.30 shows two types of heat source and how they act as a kind of cooling system. Systems using water as the heat transfer fluid are sometimes referred to as water-cooled systems, while those using air are sometimes referred to as air-cooled systems.

The heat source has a significant impact on the performance of heat pumps, so its selection is extremely important. Air or water is often used as the heat medium and atmospheric heat is often used as the heat source, but unused heat utilization, such as geothermal heat or factory waste heat, is also being promoted to achieve even more effective use of heat. As already mentioned, it is important to search for unused heat utilization at higher temperatures, as a smaller temperature difference between the heat source as low temperature side and the high temperature side can reduce the power consumption required to drive equipment.

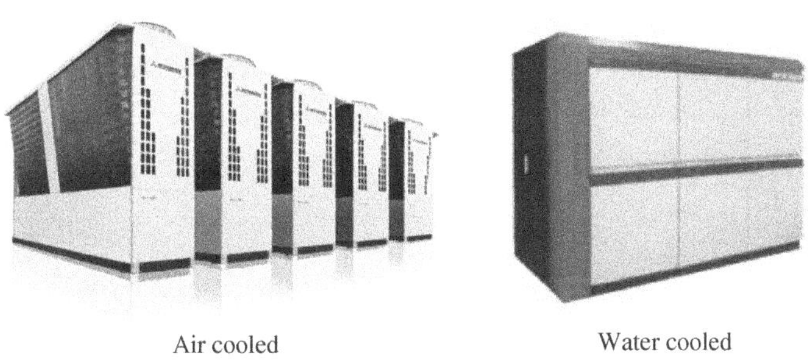

FIGURE 1.30 Air-cooled and water-cooled systems. ((Left) Based on Mitsubishi Heavy Industries Thermal Systems Ltd, "'MSV' air-cooled heat pump chiller" [29]. (Right) Based on Mitsubishi Electric Corporation, "Water-based central air conditioning" [30].)

1.4 MATERIALS SUPPORTING HEAT PUMP TECHNOLOGY

1.4.1 Thermal Conductivity of Structural Materials

In order to make the best use of the characteristics of heat pumps, it is necessary to select materials and refrigerants with excellent heat transfer properties.

In the tube walls of heat exchangers, heat is transferred by heat conduction, so the thermal conductivity of the metals used in the tube walls is important. The levels of thermal conductivity of the main metals are compared in Table 1.2. Thermal conductivity is an indicator of the ease of heat transfer; the higher this value, the easier it is for heat to be transferred. However, when selecting metals for use in heat pumps, not only those with high thermal conductivity are evaluated, but also material price, strength, workability, and whether or not the metal is corroded by refrigerant.

Generally, in products using fluorocarbon refrigerants, copper or aluminum is used for piping, which has high heat conductivity, while in the case of ammonia refrigerant, ammonia refrigerant corrodes copper and aluminum, making them unsuitable materials. Therefore, iron-based piping is used, even though it reduces thermal conductivity. When CO_2 refrigerant is used, the operating pressure is about four times higher than that of fluorocarbons. Therefore, materials and structures with higher strength are required, such as high-strength copper tubes and heat-transfer tubes with a smaller inner diameter. In addition to this, plastics will also be used extensively in air conditioners, for example, in covers.

1.4.2 Raw Materials for Refrigerants

Refrigerants used include fluorocarbon refrigerants and natural refrigerants. Fluorine is used as a raw material for fluorocarbon refrigerants, and fluorite is used as a raw material for fluorine. Lubricating oil is used to protect the rotating parts of a compressor, but a suitable lubricating oil must be selected to ensure it is compatible with next-generation refrigerants.

TABLE 1.2
Examples of Thermal Conductivity of Metals

Substance	Thermal Conductivity [W/(m-K)]
Copper	370
Aluminum	230
Steel	35–38

Source: Based on Japanese Society of Refrigerating and Air Conditioning Engineers, Advanced Standard Text Refrigeration and Air Conditioning Technology, Refrigeration Edition (4th Edition) [31].

1.4.3 Future Challenges for Structural Materials

When considering next-generation heat pump technology, not only is the performance improvement of the equipment already close to its limits, but also the performance of the next-generation refrigerant will often be reduced. As a result, the current situation is becoming such that the only way to improve equipment performance is to make the equipment even larger. The price of structural materials, copper and aluminum, are likely to continue to rise, and if equipment is made larger without sufficient progress in equipment recovery, this could lead to a significant increase in unit prices.

Significant global restrictions could also be placed on the use of plastics. The fluorite needed to produce refrigerants is also likely to be finite, and if it continues to be used without further recovery and recycling, significant shortages and price increases are also possible. From this perspective, the selection and effective use of the various materials used in the equipment will be a major factor in making next-generation heat pump technology sustainable.

1.5 TYPES OF REFRIGERANTS USED IN HEAT PUMP TECHNOLOGY

1.5.1 History of Refrigerants

As mentioned above, heat pumps can be used in various temperature ranges. The thermodynamic properties of the refrigerants suitable for these temperature ranges, such as working pressure, latent heat, and density, vary considerably, and a great number of refrigerants have been proposed (see Appendix 1 for details).

As refrigerants are originally found in nature, they started as low-temperature applications, mainly with natural refrigerants such as ammonia, which is considered to be environmentally friendly. However, ammonia is flammable and toxic and has caused many accidents, so a refrigerant conversion is called for.

Under these circumstances, CFCs, consisting of carbon, fluorocarbons, chlorine, and hydrogen, were discovered, and their molecular structures were developed one after another to suit different temperature ranges. They were hailed as dream refrigerants, and heat pumps became applicable to a wide range of temperature ranges, including not only low temperature use but also the generation of heat for air conditioning, hot water supply, and other purposes. The development of refrigerants thus left a significant mark on the progress of heat pumps.

1.5.2 Widespread Use of Refrigerants

CFCs refrigerants, which consist of chlorine, fluorine, and carbon, and hydrochlorofluorocarbons (HCFCs) refrigerants, which consist of hydrogen, chlorine, fluorine, and carbon, have good performance, are non-flammable and non-toxic, and so became widely used. However, in 1974, Dr Frank Sherwood Rowland et al. of the University of California discovered that the use of chlorine depletes the ozone layer[10] and this became a major social problem. This has led to a rapid shift to hydrofluorocarbon (HFC)-based refrigerants as chlorine-free alternatives (Figure 1.31).

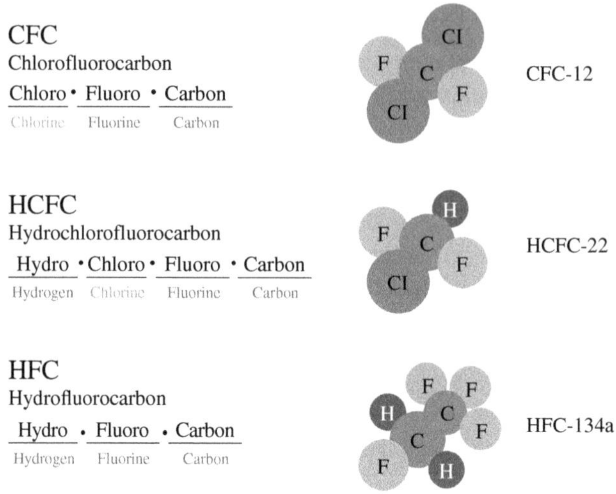

FIGURE 1.31 Basic knowledge of refrigerants. (From Saito Laboratory, Department of Mechanical Science and Aeronautics, School of Fundamental Science and Engineering, Waseda University, and International Institute for Environment and Energy Assessment, "What is a refrigerant?" [32].)

On the other hand, HFC refrigerants as this alternative refrigerant are greenhouse gases with an extremely large global warming effect, and there is a need to reduce their use.

Future refrigerants include hydrofluoroolefin (HFO) refrigerants, which are composed of carbon, fluorine, and hydrogen. These are not highly stable in the atmosphere due to the double bonds between the carbons and therefore decompose easily, and consequently have a low greenhouse effect. A return to natural refrigerants is being promoted.

1.6 HEAT PUMP TECHNOLOGY CONTRIBUTING TO DECARBONIZATION

1.6.1 The Importance of Heat Pump Technology in Decarbonization

For decades, heating was overwhelmingly based on direct combustion of fossil fuels, including in boilers, which are the main source of CO_2, and the situation is similar worldwide. Therefore, simply changing combustion equipment to heat pumps, as shown in Figure 1.33, can drastically reduce indirect CO_2 emissions, associated with electricity consumption can be significantly reduced.

In the future power generation, as the introduction of renewable energy sources will accelerate, simply converting to heat pumps powered by electricity, i.e. electrification, will also naturally lead to a reduction in CO_2 emissions.

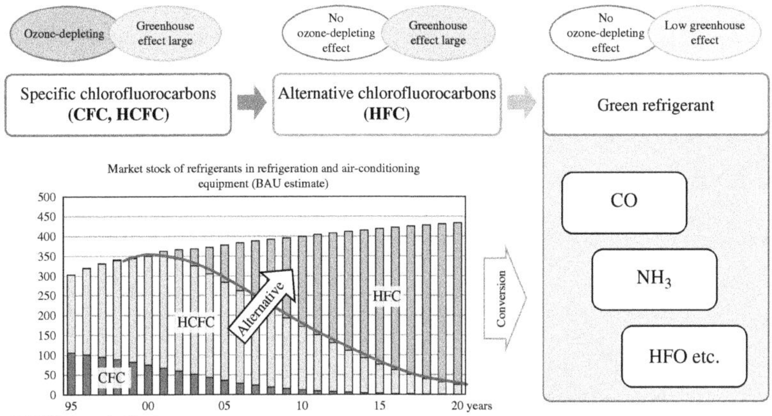

BAU: Business As Usual

The emission estimates for the fluorinated greenhouse gas sector reflect projections based on the continuation of current measures.
Source: Joint Meeting of the Subcommittee on the Examination of Post-2020 Global Warming Countermeasures of the Global Environment Committee, Central Environment Council, and the Subcommittee on Global Environmental Issues, Industrial Structure Council, Review of the Draft Commitment, Document 4

Status of Adoption of Alternative CFC Refrigerants and Green Refrigerants

*Color code: mildly flammable / Flammable

Area	Field	Current alternative refrigerants (GWP)	Green refrigerants as alternatives to HFCs	
1 Substitution is underway or is expected to proceed	Household refrigerator	(HFC-134a (1,430))	Isobutane	All new shipments have been converted to green refrigerants
	Vending machine	(HFC-134a (1,430)) (HFC-407c (1,770))	CO_2 Isobutane HFO-1234yf	
	Car air conditioner	(HFC-134a (1,430))	HFO-1234yf	Replacement is expected to proceed in the future
2 Candidate alternatives exist, but there are challenges to their widespread use	Ultra-low temperature freezer/refrigerator	HFC-23 (14,800)	Air	
	Large-capacity commercial freezer/refrigerator	HFC-404A (3,920) HFC-410A (2,090)	Ammonia, CO_2	The Ministry of the Environment provides support for implementation
	Medium-sized commercial freezer/refrigerator (freestanding display case)		CO_2	
3 Considering alternative candidates	Small commercial freezer/refrigerator	HFC-404A (3,920) HFC-410A (2,090)	(Currently evaluating alternative refrigerants)	
	Commercial air conditioners	HFC-410A (2,090) HFC-32 (675)	Development Support from the Ministry of Economy, Trade and Industry	
	Residential air conditioner	HFC-32 (675)		

FIGURE 1.32 Return to natural refrigerants and introduction of green refrigerants. (From Ministry of the Environment, Fluorocarbon Measures Office/Ministry of Economy, Trade and Industry, Ozone Layer Protection Promotion Office, "Status of alternative fluorocarbons and current initiatives" [33].)

For this reason, decarbonization is crucial, first of all, by promoting the expansion of heat pump applications and their widespread use. Some estimates suggest that the proliferation of heat pumps alone could have the effect of reducing CO_2 by 100 million tonnes per year by 2050. Although the expansion of applications in cryogenic range regions is also necessary, this is not a pressing issue in line with social needs and is therefore not specifically dealt with here.

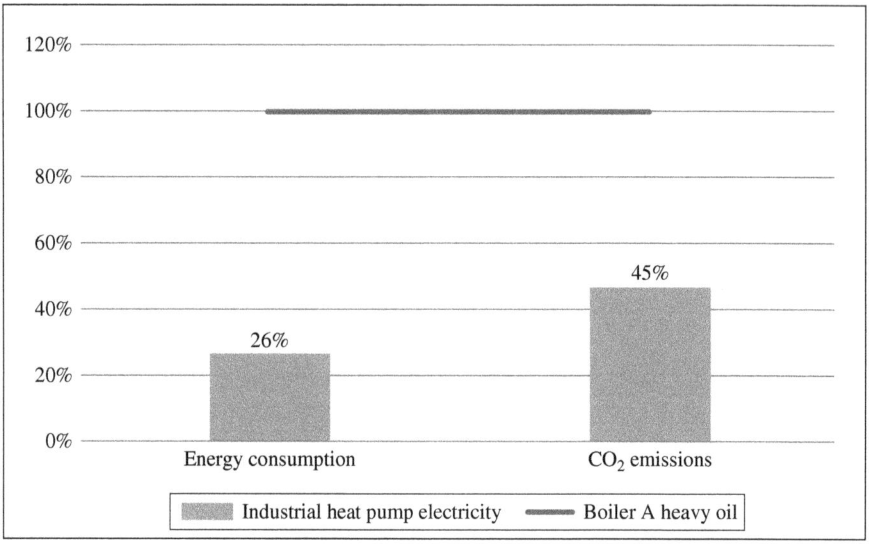

FIGURE 1.33 Comparison of CO_2 emissions between boilers and heat pumps. *Note:* 90% boiler efficiency, coefficient of performance (COP) 3.5 for industrial heat pumps, 0.0189 tCO_2/GJ emissions factor for heavy oil A, and 0.000434 tCO_2/kWh national average emissions factor for electricity (Mitsubishi Research Institute estimates).

On the other hand, in Japan, where the introduction of air conditioners (which is a representative heat pump technology) has already made steady progress, there has been a strong demand for saving energy in terms of electricity consumption, as the equipment as a whole consumes a large amount of electricity.

Air conditioners can thus be seen as a bad equipment that emits a lot of $CO_{2.}$ This may be due to the fact that air conditioners has been increasingly introduced as a cooler, but it should not be forgotten that in recent years the majority of air conditioners have been used for both cooling and heating, and that the electrification effect of heating with air conditioners is also enormous (Figure 1.34).

In the future, it is important to properly recognize heat pumps as equipment that utilizes air heat as a renewable energy source (Figure 1.34) as it is in Europe, and to promote the significant proliferation of heat pumps. This trend is beginning to be followed in the USA. It is to be hoped that such a development will also take place in Japan.

1.6.2 High Adaptability to Diverse Areas

1.6.2.1 Wide Range of Capacities and Temperature Ranges

The refrigeration and heating capacities of heat pumps are variously referred to as refrigeration capacity, cooling capacity, or heating capacity, depending on the application. The unit of this capacity is expressed as 'W'.

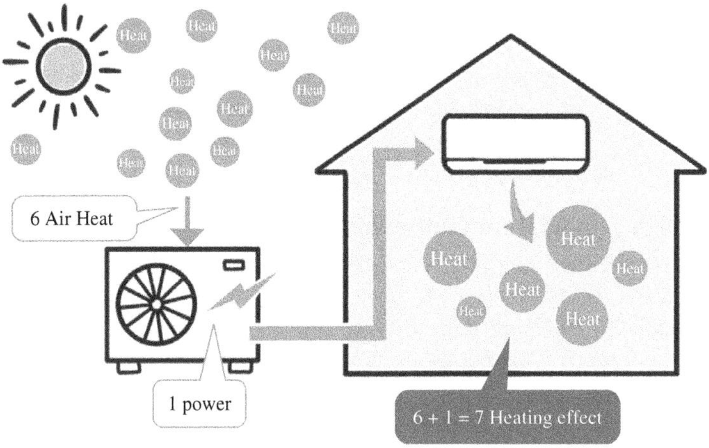

FIGURE 1.34 Heating effect of air conditioners. (From Daikin Industries Ltd, "Environment Lecture 01: Power consumption of air conditioners and global warming" [34].)

Figure 1.35 shows the range of refrigeration and heating capacities of heat pumps, with specific applications in different temperature ranges. Refrigeration and heating capacities can also be built from a few kW in the smallest applications to several thousand kW in the largest applications. Also, the applicable temperature range is very wide, from –100°C on the cooling side to nearly 200°C on the heating side.

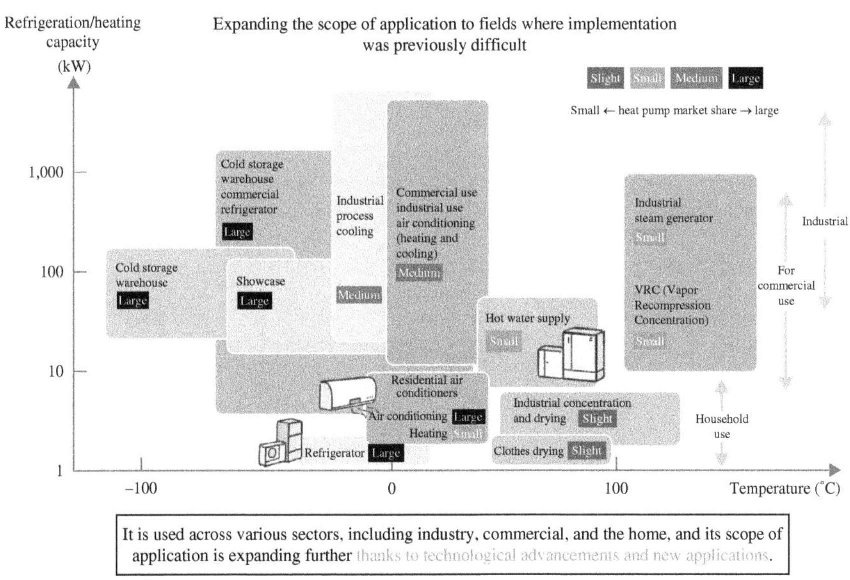

FIGURE 1.35 Capacity range of heat pumps by field of application. (From Heat Pump & Thermal Storage Technology Center of Japan, "Scope of use of heat pumps" [35].)

As heat pump technology is the only technology that can generate cooling and heating energy, its main application has been cooling and heating energy generation, but its application is expanding toward the high temperature side as well. This has created momentum to switch from combustion appliances using fossil fuels to heat pump appliances.

1.6.2.2 High Temperature Controllability

Heat pump technology not only allows the production of appliances with a wide range of capacities, as mentioned above, but also uses electricity, which can be highly variable to different capacities by means of inverter technology[11]. This means that the required temperatures can be appropriately adapted and CO_2 emissions can be minimized.

As shown in Figure 1.36, heat pump technology can generate and supply heat at the required temperature range on-site at high efficiency due to its reliable temperature controllability, whereas in the past, heat was generated and supplied intensively by boilers. In addition, when waste heat or untapped energy is available, the heat can be raised to the required temperature range with little electricity and even heat utilization can be achieved. Thus, heat pumps can generate heat at various temperatures with minimum energy use and minimum CO_2 emissions.

For example, in the future energy supply and demand system, a large amount of renewable energy will have to be used, but there will be a large gap between the time renewable energy is generated on the supply side and the time heat is utilized on the demand side. By combining heat pump technology with a relatively inexpensive thermal storage system, the time when renewable energy is used can be controlled by using renewable energy-derived power to operate the heat pump technology and storing it as heat energy in the heat storage tank. Note that renewable energy-derived electricity can be stored directly in storage batteries or other devices, but it is often not cost effective to do so.

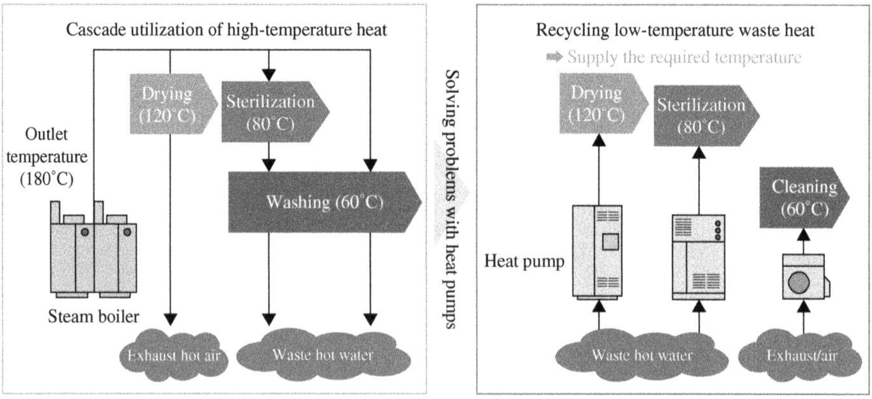

FIGURE 1.36 How heat pumps supply the required temperature for a process. (From Japan Electro-Heat Center, "Heat supply tailored to individual processes" [36].)

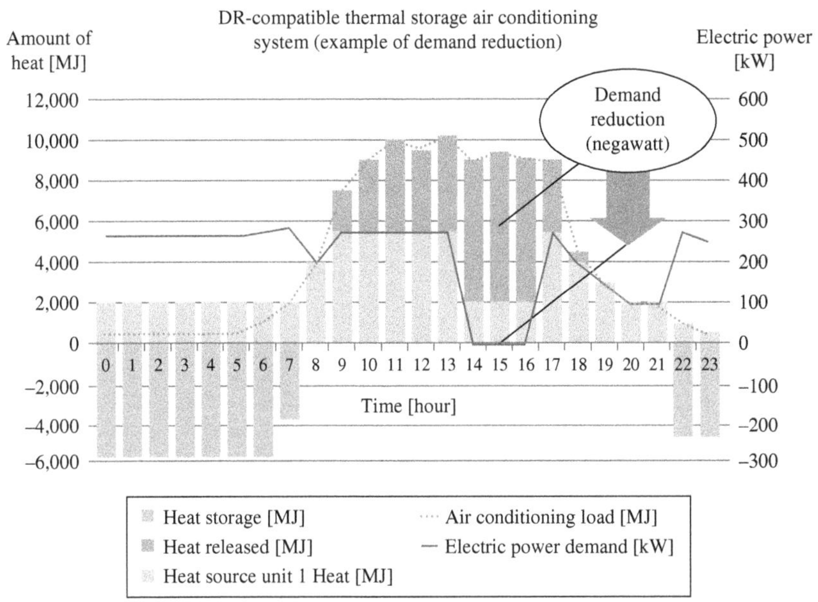

FIGURE 1.37 Time control by heat storage using heat pumps. (From Heat Pump & Thermal Storage Technology Center of Japan, "Heat pump and thermal storage systems are key equipment that contribute to energy saving and a decarbonized society—'Heat Pump and Thermal Storage Month' has started" [37].)

Until now, thermal storage systems have used cheap electricity at night for heat storage and used it for air conditioning and hot water during the day, as shown in Figure 1.37. In the future, the system is expected to play a role as a supply–demand adjustment function that contributes to the effective use of renewable energy.

As shown in Figure 1.38, generating heat on-site can reduce heat losses due to transportation. For example, heat losses and transportation power during large-scale heat transportation, such as steam piping, can be eliminated, making the total heat utilization extremely efficient.

Furthermore, equipment can be installed without spatial constraints, and if equipment installed in different locations is entirely controlled from the outside (total system) by an energy management system, it is possible to operate the equipment at higher performance. In this way, the conversion from a central method to a distributed control method facilitates heat control in a spatial manner.

1.7 DIVERSE APPLICATION EXAMPLES OF HEAT PUMP TECHNOLOGY

As mentioned above, heat pump technology has the advantage that it can be used for heating in a wide range of applications in a wide range of temperatures, with a wide range of capacities with low power consumption. Therefore, in recent years,

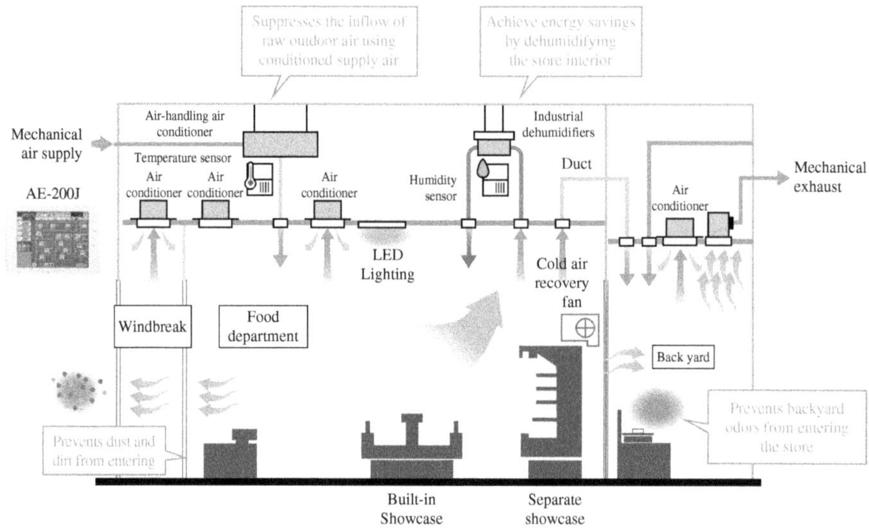

FIGURE 1.38 Decentralized arrangement and centralized control of heat pumps in a supermarket. (From Nihon Business Publishing Co. Ltd (Sendenkaigi Co. Ltd)/Kankyo Business, "Supermarket 'Life' saves 54,900 kWh of energy per year at new shop" [38].)

by combining it with various other technologies, it can be applied to diverse social purposes as life-saving technology or technology for realizing industrial development and stable energy supply.

For example, in air conditioning, air quality can be efficiently maintained and managed by combining it with other technologies, including humidity control, sterilization, disinfection, and lighting. In food systems, heat pump technology can be combined with showcase and automotive technologies to preserve food for longer periods of time, with less food waste.

It can also be used to realize heat processes in factories to enable decarbonization, and, in combination with heat storage, can be applied to demand response (DR), which will be necessary for the mass introduction of renewable energy in the future. The contribution to a stable energy supply is also expected to be significant.

This section therefore introduces specific examples of heat pump applications. Details on how these features can be developed in various social fields in the future will be presented in subsequent chapters.

1.7.1 Use in Daily Life

When heat pumps are used as air conditioning units in residential air conditioners, commercial air conditioners, and in large spaces such as concert halls, large shopping malls and airports, temperature control, humidity control, and ventilation can be combined to improve and maintain air quality and thus create comfortable spaces (Figure 1.39).

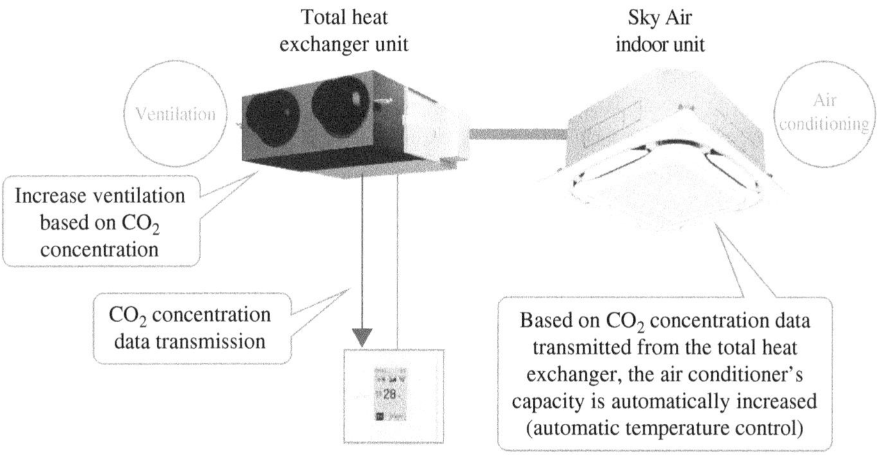

FIGURE 1.39 Example of interlocked control of ventilation and air conditioning. (From Daikin Industries Ltd, "Peace of mind with air quality monitoring" [39].)

This can lead to the mitigation of temperature shock, prevention of heatstroke, and support comfortable sleep, thereby maintaining health and improving work efficiency.

Heat pumps can easily be combined with equipment and devices that eliminate or reduce germs, viruses, and other pathogens, creating spaces that are appropriately sterile for people and food that are vulnerable to pathogens (Figure 1.40). For this reason, they are used in hospitals, retirement homes, clean rooms, plastic greenhouses, and food processing plants.

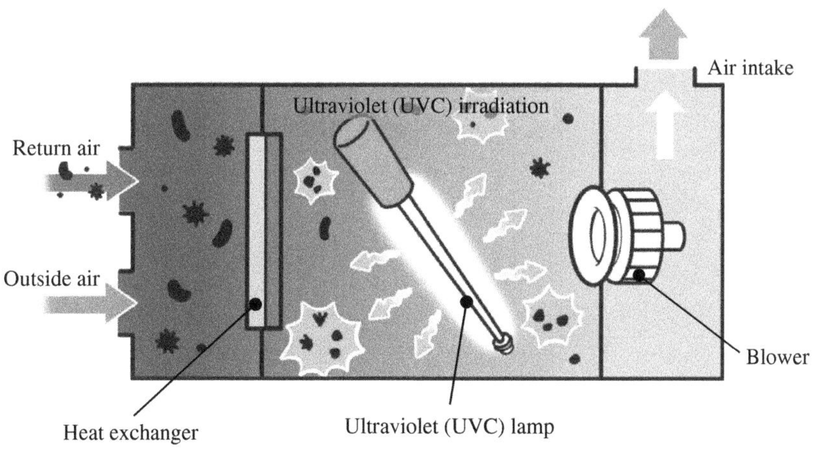

FIGURE 1.40 The removal of pathogenic bacteria using heat pumps. (From Shinko Industries Ltd, "Health air conditioning—Overview of sterilization system for facilities" [40].)

1.7.2 Use in Food Systems

1.7.2.1 Use in the Cold Chain

The use of appropriate freezing and refrigeration technologies in food processing plants, refrigerated warehouses, refrigerated transportation trucks and containers, and on fishing vessels, (including those in distant waters), can reduce the deterioration of the quality of foods that are easily damaged and have fixed expiry dates.

This results in greater food distribution and contributes to the food supply chain. In addition, food waste can be reduced.

Examples of use in freezing processes in food processing plants (Figure 1.41) and in refrigerated warehouses for freezing and cold storage (Figure 1.42) are shown below.

FIGURE 1.41 Example of a spiral freezer.

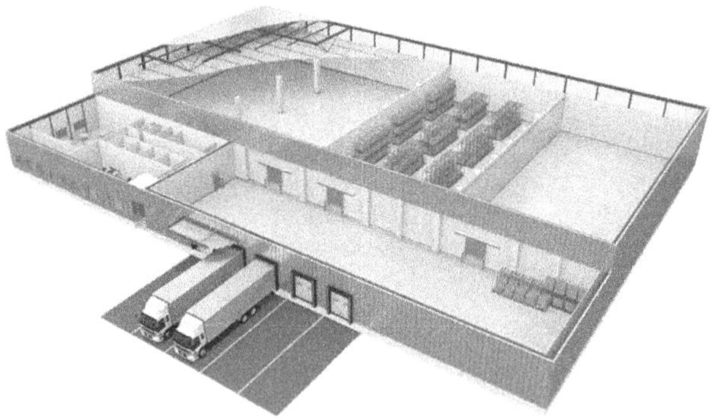

FIGURE 1.42 Example of a refrigerated/frozen warehouse. Nikkei Panel System Co. Ltd, "Freezing and refrigerating warehouses - Products comprising freezing and refrigerating warehouses" [41].)

FIGURE 1.43 Example of heat pump use in agriculture. (Based on E's Inc., "G Bazooka".)

1.7.2.2 Use in Agriculture

For plants, the heat pump system can be combined with sunlight to provide the temperature, humidity, and photosynthesis required by plants, and has become a key technology in agriculture (Figure 1.43).

1.7.2.3 Use in Aquaculture

Temperature control is also very important in aquaculture and other areas where heat pump technology is expected to be used (Figure 1.44).

FIGURE 1.44 Example of a heating system for eel ponds. (From Ministry of Agriculture, Forestry and Fisheries/Nichiman Co. Ltd, "INTERVIEW02 Project registration for introduction of eel pond heating system using heat pumps–significant CO_2 reduction achieved by wastewater heat utilization + introduction of heat pumps" [42].)

1.7.3 Use in Industrial Manufacturing Processes

In factories, the environmental conditions such as temperature and humidity in manufacturing processes need to be precisely controlled. In addition, when cooling or heating objects in manufacturing processes, the required temperature and time are strictly defined.

There is also a recent trend toward huge heat demands in some applications. In the case of semiconductor manufacturing processes, plant construction and expansion are repeated in 4- to 5-year cycles, creating heat demand in the order of 200 MWh when a new line goes into operation (Figure 1.45).

Heat pumps can be used for air conditioning and process cooling in such plants, and the high operability and controllability of the operation enables stable temperature and humidity control in production plants. This allows for the production of high-quality products and improved production efficiency.

Another feature of heat pumps is that they can utilize various types of waste heat as heat sources, such as factory waste heat, hot spring waste heat, and sewage waste heat. There are various methods of waste heat utilization, but one example is as an industrial heat pump (boiler): in the case shown in Figure 1.46, the low temperature waste heat from the boiler above is regenerated to a high temperature to supply hot air for drying at 120°C and hot water for washing at 60°C.

1.7.4 Use as a Regulating Force in the Electricity Supply and Demand System

The gap between the time for heat generation and heat utilization can be adjusted by retaining and storing the heat generated by heat pumps. Therefore, heat can also be used as a regulating force for power supply and demand by generating heat during times when there is a surplus of electricity supply, storing it in the heat storage tank, and using it during times when heat is to be utilized.

FIGURE 1.45 Example of a factory with huge heat demand (semiconductor factory). (Based on Kioxia Corporation.)

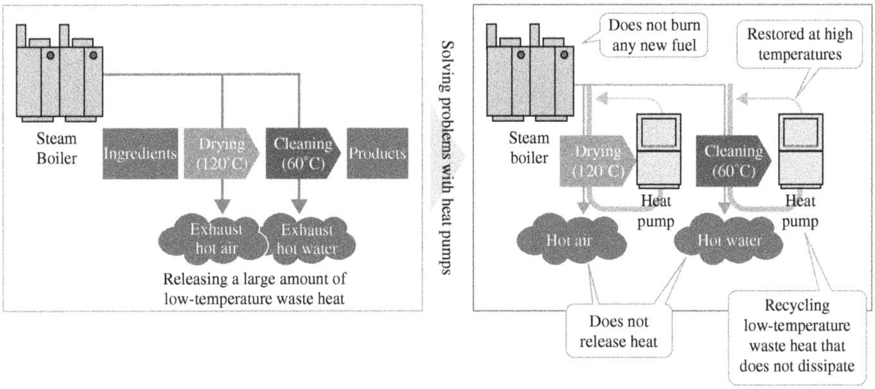

FIGURE 1.46 Example of a heat pump that recovers low temperature waste heat from a factory. (From the Japan Electro-Heat Center, "Industrial heat pump utilisation guide" [43].)

In the case of hot water supply, water has traditionally been heated using cheap night-time electricity, but recently the spread of solar power generation has been expanding, and the Ohisama Eco Cute[12] (Figure 1.47), which mainly uses surplus daytime electricity generated by household solar power generation, is being developed as a product.

1.7.5 Use in Large Facilities

The development of large-scale refrigeration equipment has made it possible to produce ice and artificial snow on a large scale (Figure 1.48). This has enabled ice skating and artificial ski slopes to be constructed outside of cold regions and utilized at event and attraction venues.

It is also used for medical equipment and facilities that require refrigeration and freezing equipment. The use of refrigeration technology in the transportation of COVID-19 vaccines is fresh in the memory.

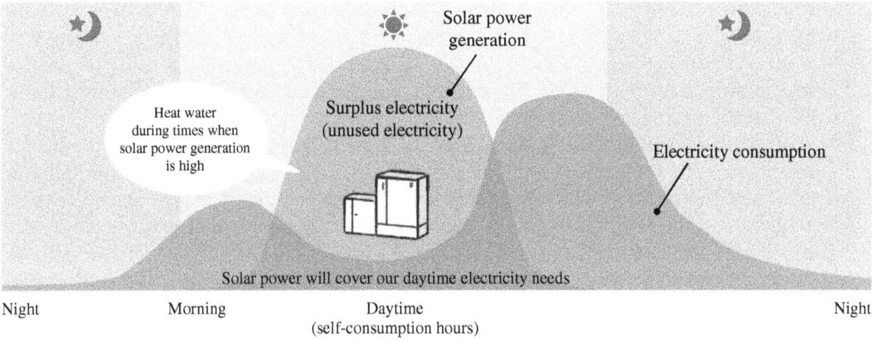

FIGURE 1.47 Example of Ohisama Eco Cute. (From Daikin Industries Ltd, "Ohisama Eco Cute" [44].)

FIGURE 1.48 Examples of an artificial snowfall machine. (From Iceman Corporation, "Product information snow accelerator" [45].)

NOTES

1. Vapor compression type: Heat cycle in which heat is transferred by electricity-driven compression and expansion of a refrigerant.
2. Absorption and adsorption types: Heat cycles in which heat is driven by heat and heat utilization is based on chemical reactions in the working fluid.
3. Phase change refers to the process of changing the state of matter between the three states (three phases) of gas, liquid, and solid.
4. Latent heat of vaporization refers to the heat required for a refrigerant to change from a liquid to a gas as heat of vaporization.
5. Superheated vapor: Superheated vapor refers to vapor that is further heated after evaporation to a temperature higher than the boiling point.
6. Capacity as a heat pump: Referred to as heating capacity or hot water heating capacity.
7. Definition of COP: In Figure 1.8, w corresponds to the change on the horizontal axis from (1) to (2), Qc from (2) to (3) for heating and heating and from (4) to (1) for cooling and cooling. Therefore, COP_h for heating/heating = COP_c for cooling/cooling$^+$ 1.
8. CO_2 emissions refers to energy-derived CO_2 generated from fossil fuels.
9. Connected to a showcase: A showcase and a refrigeration unit (condensing unit) are integrated type showcases, whereas a showcase and a refrigeration unit are prepared and combined separately, and are therefore referred to as separate type showcases.
10. Depletion of the ozone layer: Dr Rowland was awarded the Nobel Prize in Chemistry in 1995 for this discovery.
11. Inverter technology: Inverter technology that allows the speed of the motor to be freely changed and the output of the air conditioner to be freely controlled by switching the power semiconductors on and off. Energy-saving performance is significantly higher than models without inverters.

12. Ohisama Eco-Cute: "Ohisama Eco-Cute" is a product sold by Daikin Industries Ltd, Panasonic Corporation, and Corona Corporation. Ohisama Eco-Cute and Eco-Cute are registered trademarks of Kansai Electric Power Co.

REFERENCES

1. Heat Pump & Thermal Storage Technology Center of Japan, "Let's learn about heat pumps and thermal storage systems—What is a heat pump," "How heat pumps work," 2023, https://www.hptcj.or.jp/study/tabid/102/Default.aspx
2. Hitachi Global Life Solutions Inc., "I want to know how air conditioner works," 2024, https://kadenfan.hitachi.co.jp/support/ra/q_a/a19.html
3. Heat Pump & Thermal Storage Technology Center of Japan, "Highly efficient heat pump technology—Residential air conditioner efficiency APF (Annual Performance Factor): Is 7.0 or more!," 2023, https://www.hptcj.or.jp/study/tabid/104/Default.aspx
4. Yutaka Shibata, "Research and Development in High Performance of Heat Exchangers for Air Conditioners" (Japanese Society of Mechanical Engineers (JSME)), "No.07-5 Proceedings of the JSME Thermal Engineering Conference 2007," p. 25 Fig.1, 2007, https://www.jsme.or.jp/conference/tedconf07/backuphp/conf07proc/A13-shibata.pdf
5. Yutaka Shibata, "Research and Development in High Performance of Heat Exchangers for Air Conditioners" (Japanese Society of Mechanical Engineers (JSME)), No. 07-5 Proceedings of the JSME Thermal Engineering Conference 2007, p. 26 Fig.2, 2007, https://www.jsme.or.jp/conference/tedconf07/backuphp/conf07proc/A13-shibata.pdf
6. Yutaka Shibata, "Research and Development in High Performance of Heat Exchangers for Air Conditioners" (Japanese Society of Mechanical Engineers(JSME)), No.07-5 Proceedings of the JSME Thermal Engineering Conference 2007, p. 27 Fig.8, 2007, https://www.jsme.or.jp/conference/tedconf07/backuphp/conf07proc/A13-shibata.pdf
7. Japanese Society of Refrigerating and Air Conditioning Engineers, Advanced Standard Text Refrigeration and Air Conditioning Technology, Refrigeration Edition (4th Edition), p. 118, 2017.
8. Japanese Society of Refrigerating and Air Conditioning Engineers, Advanced Standard Text Refrigeration and Air Conditioning Technology (Refrigeration Edition), p. 129, 2017.
9. Japanese Society of Refrigerating and Air Conditioning Engineers, Advanced Standard Text Refrigeration and Air Conditioning Technology (Refrigeration Edition), p. 95, 2017.
10. Saginomiya Seisakusho Inc., Expansion Valve, "Temperature expansion valve/Medium type temperature expansion valve SCX," 2023, http://www.saginomiya.co.jp/auto/searchresult18.php
11. Saginomiya Seisakusho, Inc., Expansion Valve, "Electronic expansion valve VKV," 2023, https://www.saginomiya.co.jp/auto/searchresult21.php
12. Panasonic Corporation, "Cooling unit top slim type," 2023, https://panasonic.biz/appliance/cold_chain/prefab/pcu/topslim/
13. Mitsubishi Heavy Industries Thermal Systems Ltd. "Transport at refrigeration units," 2023, https://www.mhi-mth.co.jp/business/reefer-unit/
14. Panasonic Corporation, "Condensing unit/remote condensing units (R410A)," 2023, https://panasonic.biz/appliance/condense/
15. Daikin Industries, Ltd., "Residential air conditioners," 2023, https://www.ac.daikin.co.jp/roomaircon
16. Daikin Industries, Ltd., "Commercial air conditioners (store and office air conditioners) 'Sky Air' series," 2023, https://www.ac.daikin.co.jp/shopoffice
17. Daikin Industries, Ltd, "Product features of wall-mounted commercial air conditioners," 2023, https://www.ac.daikin.co.jp/shopoffice/products/indoor/wall_mounted

18. Daikin Industries, Ltd, "Floor standing type product features," 2023, https://www.ac.daikin.co.jp/shopoffice/products/indoor/floor
19. Daikin Industries, Ltd, "VRVX series high COP type," 2023, https://www.ac.daikin.co.jp/vrv/vrv_x_hicop
20. Daikin Industries, Ltd, "Commercial-use multi-air conditioner (multi-air conditioner for buildings) indoor unit," 2023, https://www.ac.daikin.co.jp/vrv/roomunit
21. Mitsubishi Heavy Industries Thermal Systems Ltd, "Car air conditioner," 2023, https://www.mhi-mth.co.jp/business/car_air_conditioning/
22. Toshiba Carrier Corporation, "Heat source system general catalogue," p. 157, 2022.
23. Mitsubishi Heavy Industries Thermal Systems Ltd, "Centrifugal chillers," 2023, https://www.mhi-mth.co.jp/business/centrifugal-chiller/turbo-freezer/
24. Mitsubishi Electric Corporation, "Mitsubishi Eco-Cute," 2023, https://www.mitsubishielectric.co.jp/home/ecocute/
25. Maekawa Manufacturing Co. Ltd, "Heat pumps," 2023, https://www.mayekawa.co.jp/ja/products/category/heat_pump/
26. Daikin Industries Ltd, "Commercial air conditioners (shop and office air conditioners) 'SkyAir' series," 2023, https://www.ac.daikin.co.jp/shopoffice
27. Daikin Industries Ltd, "Commercial multi-air conditioners (multi-air conditioner for buildings) indoor unit," 2023, https://www.ac.daikin.co.jp/vrv/roomunit
28. Mitsubishi Heavy Industries Thermal Systems Ltd, "EHP/GHP direct expansion air Han LX/GAP series," 2023, https://www.mhi-mth.co.jp/business/air-handling-unit/equipment/ehp-ghp/
29. Mitsubishi Heavy Industries Thermal Systems Ltd., "'MSV' air-cooled heat pump chiller," 2023, https://www.mhi-mth.co.jp/business/air-to-water/msv/
30. Mitsubishi Electric Corporation, "Water-based central air conditioners 'Water-based chiller for cooling only water-cooled compact cube'," 2023, https://www.mitsubishielectric.co.jp/ldg/ja/air/products/central/airchiller/lineup_03.html
31. Japanese Society of Refrigerating and Air Conditioning Engineers, "Advanced standard text refrigeration and air conditioning technology refrigeration edition," p. 18 (Table 2.1), 2017.
32. Saito Laboratory, Department of Applied Mechanics and Aerospace Engineering School of Fundamental Science and Engineering, Waseda University, and International Institute for Environment and Energy Assessment, Refrigerant Information Dissemination Website (w-refrigerant.com), "What is refrigerant?," 2019, https://w-refrigerant.com/knowledge/冷媒とは?/
33. Ministry of the Environment, Fluorocarbon Measures Office/Ministry of Economy, Trade and Industry, Ozone Layer Protection Promotion Office, "Status of alternative fluorocarbons and current measures," 2021, https://www.meti.go.jp/shingikai/sankoshin/seizo_sangyo/kagaku_busshitsu/flon_godo/pdf/010_01_00.pdf
34. Daikin Industries Ltd, "Environment lecture 01: Power consumption of air conditioners and global warming"—"Heating mechanism using heat pumps," 2023, https://www.daikin.co.jp/csr/information/lecture/lec03
35. Heat Pump and Thermal Storage Technology Center of Japan, "Scope of use of heat pumps," 2023, https://www.hptcj.or.jp/study/tabid/103/Default.aspx
36. Japan Electro-Heat Center, Comprehensive site for waste heat and unused heat utilization industrial heat pumps.com, "Heat supply tailored to individual processes," 2023, https://sangyo-hp.jeh-center.org/heatpump_heat_supply01.html
37. Heat Pump and Thermal Storage Technology Center of Japan, "Heat pump and thermal storage systems are key equipments that contribute to energy saving and a decarbonized society—'Heat pump and thermal storage month' has started," p. 4, 2022, https://www.hptcj.or.jp/Portals/0/data0/event/symposium/2022/202206271NewsReleaseInsatsuyou.pdf

38. Nihon Business Publishing Co. Ltd, Environmental Business, "Supermarket 'Life' saves 54,900 kWh of energy per year at new shop," 2021, https://www.kankyo-business.jp/news/028042.php
39. Daikin Industries Ltd, "Peace of mind through air quality mimamori," 2023, https://www.ac.daikin.co.jp/shopoffice/products/safety
40. Shinko Industries Ltd, "Health air conditioning - Overview of sterilization system for facilities," 2023, https://www.sinko.co.jp/product/health-air/
41. Nikkei Panel System Co. Ltd, "Freezing and refrigerating warehouses—Products comprising freezing and refrigerating warehouses," 2023, https://www.nikkeipanel.co.jp/products/cold-storage/
42. Ministry of Agriculture, Forestry and Fisheries, Nikkei Corporation Co. Ltd., "INTERVIEW02 Project registration for the introduction of a heating system for eel ponds using heat pumps—Significant reduction of CO_2 emissions by wastewater heat utilization + introduction of heat pumps," 2023, https://www.maff.go.jp/j/kanbo/kankyo/seisaku/climate/jcredit/jsusume/attach/pdf/top-13.pdf
43. Japan Electro-Heat Center, "Industrial heat pump utilisation guide," p. 4, 2017.
44. Daikin Industries Ltd, "Ohisama Eco Cute," 2017, https://www.ac.daikin.co.jp/sumai/alldenka/solar_ecocute_zenkoku
45. Iceman Corporation, "Product information snow accelerator," 2023, https://iceman.co.jp/product/194/

2 Social Conditions and Trends Surrounding Heat Pump Technology

OVERVIEW

As discussed in Chapter 1, heat pump technology has excellent features and has a significant role to play in realizing a sustainable society. On the other hand, there are many challenges, including the refrigerant problem.

This chapter provides an overview of the social demands that need to be addressed, mainly from a policy perspective, in order to realize next-generation heat pump technology and, furthermore, to build a sustainable society in which heat pump technology plays a central role (Figure 2.1).

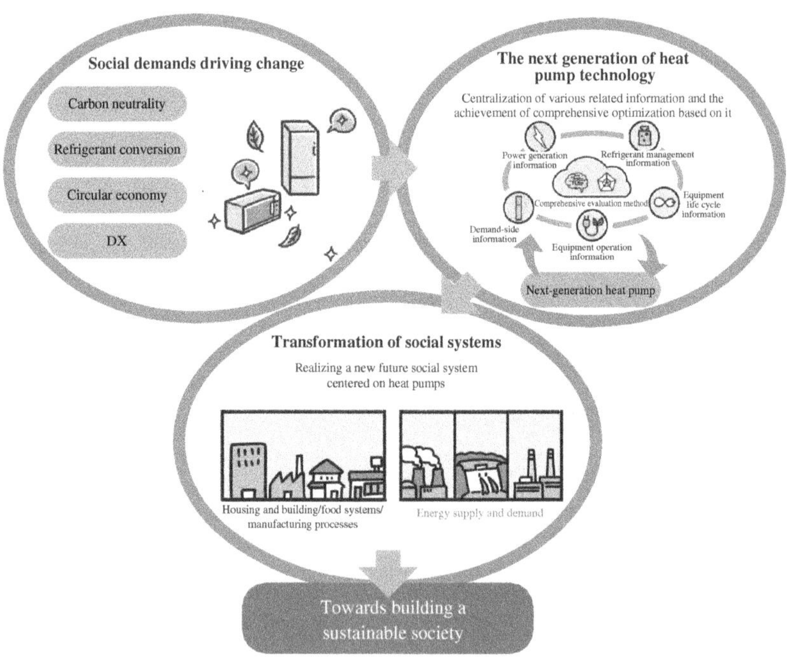

FIGURE 2.1 Toward the construction of a sustainable society.

DOI: 10.1201/9781003666950-2

2.1 CARBON NEUTRALITY AND HEAT PUMP TECHNOLOGY

2.1.1 WHAT IS CARBON NEUTRALITY?

Carbon neutrality is the goal of achieving virtually zero emissions of CO_2 and other greenhouse gases (GHGs) by balancing emissions with absorption and removal, as shown in Figure 2.2. The problem of climate change, which causes climate extremes such as heavy rainfall, heat waves, and droughts that occur frequently in many parts of the world, including Japan, is mainly caused by the massive emission of GHGs such as CO_2 generated by the combustion of fossil fuels like oil, natural gas, and coal, and the entire world is therefore working to achieve carbon neutrality Therefore, urgent action is required to become carbon neutral worldwide.

In Japan, the Government of Japan declared in October 2020 that it will aim to become carbon neutral by 2050. Following this, in April 2021, the government also raised its GHG reduction target for 2030 from 26% of the previous target for 2013 to 46%, and announced that it would continue its challenge to reach the 50% mark.

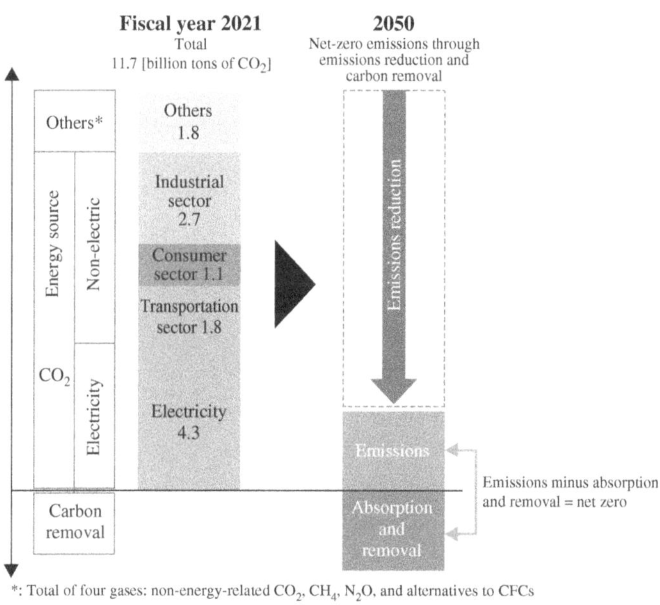

FIGURE 2.2 Japan's carbon neutrality achievement flow. *Note:* Based on actual emissions data for FY2021 from the National Institute for Environmental Studies' Greenhouse Gases Inventory Office, "Japan's Greenhouse Gases Emissions Data (FY1990–FY2021), Factual Data." (From Ministry of Economy, Trade and Industry (METI), Agency for Natural Resources and Energy, "What is 'carbon neutral'? (Part 1)—When and by whom will it be achieved?" [1].)

2.1.2 International Treaties for the Realization of Carbon Neutrality

As a response to the problem of climate change caused by GHGs, the United Nations Framework Convention on Climate Change (UNFCCC), adopted in 1992, agreed that the entire world should work toward stabilizing the atmospheric concentration of GHGs as the ultimate goal. Based on the Convention, the United Nations Framework Convention on Climate Change COP (Conference of the Parties) has been held annually since 1995 as a meeting where Member States regularly discuss specific policies on global warming. Member States are also obliged to report their specific policies and GHG emissions to the COP.

In November 2015, the COP21 adopted the Paris Agreement as a framework for all countries to participate in, entering into force in 2016. The long-term global target was set to limit the temperature increase from pre-industrial times to 2°C, and all countries were required to submit and renew their reduction targets every 5 years. The Glasgow Climate Accord at COP26 in Glasgow, UK in November 2021 raised this target to 1.5°C. In order to achieve the 1.5°C target, global CO_2 emissions need to be reduced to 45% below 2010 levels by 2030 and to virtually zero by around 2050. As shown in Figure 2.3, by the end of COP26, more than 150 countries, including Japan, pledged to achieve carbon neutrality by 2050.

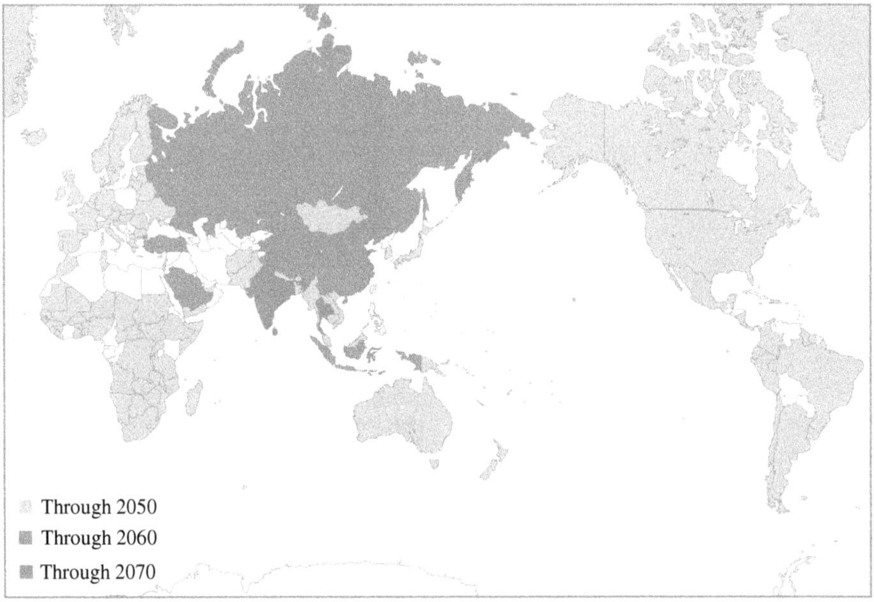

FIGURE 2.3 Countries and regions that have pledged to become carbon neutral (at the end of COP26). (From METI, Agency for Natural Resources and Energy, "Annual Report on Energy in 2021 (Energy White Paper 2022)" [2].)

2.2 DOMESTIC TRENDS TOWARD ACHIEVING CARBON NEUTRALITY

2.2.1 Japan's Measures toward Decarbonization

Japan's GHG emissions in FY2021 amounted to 1.17 billion tonnes in CO_2 equivalent, with energy-based CO_2 accounting for the largest breakdown by type at approximately 85% of the total, followed by non-energy-based CO_2 and hydrofluorocarbons (HFCs). The breakdown of CO_2 (energy- and non-energy-derived) emissions by sector is shown in Figure 2.4, with the industrial sector accounting for the largest share, especially in the non-power sector (heat and fuels).

In Japan as a whole, heat and fuels other than electricity account for about 70% of final energy consumption, with the industrial sector in particular accounting for about 80%. The combustion of fuels for the generation of steam, heating steam, or direct heating at the point of demand always results in energy-derived CO_2 emissions, so it is important to promote decarbonization of heat and fuel demand, which accounts for the majority of energy consumption, in order to achieve carbon neutrality.

Based on the current status of emissions, since the declaration of carbon neutrality by 2050 in October 2020, Japan has developed various strategies and plans to realize this goal, such as the Green Growth Strategy, which was further realized in June 2021.

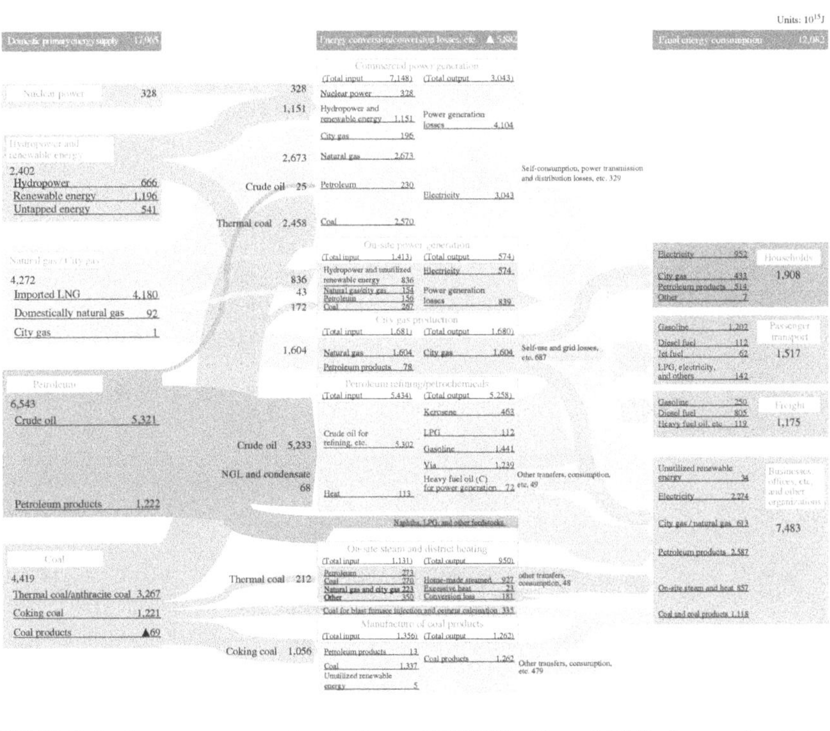

FIGURE 2.4 Energy consumption in Japan (FY2020). (From METI, Agency for Natural Resources and Energy, Annual Report on Energy in FY2021 (Energy White Paper 2022) [3].)

In addition, based on the new 2030 reduction target, the Sixth Basic Energy Plan and the Plan for Global Warming Countermeasures were approved by the Cabinet in October 2021, setting out the direction of specific energy and climate change policies. Furthermore, in May 2022, the Interim Clean Energy Strategy was compiled as a position paper to draw a feasible direction, linking the targets for 2030 and 2050.

2.2.2 Plans for Decarbonization in Different Sectors

As described above, various strategies and plans have been developed in a bid to decarbonize, but the basic path and direction of thinking toward achieving carbon neutrality in 2050 is common to all of the strategies and plans.

Specifically, as shown in Figure 2.5, the power sector will pursue innovations such as hydrogen and ammonia power generation and thermal power generation based on carbon capture and storage (CCS) and carbon storage and reuse (CCS/CCUS) through carbon recycling, while decarbonizing power sources through the maximum introduction of renewable energy. The following are some of the key issues to be addressed.

In the non-power sector (industrial, commercial and residential, transportation), electrification through thorough energy saving and decarbonized electricity will be promoted, while in sectors and areas where electrification is difficult (e.g. high temperature areas of heat demand in the industrial sector), decarbonization will be pursued through the use of hydrogen, synthetic methane, and synthetic fuels, while the remaining CO_2 emissions are to be reduced to practically zero through afforestation, direct air carbon capture and storage (DACCS)[1] etc.

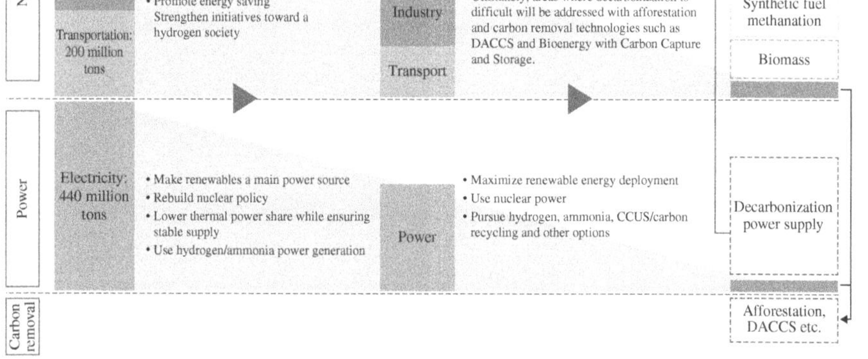

FIGURE 2.5 Image of transition to carbon neutrality. *Note:* The figures represent energy-derived CO_2 emissions only, but Japan's goal of carbon neutrality covers all GHGs, including methane, N_2O, fluorocarbons (FCs), and other gases. (From METI, "Green growth strategy associated with carbon neutrality in 2050" [4].)

2.2.3 Need for Demand-Side and Supply-Side Electrification

As described above, the main premise for achieving carbon neutrality is that decarbonization of supply-side power sources and electrification of the demand side should be pursued in both directions, and the policy is to use decarbonized fuels among other things in areas that are difficult to deal with through these means.

On the supply side, it is essential to expand the use of renewable energy, but this is likely to lead to greater fluctuations in output from naturally variable power sources such as solar and wind power. In order to ensure the stable operation of the power system, it is necessary to increase its flexibility by, for example, securing regulating power.

On the demand side, the policy directions (shown in Table 2.1) are based on sector-specific energy consumption characteristics. In all sectors, energy saving, electrification, and other energy conversions are positioned at the center of the policy, and heat

TABLE 2.1
Direction of Demand-Side Policies to Achieve Carbon Neutrality

Sector	Main Policies toward Carbon Neutrality	
Industry	Energy saving	Support for technology development and introduction to increase the diffusion of technologies with high energy-saving potential
		Review of institutional measures related to energy efficiency and conservation (e.g. benchmarking system, energy conservation law)
	Energy conversion	Low temperature area: Promote introduction of **industrial heat pumps**
		Medium to high temperature area: Support for development, demonstration, and introduction of decarbonized fuels (hydrogen, ammonia, synthetic fuels, etc.)
	Manufacturing process conversion	Support for development, demonstration, and introduction of innovative manufacturing processes, e.g. hydrogen-reduced ironmaking in the steel industry and artificial photosynthesis in the chemical industry
Consumer	Energy saving	Strengthening of energy efficiency and conservation regulations for housing and buildings (obligatory compliance with energy efficiency and conservation standards for small-scale buildings and housing, and gradual increase in standards to net zero energy building (ZEB)/net zero energy housing (ZEH)
		Further performance improvement and promotion of building materials and equipment
		Technical development, demonstration support, and institutional review to promote energy efficiency and conservation through digitalization
	Energy conversion	Development and promotion of technologies that contribute to reducing the cost of **electrification** and hydrogenation equipment and eliminating installation restrictions

(Continued)

TABLE 2.1 *(Continued)*
Direction of Demand-Side Policies to Achieve Carbon Neutrality

Sector	Main Policies toward Carbon Neutrality	
Transport	Energy saving	Steady implementation of fuel efficiency standards
		Review of the shippers' system under the energy conservation law
		Improvement of overall logistics efficiency through the use of digital technology
	Energy conversion	Promotion of the introduction of electric vehicles and infrastructure with the target of 100% electric vehicles in new car sales by 2035
		Support for large-scale and technological development of synthetic fuels

Source: Based on the METI, Agency for Natural Resources and Energy, "Study on the realisation of carbon neutrality in 2050" [5].

pump technology can be said to be a core technology in this context. In addition, the energy-saving performance of facilities and equipment other than heat pump technology, as well as of houses and buildings themselves, will also continue to improve, and heat pump technology will be required to work together and co-operate with them to achieve further energy saving.

2.2.4 Heat Pump Technology Measures for Decarbonization

The Sixth Basic Energy Plan, approved by the Cabinet in October 2021, sets out specific energy policy responses to realize the new GHG reduction targets for 2030 expressed in April 2021, based on the aforementioned direction.

In this context, heat pump technology should be introduced to promote energy saving and electrification in the non-power sector. It also indicates a direction to utilize heat pumps and other power-using equipment on the demand side in demand response (DR) as one of the regulating forces to increase the flexibility of the power system, in order to expand the introduction of renewable energy in the power sector.

The Sixth Basic Energy Plan also presents an ambitious supply–demand outlook for the FY2030, assuming that the various issues in energy supply and demand have been resolved through these policy measures (Figure 2.6). It aims to increase the share of renewable energy in the primary energy supply to 22–23% by FY2030, while the amount of energy saved on the demand side (reduction from before the measures) is to be 62 million kL in FY2030.

Like the Sixth Basic Energy Plan, the Global Warming Countermeasures Plan approved by the Cabinet in October 2021 was revised in response to the new reduction target for FY2030 expressed in April 2021. In this plan, reduction targets were set for each type of GHG (and for energy-based CO_2 by sector).

Table 2.2 shows the reduction targets by type of GHG, with the energy-derived CO_2, which accounts for approximately 80% of total GHG emissions, targeted to be

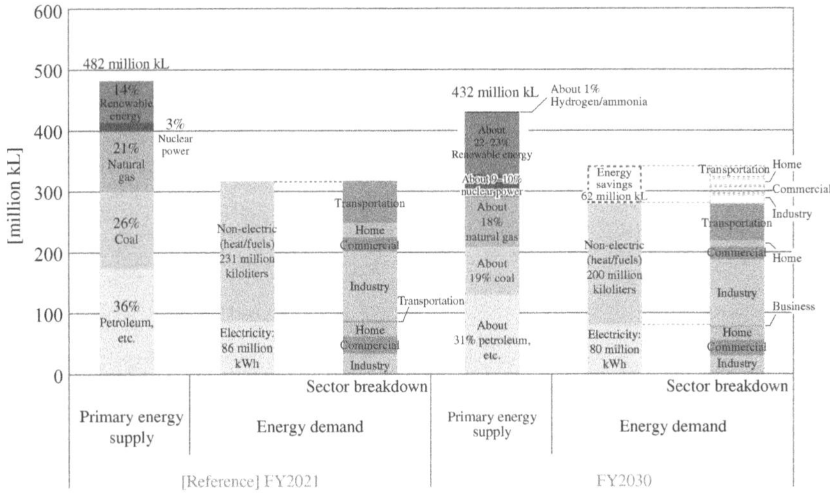

FIGURE 2.6 Outlook for energy supply and demand in FY2030 in the basic energy plan. (From METI, Agency for Natural Resources and Energy, "Comprehensive Energy Statistics Energy Balance Table (FY2021)" [6] and "Outlook for Energy Supply and Demand in FY2030 (Related Data)" [7].)

reduced by 46% compared to FY2013, with the residential and commercial sectors achieving 66% and 51% respectively, showing that all sectors have been strengthened compared to previous targets.

The Global Warming Countermeasures Plan also presents a list of measures and policies to support the new reduction targets mentioned above, which includes the promotion of the introduction of heat pumps as one of the measures and policies to reduce energy-based CO_2. Table 2.3 shows the specific prospects for the introduction of heat pumps, which are expected to increase significantly in each sector.

2.3 OVERSEAS TRENDS TOWARD ACHIEVING CARBON NEUTRALITY

As mentioned above, many COP Member States have declared themselves carbon neutral, and policies to realize this have been actively pursued in each country. This section describes specific decarbonization policy trends in European Union (EU) and the USA, as well as the position of electrification and heat pumps in this context.

2.3.1 Europe's Carbon Neutral Policies

As shown in Table 2.4, in EU, relevant legislation is being amended to achieve a GHG reduction of at least 55% by 2030 compared to 1990 levels in order to achieve the target of achieving climate neutrality by 2050.

In May 2022, the REPowerEU, a plan to break away from dependence on Russian fossil fuels, partly due to the invasion of Ukraine, was announced, based

TABLE 2.2
GHG Reduction Targets in the Global Warming Countermeasures Plan (Unit: million t-CO_2)

	FY2013 actual	FY2019 actual (compared to FY2013)	Targets/guidelines for FY2030 (compared to FY2013)
Greenhouse gas emissions/sequestration	1,408	1,166(▲17%)	760(▲46%)
Energy-derived carbon dioxide (CO_2)	1,235	1,029(▲17%)	677(▲45%)
Industrial sector	463	384(▲17%)	289(▲38%)
Commercial and other sectors	238	193(▲19%)	116(▲51%)
Residential sector	208	159(▲23%)	70(▲66%)
Transportation sector	224	206(▲8%)	146(▲35%)
Energy conversion sector	106	89.3(▲16%)	56(▲47%)
Non-energy origin CO_2	82.3	79.2(▲4%)	70.0(▲15%)
Methane (CH_4)	30	28.4(▲5%)	26.7(▲11%)
Dinitrogen monoxide (N_2O)	21.4	19.8(▲8%)	17.8(▲17%)
4 gases such as alternative fluorocarbons (FCs)	39.1	55.4(+42%)	21.8(▲44%)
Hydrofluorocarbons (HFCs)	32.1	49.7(+55%)	14.5(▲55%)
Perfluorocarbons (PFCs)	3.3	3.4(+4%)	4.2(+2.6%)
Sulphur hexafluoride (SF_6)	2.1	2.0(▲4%)	2.7(+27%)
Nitrogen trifluoride (NF_3)	1.6	0.26(▲84%)	0.5(▲70%)
Greenhouse gas sinks	–	▲45.9	▲47.7
Joint Crediting Mechanism (JCM)	Aim to achieve international emission reductions and absorption of around 100 million t-CO_2 cumulatively by FY2030 through public–private partnerships. The credits earned by Japan will be counted appropriately for the achievement of Japan's nationally determined contribution		

Source: Based on the Ministry of the Environment's, "Global Warming Countermeasures Plan" [8].

TABLE 2.3
Prospects for the Introduction of Heat Pumps in the Annex to the Global Warming Countermeasures Plan

Sector		Heat Pumps	Cumulative Introduction Forecast	
			FY2013	FY2030
Industry	Across industries	Industrial heat pumps	11,000 kW	1,673,000 kW
	Facility horticulture	Heat pumps for institutional horticulture	–	+26.7,000 units
Commercial		Commercial heat pump water heaters	29,000 units	140,000 units
Residential		Heat pump water heaters	4.22 million units	15.9 million units

Source: Based on the Ministry of the Environment's Global Warming Countermeasures Plan [8].

TABLE 2.4
Decarbonization Policies, Electrification, and Heat Pump Policy Trends in Europe

Policies and key measures for decarbonization	**European Green Deal** Proposed as a policy document to realize "A Clean Planet for All" with the aim of achieving climate neutrality by 2050.
	Fit for 55 package Adopted as a comprehensive climate change policy package to achieve the 2030 and 2050 targets. Broadly divided into carbon pricing, target updates, rule revisions and support measures, strengthening and amending existing instruments such as the EU Emission Trading Scheme and introducing new policies such as border carbon adjustment measures.
	REPowerEU Presented as a plan to move away from dependence on Russian fossil fuels, in light of the global energy market turmoil caused by Russia's invasion of Ukraine. Additional measures for energy saving, diversification of energy supply, and accelerated diffusion of renewable energy are presented.
Key electrification and heat pump-related policies	**Renewable Energy Directive** Defines atmospheric heat utilized by heat pumps, etc. for heating, hot water supply, and cooling applications as renewable energy, but a revised directive, which stipulates a method for calculating the amount of renewable energy during cooling in addition to the existing method for calculating the amount of renewable energy during heating, will enter into force in June 2022.
	Energy Performance of Buildings Directive (EPBD) The directive sets out the direction of progress in replacing fossil fuel heating equipment with heat pumps and other renewable energy-based technologies, and is expected to promote decarbonization and electrification of both new and existing buildings following the development of the revised directive.
	Comprehensive energy system strategy As part of the European Green Deal, the strategy sets out integrated measures for all sectors of the energy system, from the supply side to the demand side, toward efficient decarbonization, with electrification as a core measure among the specific measures outlined. Electrification through the deployment of heat pumps for heating and cooling in buildings will play a central role as one of the concrete measures, and heat pumps are positioned as a key technology, as they are expected to be used for district heating and cooling and for industrial demand for low temperature heat.
	REPowerEU As part of the above-mentioned measures, subsidies for fossil-fueled boilers in buildings will be phased out by Member States by 2025, and incentive support for heat pumps will be encouraged instead to double the current installation rate of heat pumps and to achieve a target set to reach a cumulative total of 10 million units.

Source: Based on New Energy and Industrial Technology Development Organization (NEDO)/Technology Strategy Center (TSC)/Global Technology Research Unit, "TSC Foresight: Trends in Major Countries and Regions (the US, China, the EU, the UK, Germany, Indonesia, Egypt, India) Regarding Carbon Neutrality in the Run-Up to COP27—From the Perspectives of Renewable energy and Fossil Energy" [9]. Heat Pump & Thermal Storage Technology Center of Japan, "Related policies and diffusion status of heat pumps in Europe and the United States" [10].

on the policy package "Fit for 55" adopted in July 2021, energy saving, diversification of energy supply, and renewable energy promotion. Additional measures were proposed to accelerate the process. As part of this, subsidies for fossil-fueled boilers in buildings will be phased out in Member States by 2025. Instead, incentive support for heat pumps was to be encouraged, with the aim of doubling the current installation rate of heat pumps to a cumulative total of 10 million units over 5 years.

Thus, since Russia's invasion of Ukraine, Europe's policy of focusing on the expansion of renewable energy and energy saving and electrification in order to break away from dependence on Russian fossil fuels has become more clearly defined. In this context, heat pumps have become even more important and are actively supported.

In addition, Europe's other heat pump-related policies clearly state that the definition of renewable energy includes atmospheric heat used by heat pumps, as well as specific calculation methods for the amount of renewable energy utilized. This positive evaluation of heat pumps as a renewable energy utilization technology is considered to be one of the factors driving their widespread use.

In Japan, the definition of renewable energy sources in the Act on the Upgrading of the Energy Supply Structure also includes atmospheric heat utilized by heat pumps. However, unlike in Europe, this is not assessed as renewable energy use, as the method for its calculation has not been defined. In addition, non-fossil energy sources were also added to the scope of the Energy Conservation Law with the 2022 amendment, and its use is being promoted to businesses operators. However, natural heat subject to quantitative assessment is limited to supply-constrained solar, geothermal, hot spring, and snow and ice thermal energy, while atmospheric heat and river water heat are not covered.

2.3.2 USA's Carbon Neutral Policy

The US withdrew from the COP's Paris agreement under the Trump administration, but returned in 2021 under the Biden administration, submitting an updated version of its nationally determined contribution (NDC[2]) to the UN in April 2021, which calls for a 50–52% reduction in GHG emissions in 2030 compared to 2005 levels, and net zero is to be achieved by 2050. As a long-term strategy to achieve the 2050 target, the power sector is to decarbonize (achieving zero CO_2 emissions from power generation) by 2035, while the industrial sector is to be electrified. In the building sector, the strategy calls for improving the energy efficiency of buildings and promoting the use of electric products such as heat pump air conditioners and heat pump water heaters.

One of the specific measures being implemented is the Inflation Reduction Act, which aims to increase the spread of so-called building electrification by including tax credits and rebates to encourage the purchase of energy efficiency equipment such as heat pumps instead of gas and oil heating and hot water supply. The Infrastructure Investment and Jobs Act also provides a lot of funding for energy infrastructure and decarbonization-related activities (Table 2.5). In California, a leading state in this area, there has been some discussion of banning the use of gas furnaces and gas water heaters.

TABLE 2.5
Decarbonization Policies in the USA: Electrification and Heat Pump Policy Trends

Policies and key measures for decarbonization	**Inflation Reduction Act** Includes a wide range of measures, including a $60 billion US domestic manufacturing fund for wind turbines, solar panels and electric vehicles (EVs), a regional electricity decarbonization fund and loans, $73.6 billion in government purchases of clean products, $27 billion in conversion to climate-smart agriculture, and tax credits for consumers' purchases of eco-friendly heating and cooling equipment, solar equipment and other products, which declares that this will fund a 40% reduction in carbon emissions by 2030.
	Infrastructure Investment and Jobs Act Invests a total of $88 billion in energy infrastructure and decarbonization, in addition to transport, digital, and water infrastructure. Allocates much of the money to making power infrastructure stronger and smarter, building hydrogen hubs, battery storage and vehicle electrification, zero emissions, nuclear technology, and develop nuclear technology.
Key electrification and heat pump-related policies	Long-term strategy toward net zero GHG emissions by 2050 specifies the importance of dramatically improving energy efficiency in buildings and increasing the sales share of electrical products such as heat pump air conditioners, heat pump water heaters, induction cookers, and electric clothes dryers.

Source: Based on New Energy and Industrial Technology Development Organization (NEDO)/ Technology Strategy Center (TSC)/Global Technology Research Unit, "TSC Foresight: Trends in Major Countries and Regions (the US, China, the EU, the UK, Germany, Indonesia, Egypt, India) Regarding Carbon Neutrality in the Run-up to COP27—From the Perspectives of Renewable energy and Fossil Energy" [9]. Heat Pump & Thermal Storage Centre of Japan, "Related policies and diffusion status of heat pumps in Europe and the United States" [10].

2.3.3 China's Carbon Neutral Policy

China announced in a speech to the UN General Assembly in September 2020 that President Xi Jinping will peak out CO_2 emissions by 2030 and achieve net zero GHG emissions by 2060. "China's National Voluntary Contribution Effectiveness, New Targets and New Measures" submitted in October 2021. "China's Mid-Century Long-term Low Greenhouse Gases Emissions Development Strategy," officially stated in the Strategy for the Development of Long-term Low Greenhouse Gases, updated the NDC and implementation measures.

Measures in China have previously focused on reducing coal consumption and increasing the use of non-fossil energy sources, including renewable energy. The NDC also states, among other things, that the share of non-fossil fuels in primary energy consumption should be 25% by 2030. In addition, in October 2021, an Action Plan for Carbon Peak by 2030 was formulated, setting specific measures and numerical targets for 10 cross-cutting areas. One of the 10 areas, for example, is the Action

Plan for Carbon Peak in the Industrial Area, in which electrification of industrial energy is listed as one of the priority initiatives.

2.3.4 Carbon Neutral Policies in Southeast Asia

Southeast Asian countries are generally set to achieve carbon neutrality of GHG emissions by 2050 or as soon as possible, with GHG emissions peaking in 2030 (Table 2.6).

Energy consumption in Southeast Asia has increased significantly due to large population growth and rapid economic growth. Cooling energy consumption has also increased significantly and is expected to continue to do so in the future due to economic growth and the associated increase in incomes, as well as the effects of rising temperatures. Analysis by the International Energy Agency (IEA) estimates that the number of air conditioning units in the Association of Southeast Asian Nations (ASEAN) countries will reach 3 billion units by 2040, up from 40 million

TABLE 2.6
GHG Reduction Targets for Major Countries in Southeast Asia (as of August 2023)

Country	Nationally Determined Contribution (NDC)	Carbon Neutral/Net Zero Target Deadline
Indonesia	43.2% reduction compared to BAU ratio*1 conditional*2 by 2030	2060
Cambodia	41.7% reduction compared to BAU ratio*1 by 2030	2050
Singapore	Reduce emissions to approx. 60Mt-CO_2 by 2030	2050
Thailand	40% reduction compared to BAU ratio*1 conditional*2 by 2030	Carbon neutral by 2050, net zero by 2065*3
Philippines	75% reduction compared to BAU ratio*1 conditional*2n by 2030 (72.29% for conditionalities)	–
Brunei	20% reduction compared to BAU ratio*1 by 2030	–
Vietnam	43.5% reduction compared to BAU ratio*1 conditional*2 by 2030	2050
Malaysia	Reduce emissions per GDP by 45% of 2005 levels by 2030	2050

*1 BAU ratio is a concept that refers to the effect compared to the natural case (business as usual) where no special measures are taken, and means the emissions ratio in the case where no additional measures are taken.

*2 Conditional refers to cases where financial, technical, or other support is obtained from developed countries.

*3 In Thailand, carbon neutral is a target to achieve virtually zero CO_2 emissions, while net zero is a target to achieve virtually zero GHG emissions overall.

Source: Based on UNFCCC, "NDC registry" [11] and "Long-term strategies portal" [12].

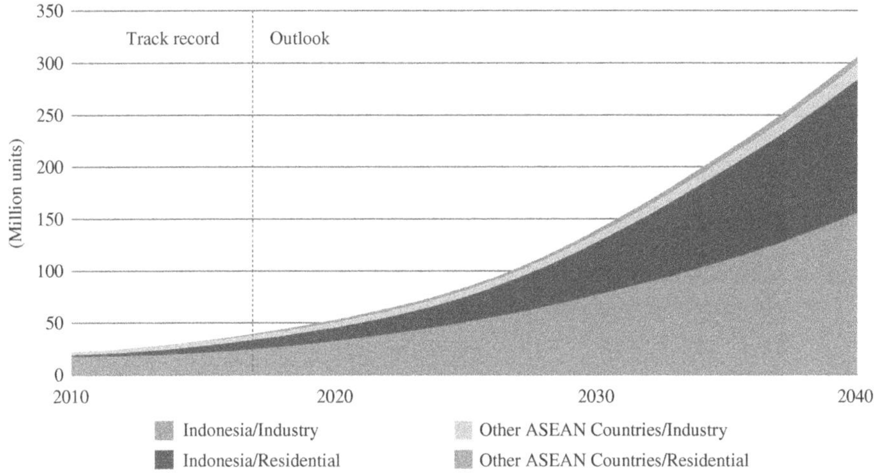

FIGURE 2.7 Outlook for air conditioner stock in ASEAN countries. (From IEA, "The future of cooling in Southeast Asia" [13].)

in 2017, and that the amount of energy required for cooling will accordingly increase from 75 TWh in 2017 to 300 TWh in 2040 (Figure 2.7).

Thus, energy saving through improved air conditioner efficiency and increased use of renewable energy to decarbonize the electricity used to meet this cooling demand are crucial to achieving decarbonization in Southeast Asia.

As described above, the main prerequisites for achieving carbon neutrality in any country or region are the expansion of renewable energy on the supply side, and energy saving and electrification on the demand side. Heat pump technology is one of the core technologies.

On the other hand, measures against refrigerants are important as a prerequisite for GHG reduction through heat pump technology, and as unique regulations have been introduced, these must be addressed.

2.4 THE NEED FOR ENVIRONMENTALLY FRIENDLY REFRIGERANT CONVERSION

2.4.1 Refrigerant Conversion to Achieve Carbon Neutrality

The current social requirement for refrigerant conversion is to convert those with a high environmental impact to those with a lower environmental impact. Initially, conversion was promoted to avoid the ozone layer depletion effects of using FC refrigerants, but now alternative FCs are used that do not deplete the ozone layer. On the other hand, it has been found that refrigerants have a significant impact on global warming, and there is a renewed focus on reducing the use of these alternative refrigerants.

Ironically, alternative refrigerants are being sought to replace HFC refrigerants, which are currently the mainstream alternative refrigerants, with the ongoing conversion to natural refrigerants or HFO refrigerants[3]. It is also a reminder of the difficulties of dealing with chemicals.

This demand for refrigerant conversion can also be seen as part of the prevention of global warming and move toward carbon neutrality. The refrigerant issue is also one of the biggest problems of heat pump technology.

2.4.2 Domestic and International Trends on Refrigerant Conversion

2.4.2.1 International Treaties: From Ozone Layer Protection to Global Warming Prevention

FCs are a type of GHG and are used as refrigerants in heat pump technology. They are also ozone layer-depleting substances. Their production, consumption, and trade are regulated under the Vienna Convention for the Protection of the Ozone Layer—an international treaty adopted in 1985.

Specifically, the Montreal Protocol, adopted in 1987 on the basis of the Vienna Convention, stipulates that CFCs used as refrigerants in air conditioner have already been phased out and that HCFCs, which replaced CFCs, are to have been phased out by 2020 in developed countries and by 2030 in emerging countries. Furthermore, HFCs, which have been used as a substitute for HCFCs, have recently become an issue of global warming impact, and it was decided in October 2016 to establish an obligation to reduce their production and consumption within the framework of the Montreal Protocol. This decision, known as the Kigali Amendment, entered into force in January 2019.

The reduction schedule for production and consumption of HFC reductions in the Kigali Amendment is shown in Table 2.7. For the group of developed countries to which Japan belongs, the target is to achieve an 85% reduction, compared to the base amount, by 2036. Japan's FC reduction targets set out in the aforementioned global warming countermeasures plan and the growth strategy based on the Paris Agreement are based on the assumption that this Kigali Amendment will be steadily implemented.

TABLE 2.7
HFC Reduction Schedule under the Kigali Amendment

	Developed Countries	Developing Countries Group 1 (e.g. China)	Developing Countries Group 2
Base year	2011–2013	2010–2022	2024–2026
Reduction schedule	2019: ▲10%	2029: ▲10%	2032: ▲10%
	2024: ▲40%	2035: ▲30%	2037: ▲20%
	2029: ▲70%	2040: ▲50%	2042: ▲30%
	2034: ▲80%	2045: ▲80%	2047: ▲85%
	2036: ▲85%		

Source: Based on the METI, Ozone Layer Protection Promotion Office, "Summary of the Montreal Protocol and Kigali Amendment" [14].

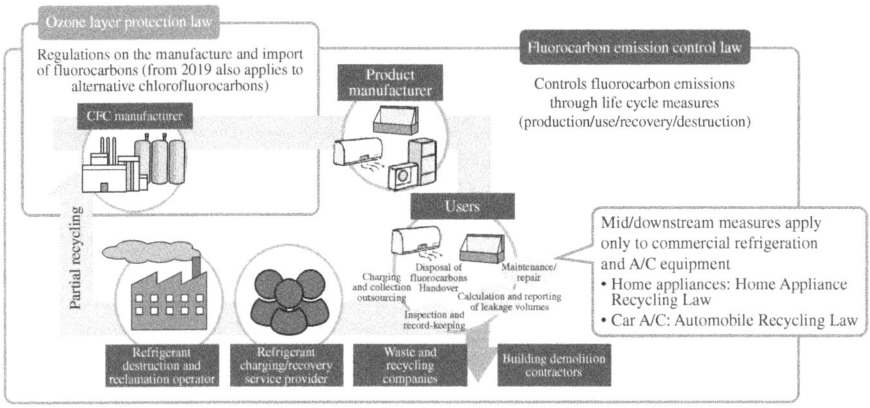

FIGURE 2.8 Overview of fluorocarbon measures in Japan. (From Ministry of the Environment and METI, "Overview of trends surrounding fluorocarbons and the revised Fluorocarbons Emissions Control Law" [15].)

2.4.2.2 Japanese Policy: Fluorocarbon Emissions Control Law

In Japan, the Ozone Layer Protection Law to reduce the production and consumption of CFCs, based on the Montreal Protocol and the Fluorocarbons Emissions Control Law to control the emissions of CFCs throughout their life cycle, is in force. In this context, the law restricts the use of high-global warming potential (GWP) refrigerants, which have a significant greenhouse effect, and encourages refrigerant conversion (Figure 2.8).

Under the Ozone Layer Protection Law, the METI authorizes the manufacture and import of controlled FCs within the annual limits set under the Montreal Protocol.

In addition, the Fluorocarbons Emissions Control Law requires manufacturers[4] to achieve a target GWP value[5] for designated products in a target year in order to promote low GWP and non-fluorocarbon products (Figure 2.9, Table 2.8).

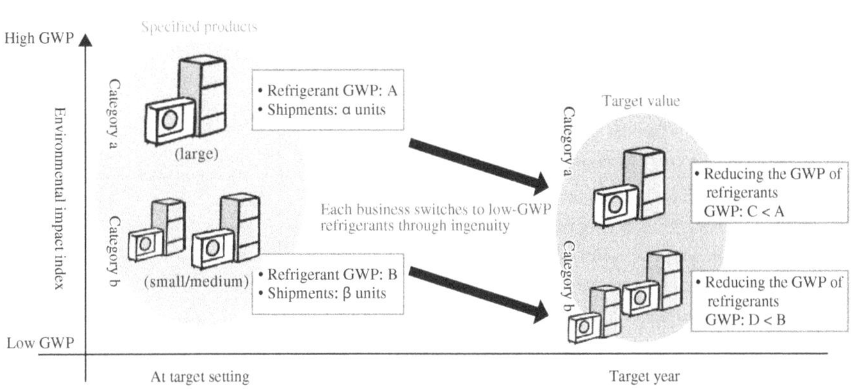

FIGURE 2.9 Overview of the designated product system in the Fluorocarbons Emissions Control Act. (From METI, "Regulations for designated products manufacturers" [16].)

TABLE 2.8
Products Subject to the Designated Product System under the Fluorocarbons Emissions Control Act (Refrigeration and Air Conditioning Equipment) (2023)

Classification of Designated Products	Major Refrigerants Currently Used and GWP	Target Environmental Impact	Target Year
Residential air conditioners (excluding through-the-wall type, etc.)	R410A(2090) R32(675)	750	2018
Air conditioning equipments for shops and offices			
1. Those with a legal refrigeration capacity of less than 3 tonnes, excluding floor-mounted types, etc.	R410A(2090) R32(675)	750	2020
2. Those with a legal refrigeration capacity of 3 tonnes or more, excluding floor-standing type, etc., and excluding (3) to (6) above	R410A(2090)	750	2023
3. Central air conditioning equipments that use centrifugal chiller	R134a(1430) R245fa(1030)	100	2025
4. Central air conditioning equipments that use positive displacement compressor chillers (chilling units for air conditioning)	R410A(2090)	750	2027
5. Multi-air conditioner for buildings (limited to those involving the installation of new equipment and renewal of refrigerant piping; excluding those of simultaneous cold/heat operation type and those for cold climates)	R410A(2090)	750	2025
6. Gas engine driven heat pump air conditioners (limited to those involving new installation and renewal of complete refrigerant piping, excluding those of simultaneous cold/heat operation and those for cold climates)	R410A(2090)	750	2027
7. Air conditioners for installations (limited to those involving new installation and renewal of complete refrigerant piping systems; excluding equipment for specific purposes such as for computers, medium-temperature use, integrated type, etc.)	R410A(2090)	750	2027

(Continued)

TABLE 2.8 *(Continued)*
Products Subject to the Designated Product System under the Fluorocarbons Emissions Control Act (Refrigeration and Air Conditioning Equipment) (2023)

Classification of Designated Products	Major Refrigerants Currently Used and GWP	Target Environmental Impact	Target Year
Air conditioning systems for vehicles			
Limited to those installed in passenger vehicles (except those with a seating capacity of 11 or more)	R134a(1430)	150	2023
Limited to those installed in trucks (used for transportation of goods) and buses (with a seating capacity of 11 or more)	R134a(1430)	150	2029
Condensing units and stationary refrigeration and freezing units (excluding, for example, those with a compressor rated output of 1.5 kW or less)	R404A(3920) R410A(2090) R407C(1770) CO_2 (1)	1500	2025
Commercial integrated type refrigeration and freezing equipment (built-in, small refrigeration equipment)			
Commercial refrigeration-freezers (excluding those where the lower limit of the evaporation temperature of the refrigerant in the evaporator is less than −45°C)	R404A(3920) R410A(2090) R407C(1770) CO_2 (1)	150	2029
Showcases (limited to compressors with a rated output of 750 W or less)		150	2029
Refrigeration and freezing equipment with rigid polyurethane foam	HFC-245fa(1030) HFC-365mfc(795)	100	2024
Vending machines with refrigeration or freezing functions using rigid polyurethane foam	HFC-245fa(1030) HFC-365mfc(795)	100	2024
Central system refrigeration and freezing equipment (limited to those shipped for new refrigerated warehouses with an effective volume of 50,000 m^3 or more)	R404A(3920) Ammonia (single digit)	100	2019

Source: Based on the METI, "Regulations for designated product manufacturers, etc." [16].

Furthermore, as indicated in the previous Section 2.2.4, in order to reduce FC emissions, the global warming countermeasures plan sets a reduction target for HFCs of approximately 55% of the FY2013 level by FY2030.

On the other hand, safety considerations need to be taken into account in refrigerant conversions. Refrigerants become high-pressure gases in compressors and are therefore subject to the High Pressure Gas Safety Act in order to ensure safety.

The High Pressure Gas Safety Act requires the deployment of refrigeration safety officers and the implementation of safety inspections, depending on the hazardous nature of the high-pressure gas. It sets separate categories and actions for safety standards[6] depending on whether the gas is an inert gas, flammable gas[7], or toxic gas[8] (Figure 2.10).

As the use of flammable gases has increased with refrigerant conversions, the less flammable gases are separated from flammable to mildly flammable gases after safety assessment. This is the so-called A2L refrigerant[9]. Although standards have been partially relaxed for mildly flammable gases such as HFCs, flammable gases such as propane are the most strictly regulated.

2.4.2.3 European Policy: F-Gas Regulation

In Europe, the F-Gas Regulation (Regulation (EU) No. 517/2014), which started to apply in 2015 with the aim of protecting the environment by reducing GHG (F-gas) emissions, provides for refrigerant regulation: to regulate the total amount of HFCs by weighted addition according to GWP values in the region.[10] A quota system for manufacturers and importers was introduced to allow the sale of only the quota amount. The total volume will be phased down every 3 years.

In April 2022, the European Commission proposed new regulations targeting the real phase-out of F-gas emissions toward carbon neutrality in 2050. This includes a revised phase-out reduction and a limit on the GWP value of the refrigerant to be enclosed in each product group. Deliberations are currently underway on the content of the proposal, with the European Environment Committee taking the lead in consolidating opinions, and a policy is expected to be finalized by the end of 2023.

In addition, restrictions on chemical substances that affect human health have been strengthened and, in particular, restrictions on fluorine-containing compounds (Perfluoroalkyl and Polyfluoroalkyl Substances (PFAS)) have been established in the EU as the Restriction of the Use of Certain Hazardous Substances in Electrica and Electronic Equipment (RoHS) regulations and the Registration, Evaluation, Authorisation and Restriction of Chemicals (REACH) regulations. In the EU, there is a move toward comprehensive regulation of PFAS within the REACH regulations [18].

2.4.2.4 US Policy: The SNAP Program

In the USA, the US Environmental Protection Agency (EPA) in 2021, when it returned to the Paris Agreement, issued a final rule establishing a comprehensive program to limit and phase out the production and consumption of climate change-causing HFCs in the USA. The rule will phase out the production and consumption of HFCs in the US by 85% over the next 15 years, as required by the American Innovation and Manufacturing (AIM) Act, which was passed in December 2020.

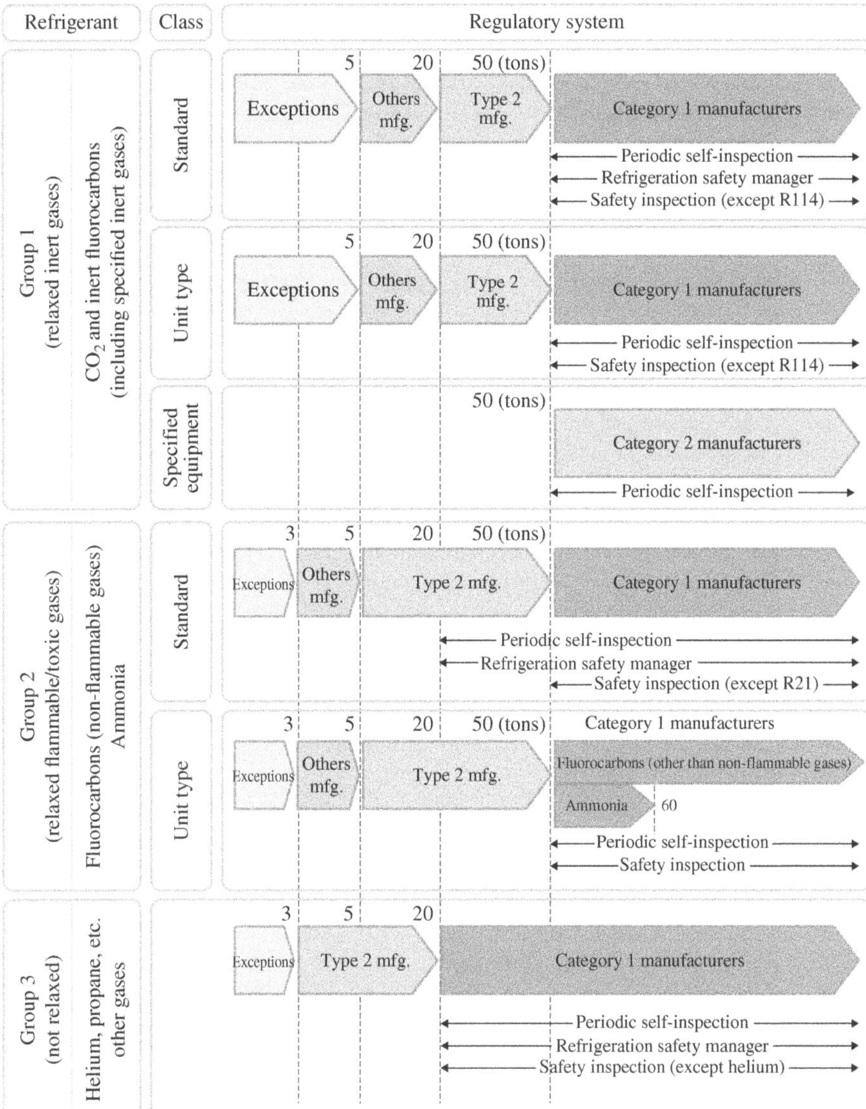

FIGURE 2.10 List of regulation systems by refrigerant gas type (after deregulation). (From Japan Refrigeration and Air Conditioning Industry Association, "About JRA GL-20" [17].)

In addition, the EPA has added HFC refrigerants to the Significant New Alternatives Policy (SNAP) program for alternative substance applications under the Clean Air Act, which establishes regulations for the use of refrigerants for each product group.

The SNAP program sets application dates for each state, and in California, the lead state, the California Air Resources Board was planning to implement a GWP of 750 or less for HFCs in all air conditioning equipment starting in January 2023. However, this has been postponed to January 2025 for separate type air conditioners and January 2026 for variable refrigerant flow (VRF), as it was determined that amendments to the Building Code on the use of flammable refrigerants would not be met in time.

In addition, the Toxic Substances Control Act (TSCA) regulation has been established as the regulation for compound PFAS.

2.5 HEAT PUMP EQUIPMENT AND RESOURCE CIRCULATION

2.5.1 Resource Circulation and Circular Economy

Resource circulation refers to the recycling of resources through initiatives such as reducing the generation of waste, reusing and recycling products, and using recycled materials in products, considering products and raw materials that have previously been discarded after being used in economic activities as resources.

In recent years, the idea of a "circular economy" has spread globally in a shift called for in order to realize resource circulation.

As shown in Figure 2.11, the circular economy is an economic activity that, in addition to conventional 3R (reduce/reuse/recycle) initiatives, generates added value through servicing and other means while reducing resource input and consumption, and making effective use of stock. Furthermore, it aims to maximize the value of resources and products, minimize resource consumption, and prevent the generation of waste, with the ultimate long-term goal of zero new resource input (excluding renewable resources).

The mass-production and mass-consumption economic and social activities of the past have formed a mass-waste society. However, in addition to impeding healthy material circulation, these activities are also closely linked to various

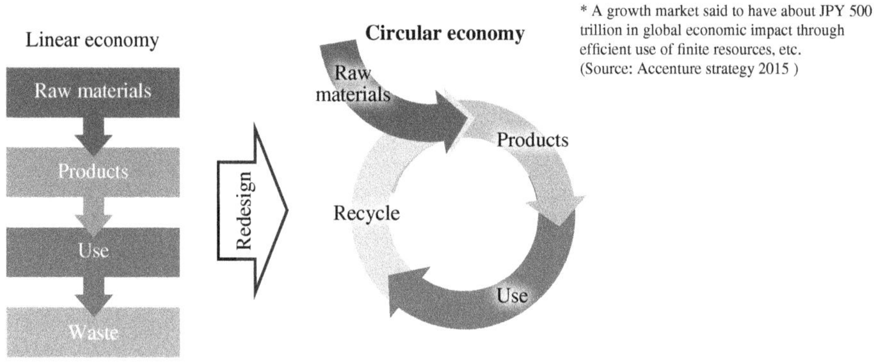

FIGURE 2.11 What is the circular economy? (From Ministry of the Environment, 2021 White Paper on the Environment, Circular Society and Biodiversity [19].)

environmental problems, such as climate change, the depletion of natural resources, and the destruction of biodiversity due to large-scale resource extraction. The increasing demand for resources, energy and food, and the increase in waste generation are becoming more serious throughout the world, and the global trend is toward a shift from one-way economic and social activities to a circular economy that uses resources in a sustainable manner.

The transition to a circular economy also has the potential to become a new source of competitiveness for companies, as it increases the sustainability of their business activities, with new business models currently being developed both domestically and internationally.

2.5.2 Measures for the Transition to a Circular Economy

2.5.2.1 Japan's Approach to the Circular Economy

In Japan, initiatives related to the 3Rs and circular economy have been promoted so far through the collaboration of various actors, including the government, the business community, and the public, under various systems to promote the formation of a circular society. In recent years, discussions on the transition to a circular economy have accelerated against the backdrop of growing concerns about climate change and the risk of resource supply disruptions. Table 2.9 shows the most recent major policy trends.

The circular economy roadmap published by the Ministry of the Environment in September 2022 (Figure 2.12) provides direction for each material and each product.

TABLE 2.9
Trends in Measures Related to the Circular Economy in Japan

Next Period	Trend
May 2020	The METI publishes its Circular Economy Vision 2020, which calls for a shift to highly circular business models
Jan 2021	The METI and the Ministry of the Environment publish the "Disclosure and Dialogue Guidance for Promoting Sustainable Finance in the Circular Economy," which provides guidance to promote smooth dialogue between companies specializing in the circular economy and investors
Mar 2022	The Circular Economy Partnership (J4CE) is established by the METI, the Ministry of the Environment, and Nippon Keidanren (Japan Business Federation)
Apr 2022	Law on Promotion of Resource Circulation of Plastics (Plastic Resource Recycling Promotion Law) comes into force
Sept 2022	The Ministry of the Environment publishes the Circular Economy Roadmap, which outlines the direction of the circular economy to be taken by 2050 and measures to be taken by 2030
Mar 2023	The METI formulates a "Growth-oriented Resource-autonomous Circular Economy Strategy" based on discussions in the design study group launched in October 2022

Source: Based on Keidanren, "Proposal for the realisation of a circular economy" [20].

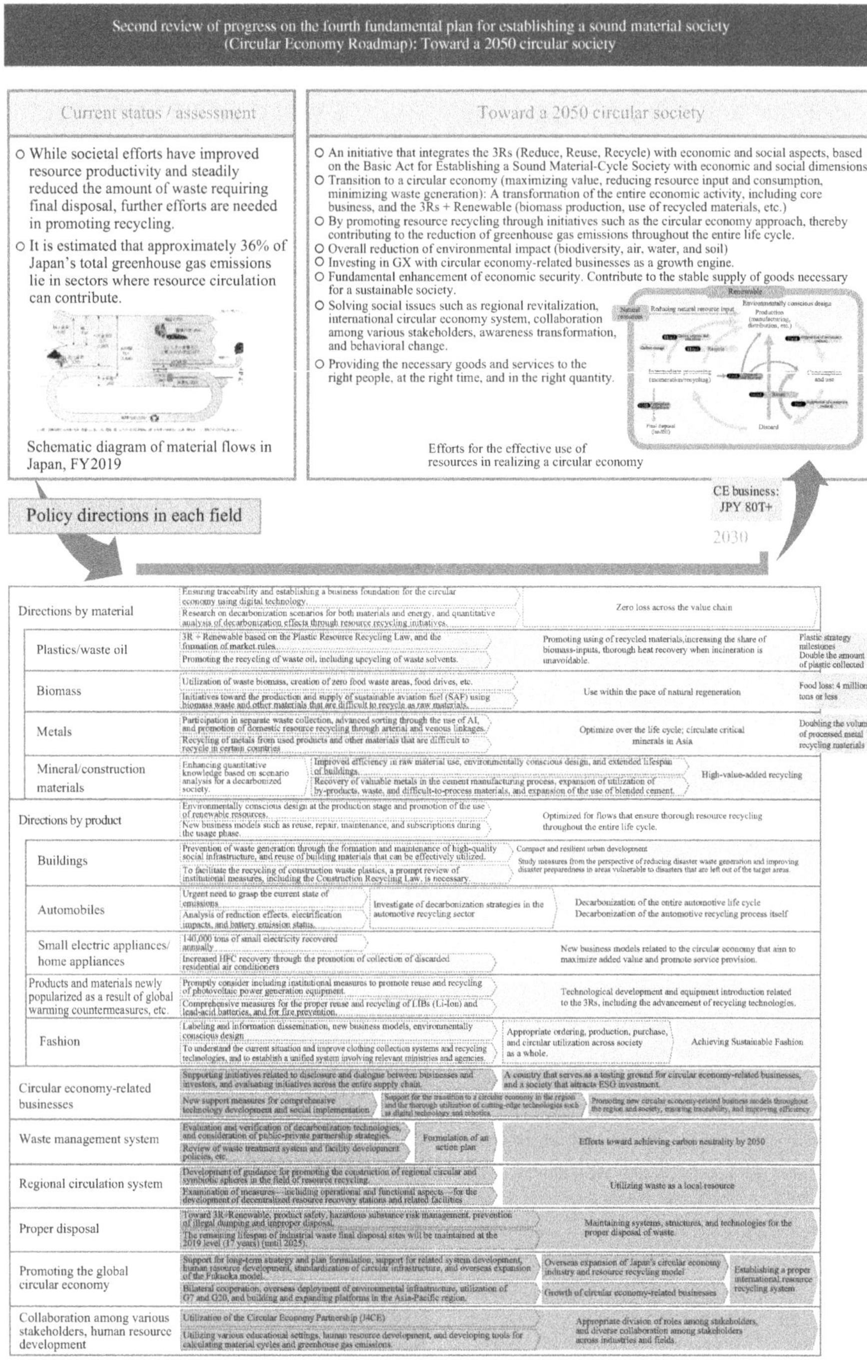

FIGURE 2.12 The Ministry of the Environment's circular economy roadmap. (From Ministry of the Environment, Central Environmental Council, Circular Society Subcommittee, "Second review of the progress of the Fourth Basic Plan for the Establishment of a Circular Society (Circular Economy Roadmap) towards a circular society in 2050" [21].)

For example, with regard to metals used in heat pump technology, the Ministry intends to double the processing volume of recycled raw materials for metals by FY2030 by promoting the recovery of recyclable resource circulation from both domestic and foreign sources, thereby ensuring a stable supply of metals contained in end-of-life products, etc. as recycled resources.

In addition, for residential air conditioners employing heat pump technology, by strictly enforcing the Home Appliance Recycling Law and thoroughly raise public awareness to promote the collection of air conditioner units and strive to improve the recovery rate and promote the recycling of HFCs (hydrofluoric acid) used as refrigerants in air conditioners.

A new recovery rate target for residential air conditioner was set in the Report on the Evaluation and Examination of the Implementation Status of the Home Appliance Recycling System, compiled separately by the METI and the Ministry of the Environment in June 2022, which aims to raise the rate to 53.9% or more by 2030; however, the recovery rate in FY2020 was 37.6%, a large gap from the target value. The gap between the target and the recovery rate remains quite large.

Commercial refrigeration and air conditioning equipment is subject to the Fluorocarbons Emissions Control Law, which requires appropriate refrigerant management of the equipment. The Industrial Structure Council and the Central Environment Council compiled the "Direction of measures to improve the recovery rate of fluorocarbons at the time of disposal" in February 2019, which states that the low refrigerant recovery rate at the time of equipment disposal is due to the fact that the recovery of FCs is not carried out, especially for models such as package air conditioners, showcases, and chilling units. It was pointed out that many appliances are disposed of without CFC recovery work being carried out.

As described above, both residential air conditioners and commercial refrigeration and air conditioning equipment have inadequate recovery of both the discarded equipment and the refrigerant used. Further measures are required from the perspective of ensuring the reuse and recycling of resources such as metals, resins, and refrigerants by collecting all used equipment, which is one of the premises of the circular economy.

2.5.2.2 European Approaches to the Circular Economy

This section describes developments in Europe, where efforts to transition to a circular economy are particularly active.

In Europe, various measures are being developed, including the Circular Economy Action Plan adopted in December 2015, as shown in Table 2.10.

Of these, the Draft Ecodesign for Sustainable Products Regulation, which is core element of the new circular economy-related policy package (Phase 1) published by the European Commission in March 2022, includes a new obligation to introduce a Digital Product Passport (DPP). The new regulation includes a mandatory introduction of the DPP.

DPP is a generic term for digitalized data on the origin, materials used, recyclability, and dismantling method of a product, with the aim of ensuring transparency and traceability of information on product sustainability.

The DPP makes this information readily available to actors involved in the product value chain, such as consumers (users), manufacturers, importers, distributors,

TABLE 2.10
Trends in Measures Related to the Circular Economy in Europe

Time Period	Trend
Dec 2015	The European Commission publishes the Circular Economy Action Plan
Dec 2019	The European Commission publishes its growth strategy, the European Green Deal, which positions the transition to a circular economy as a core policy
Mar 2020	The European Commission publishes the Second Circular Economy Action Plan
Mar 2022	The European Commission publishes the first policy package (Phase 1) related to the circular economy
Nov 2022	The European Commission publishes the second policy package (Phase 2) related to the circular economy

Source: Based on Japan External Trade Organization (JETRO), "EU Circular Economy Policy (Phase 1)—What the EU aims to achieve in the first policy package of 2022" [22] and "EU circular economy policies (Phase 2)—Second policy package of 2022 focusing on the draft Packaging and Packaging Waste Regulation" [23].

and re-manufacturers. It also allows for measures such as high tariffs on products that do not comply with EU environmental standards, which could force unsustainable products out of the EU market.

As part of the DPP, the Battery Passport for storage batteries was expected to be gradually introduced from 2024, and business associations and organizations in related industries are actively involved in the development, demonstration, and standardization of the Battery Passport.

It is expected to be introduced gradually from finished products such as electronics, IT equipment, textiles, and furniture, as well as intermediate products such as steel, cement, and chemicals.

In the future, other countries and regions promoting the transition to a circular economy, including Japan, may also make progress toward developing an infrastructure for the centralized management of information throughout the entire product life cycle to promote sustainable manufacturing, product management, and consumption.

2.6 DX-DRIVEN DIGITALIZATION OF INDUSTRY

2.6.1 What is Digital Transformation (DX)?

One of the most common definitions of DX is that it is the transformation of a company's products, services and business models based on the needs of customers and society, as well as the transformation of operations themselves, organizations, processes, corporate culture and climate in response to rapid changes in the business environment, using data and digital technology. It is often referred to as "transforming operations themselves, organizations, processes, corporate culture and culture, and establishing a competitive advantage".

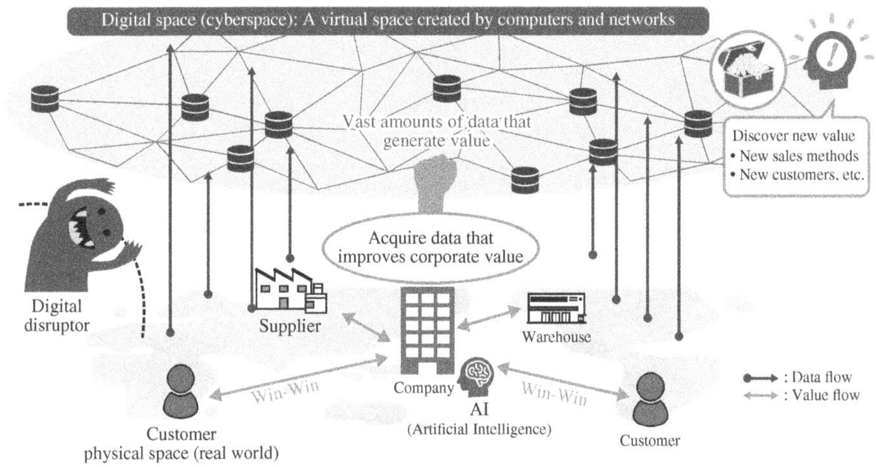

FIGURE 2.13 Image of DX. (From METI, WG1 Overall Report [24].)

This definition was compiled by the METI in 2018 in its Guidelines for Promoting Digital Transformation (DX), which led to the concept becoming widely accepted in Japan.

In recent years, as shown in Figure 2.13, a vast amount of information pertaining to physical events in physical space (real world) has been accumulated in cyberspace (virtual world)[11] and can be reproduced in digital data. Furthermore, in cyberspace, this big data can be analyzed using AI and other digital technologies, and fed back into physical space as information that generates new value.

The source of value creation in business is shifting to the use of such data and digital technologies, and it is becoming increasingly difficult for companies that cannot make good use of these technologies, and reform their corporate culture and business models to survive the digital competition of the future.

Key technologies supporting the realization of DX include Internet of Things (IoT), artificial intelligence (AI), digital twin, Cyber-Physical System (CPS), and 5G/Beyond 5G (6G), as shown in Figure 2.14. An overview of each technology is given below.

2.6.1.1 Internet of Things (IoT)

IoT is a product technology that allows all objects in the world, including electric appliances, to have communication functions and to connect to the internet and communicate with each other. In order to create a highly accurate cyberspace, a lot of data is needed, and it is important to keep collecting data from all kinds of objects through IoT.

2.6.1.2 Artificial Intelligence (AI)

AI refers to programs that operate in the same form as human thought processes. In addition to improvements in AI's own information-processing capabilities, the aforementioned increase in the amount of data due to the development of the IoT is also helping to achieve more accurate predictions of the future.

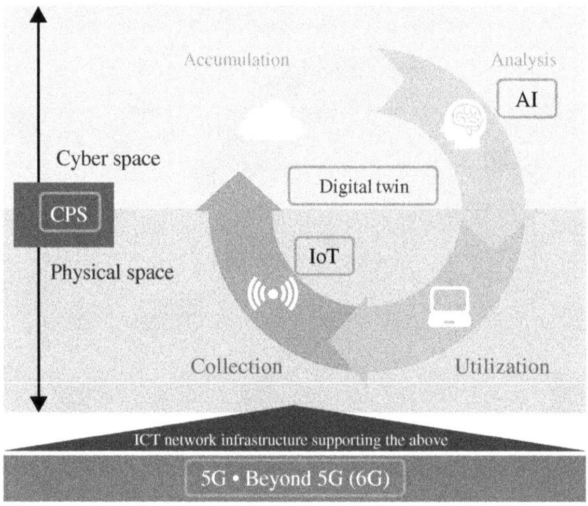

FIGURE 2.14 Key technologies supporting the realization of Society 5.0 and DX. (Based on various documents.)

2.6.1.3 Digital Twin

Digital twin is a technology that collects and accumulates information in physical space (real world) using IoT, etc. and reproduces physical space in cyberspace (virtual world) based on this data. It is called a "digital twin" because the behavior of the physical world is mirrored in the digital space like a virtual double.

Unlike conventional cyberspace, digital twin is characterized by its ability to reproduce a more realistic physical space in real time. Behind this are the aforementioned advances in IoT and AI, which enable the reproduction of real-time physical space by transmitting various data acquired by IoT to servers in the cloud in real-time, which are analyzed and processed by AI.

2.6.1.4 Cyber-Physical System (CPS)

CPS refers to a system that takes information from physical space into cyberspace, analyzes it, and uses the analysis results to solve problems in physical space and create new value.

While the aforementioned digital twin focuses on technology to reproduce physical space, CPS refers to a series of cycles from information collection, storage, analysis, and feedback.

2.6.1.5 5G and Beyond 5G (6G)

5G refers to the fifth-generation mobile communication system (5Generation), a technology that is expected to be highly effective in reflecting data in cyberspace in real time, as it will enable the transmission and reception of large volumes of data at very high speed and very low latency.

On the other hand, as in the aforementioned CPS, Beyond 5G (6G), which is a more advanced communications infrastructure than 5G, is required to realize advanced data synchronization across physical and cyberspace in any location, safely and reliably.

2.6.2 Society 5.0 to Solve Economic Growth and Social Challenges

Society 5.0 is a concept similar to DX[12], proposed in the Fifth Science and Technology Basic Plan as an image of a future, sought after society. It is defined as a "super-smart society" that combines economic growth and solutions to social issues through systems that highly integrate physical and cyberspace.

While DX refers to the initiatives of companies, etc., Society 5.0 is a concept that represents the image of future society: companies and others utilizing data and digital technology based on the DX concept will create better products, services, and business models, which will ultimately lead to the realization of Society 5.0.

2.7 DX Initiatives and Measures

2.7.1 Examples of DX Use in Manufacturing and Distribution

The DX initiatives using digital technology described in Section 2.6 are already spreading in various fields. For example, as shown in Figure 2.15, examples of initiatives toward the construction of digital twin technologies that incorporate manufacturing data from physical space (manufacturing sites) into cyberspace, derive

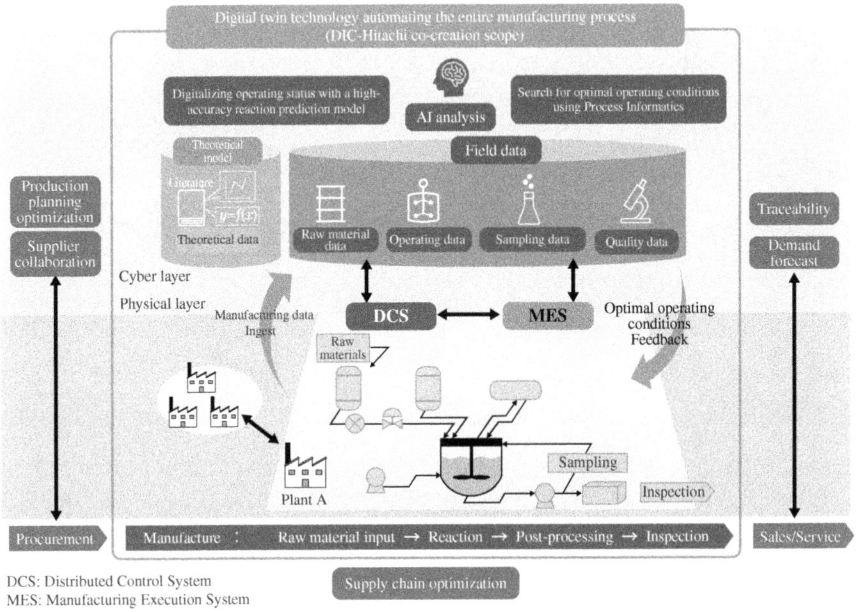

FIGURE 2.15 A manufacturing plant automated by digital twin technology. (From DIC Corporation, "DIC and Hitachi Ltd. commence full-scale collaboration to realise next-generation plants in resin manufacturing" [25].)

optimum operating conditions through AI-based analysis and automate manufacturing processes are beginning to appear.

Furthermore, AI and digital twin technologies are being used to accurately predict the flow of goods, money, and information throughout the supply chain, including not only the manufacturing site, but also the procurement of raw materials and the transportation and sale of manufactured products. Services that automate planning in response to demand fluctuations are already available and are expected to streamline and improve the efficiency of the entire manufacturing process and supply chain in the future.

2.7.2 Examples of DX Use in Energy Management

The use of digital technology is also becoming essential in realizing the transition to carbon neutrality, as discussed in Chapter 2.1, and to a circular economy, as discussed in Chapter 2.5.

For example, the construction of advanced energy management systems (EMS) is becoming increasingly important for companies and other stakeholders in decarbonization. Specifically, as shown in Figure 2.16, this is the construction of EMS that utilize AI and digital twins. By precisely predicting and managing the power supply and demand balance, and controlling equipment and other facilities without reducing productivity or quality, it will be possible to cover all electricity use with renewable energy.

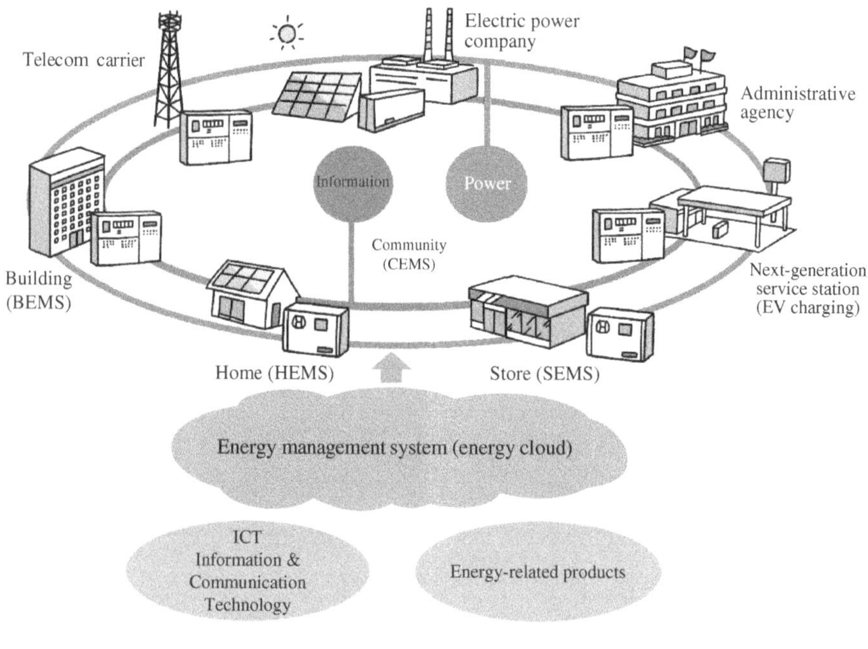

FIGURE 2.16 An energy management system using digital technology. (From NEC Corporation Corporation, "Energy management system (xEMS)" [26].)

2.7.3 Examples of DX Use in the Circular Economy

As mentioned in Section 2.5, a mechanism for centralized management of information[13] for the entire life cycle of individual pieces of equipment is required to realize a circular economy. These digital technologies will be indispensable in the construction of such a mechanism.

In Europe, DPPs are planned to be introduced to record information on the entire product value chain and each stage of the life cycle on a unified digital platform. First of all, the Battery Passport (Figure 2.17), which covers storage batteries, is expected to be gradually introduced from 2024 (see Chapter 2.5).

In response to these trends, initiatives related to the development, demonstration, and standardization of the Battery Passport by business alliances and associations in related industries are gaining momentum. Digital technologies such as blockchain and digital twin are being used in this context.

2.7.4 Toward the Realization of DX

The evolution of the underlying communications infrastructure is essential for the promotion of DX and the consequent realization of Society 5.0.

In June 2020, the Ministry of Internal Affairs and Communications compiled the Beyond 5G Promotion Strategy, which, as shown in Figure 2.18, sets out a policy to enhance 5G functions and create a Beyond 5G-ready environment during the 2020s.

Along with the development of such communications infrastructure, the development and advancement of key technologies required for DX going into the 2030s is also expected to accelerate.

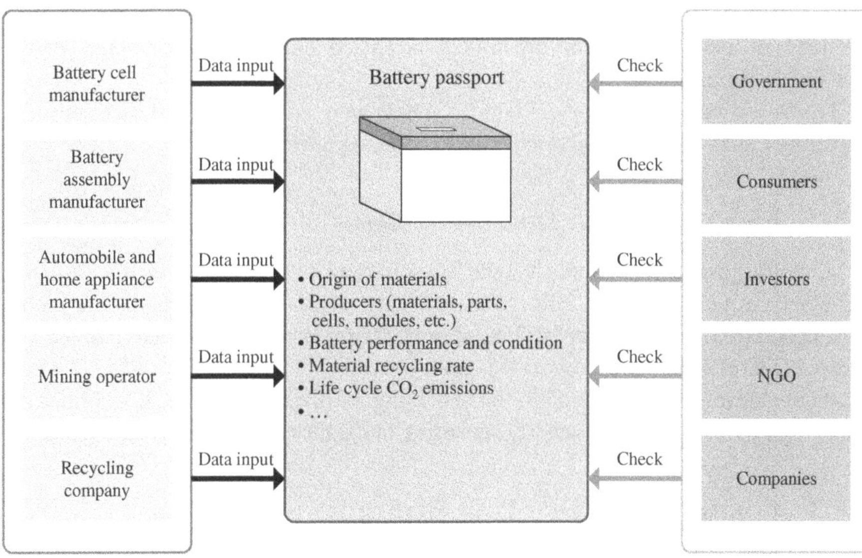

FIGURE 2.17 The battery passport. (From Global Battery Alliance, "Battery Passport Pilots" [27] and Battery Pass, "Battery Passport Content Guidance" [28].)

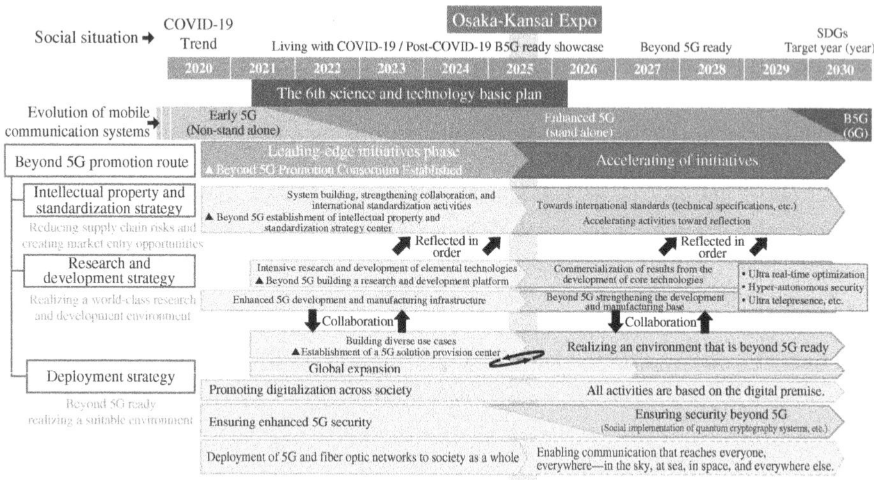

FIGURE 2.18 Overview of the "Beyond 5G promotion strategy" roadmap overview. (From Ministry of Internal Affairs and Communications, "Beyond 5G Promotion Strategy (Overview)" [29].)

2.8 THE ROLE OF HEAT PUMP TECHNOLOGY IN REALIZING A SUSTAINABLE SOCIETY

2.8.1 What Are the Sustainable Development Goals Worldwide?

The term "sustainable society" is defined as a society in which the global and natural environment is properly preserved and development is carried out in such a way that the needs of future generations are not compromised and the demands of the present generation are also met….

The Sustainable Development Goals (SDGs) were adopted at the UN Summit in September 2015 as an international call to action (Figure 2.19).

2.8.2 Japan's Sustainable Development Goals

In Japan, the government established the Headquarters for the Promotion of Sustainable Development Goals in May 2016, and various government-led initiatives are underway. In addition, in November 2017, Keidanren (Japan Business Federation) revised its Charter of Corporate Behaviour based on the concept of "Society 5.0 for SDGs," which aims to achieve the SDGs through the realization of Society 5.0. The SDG initiatives are being promoted and are steadily spreading in the business scene (Figure 2.20).

2.8.3 Contribution of Heat Pump Technology to Sustainable Development Goals

The SDGs themselves are set out to be achieved by 2030. However, how to create a sustainable society in the face of very stringent social demands, including carbon neutrality, is an issue that should continue to be considered beyond 2030. Many of

Goal 1: Poverty End poverty in all its forms, everywhere.	**Goal 10: [Complaints, etc.]** Reduce inequality within and among countries
Goal 2: [Hunger] End hunger, achieve food security and improved nutrition, and promote sustainable agriculture	**Goal 11: [Sustainable Cities]** Make cities and human settlements inclusive, safe, resilient and sustainable
Goal 3: [Health Care] Ensure healthy lives and promote well-being for all at all ages	**Goal 12: [Sustainable Consumption and Production]** Ensure sustainable consumption and production patterns
Goal 4: [Education] Ensure inclusive and equitable quality education and promote lifelong learning opportunities for all	**Goal 13: [Climate Change]** Take urgent action to combat climate change and its impacts
Goal 5: [Gender] Achieve gender equality and empower all women and girls	**Goal 14 [Marine Resources]** Conserve and sustainably use the oceans, seas and marine resources for sustainable development
Goal 6: [Water and Hygiene] Ensure availability and sustainable management of water and sanitation for all	**Goal 15: [Land Resources]** Protect, restore and promote sustainable use of terrestrial ecosystems; sustainably manage forests; combat desertification; halt and reverse degradation; and halt biodiversity loss
Goal 7: [Energy] Ensure access to affordable, reliable, sustainable, and modern energy for all	**Goal 16: Peace** Promote peaceful and inclusive societies for sustainable development, provide access to justice for all, and build effective, accountable, and inclusive institutions at all levels
Goal 8: [Economic Growth and Employment] Promote inclusive and sustainable economic growth, full and productive employment and decent work for all	**Goal 17: [Means of Implementation]** Strengthen the means of implementation and revitalize the global partnership for sustainable development
Goal 9: [Infrastructure, Industry, and Innovation] Promote inclusive and sustainable economic growth, full and productive employment and decent work for all	

FIGURE 2.19 The 17 SDGs. (From Ministry of Foreign Affairs of Japan, "Sustainable Development Goals (SDGs) and Japan's efforts" [30].)

the directions of the goals set out as SDGs are not something that can be solved in a short period of time.

In this context, heat pump technology also faces a number of problems, such as the global warming problem caused in part by refrigerant leakage and the lack of progress in resource circulation of equipment and refrigerants. For this reason, heat pump technology itself needs to evolve into a sustainable technology. The technology that achieves this is the next-generation heat pump technology.

Figure 2.21 shows that if next-generation heat pump technology that has evolved to meet social demands such as carbon neutrality, DX, and circular economy can be realized, it will contribute to many of the SDGs.

Next-generation heat pump technology has an extremely wide range of applications in diverse societies, as well as contributing to Goal 13: "Climate change," due to its inherent energy efficiency and decarbonization, and its excellent properties of wide temperature, time, and spatial controllability.

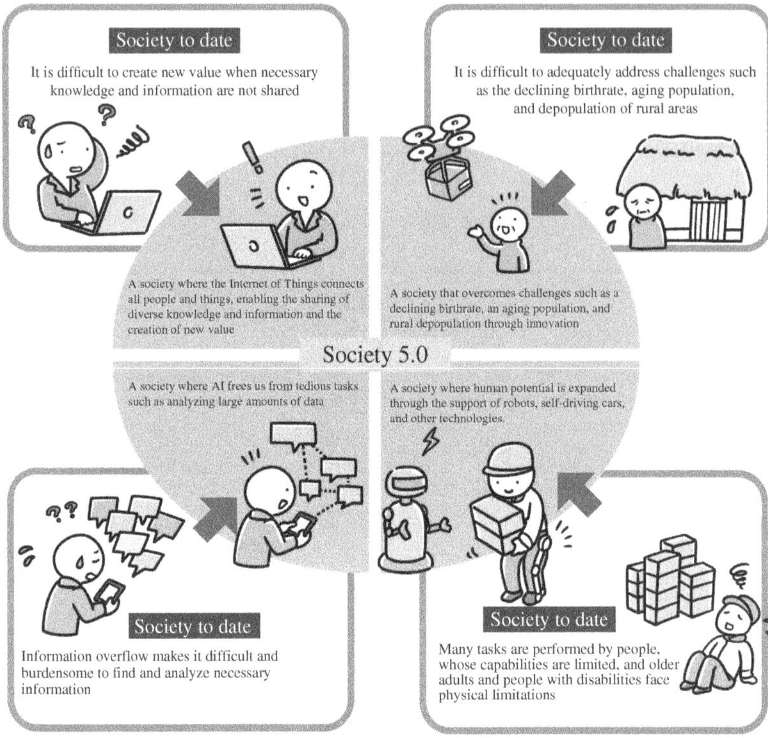

FIGURE 2.20 Society to be realized by Society 5.0. (From Ministry of Education, Culture, Sports, Science and Technology, "Grand design of higher education towards 2040 (Report)" [31].)

The value of next-generation heat pump technology	SDGs for which contributions are expected					
	2 Zero Hunger	3 Health and well-being for all	7	9 Let's build a foundation for industry and technological innovation	11	13 Take concrete measures to combat climate change
[Common goals] Improving energy efficiency (energy saving) and effectively utilizing renewable and untapped energy sources			○		○	○
[Housing and buildings] Improving the health and comfort of living environment		○			○	
[Food] Improving the sustainability and efficiency of food systems (cold chain, horticultural facilities, etc.)	○					
[Industry] Manufacturing process reform and productivity improvement				○		
[Electricity supply and demand] Stabilizing the electricity supply-demand system through the use of DR, VPP, and other technologies (provision of balancing capacity)			○			○

FIGURE 2.21 SDGs to which heat pump technology is expected to contribute.

For example, it will contribute to the realization of various future societies by bringing values such as improved health and comfort in the living environment, improved sustainability and efficiency in the food system, manufacturing process reform and productivity improvement, and stabilization of the electricity supply and demand system. These are discussed in detail in Chapters 5 and 6.

NOTES

1. DACCS: Direct Air Carbon Capture and Storage, a technology for the direct capture and storage of carbon dioxide.
2. NDC: Nationally Determined Contribution, e.g. GHG emission reduction targets that countries must prepare, report, and maintain under the Paris Agreement.
3. HFO refrigerants: Hydrofluoroolefins (HFOs) are also known as green refrigerants due to their low toxicity and low impact on the environment and human health.
4. Manufacturers who handle Class I specified products that use fluorocarbons as refrigerants. This applies to commercial air conditioning equipment, commercial refrigeration and freezing equipment, etc.
5. GWP value as a weighted average of the number of products shipped.
6. Actions include applications and notifications. For example, for inert gases, notification is not required if the legal refrigeration capacity is less than 20 refrigeration tonnes, but for flammable gases (e.g. propane (R290)), notification is required from 3 refrigeration tonnes and above.
7. Flammability is classified into four categories based on the presence or absence of flame spread and burning speed: Category 1 (non-flammable), 2L (mildly flammable), 2 (flammable), and 3 (highly flammable).
8. Toxic gases: Toxicity is classified as A (low toxicity) or B (high toxicity) according to the chronic toxicity index.
9. A2L, a mildly flammable refrigerant, is a combination of A (low toxicity) and 2L (lower flammability) in the safety classification defined by the American Society of Heating, Refrigerating and Air-Conditioning Engineers (ASHRAE). It includes HFC refrigerants, hydrofluoroolefins (HFOs) and their mixtures, which are characterized by mild flammability, low toxicity and low global warming potential (GWP).
10. Total HFC limits: A 79% phase-out of HFCs by 2030 compared to 2015, which is far higher than the reductions applicable to developed countries under the Kigali Amendment.
11. Cyberspace (virtual world): A virtual space constructed by computers and networks, using sensors and IoT.
12. Society 5.0: Society 1.0 is a hunting society, 2.0 an agricultural society, 3.0 an industrial society, and 4.0 an information society.
13. Information on the entire life cycle: Information on the materials used, performance information, usage status, recovery and disposal status of equipment after use, etc.

REFERENCES

1. Agency for Natural Resources and Energy, Ministry of Economy, Trade and Industry, "What is 'carbon neutral'? (Part 1)—When and by whom will it be achieved?," 2021, https://www.enecho.meti.go.jp/about/special/johoteikyo/carbon_neutral_01.html
2. Agency for Natural Resources and Energy, Ministry of Economy, Trade and Industry, "Annual Report on Energy 2021 (Energy White Paper 2022)," "Section 1: Overview of Energy Supply and Demand," "[211-1-3] Overview of Energy Balance Flows in Japan (FY2020)," 2022, https://www.enecho.meti.go.jp/about/whitepaper/2022/html/2-1-1.html
3. Agency for Natural Resources and Energy, Ministry of Economy, Trade and Industry, "Annual Report on Energy in FY2021 (Energy White Paper 2022)," p. 2, 2022, https://www.enecho.meti.go.jp/about/whitepaper/2022/pdf/2_1.pdf

4. Ministry of Economy, Trade and Industry, "Green growth strategy associated with carbon neutrality in 2050," p. 6, 2021, https://www.meti.go.jp/policy/energy_environment/global_warming/ggs/pdf/green_gaiyou.pdf
5. Agency for Natural Resources and Energy, Ministry of Economy, Trade and Industry, "36th Meeting of the Strategic Policy Committee, Advisory Committee for Natural Resources and Energy", Document 2, "Study for the realisation of carbon neutral by 2050," 2021, https://www.enecho.meti.go.jp/committee/council/basic_policy_subcommittee/036/036_005.pdf
6. Agency for Natural Resources and Energy, Ministry of Economy, Trade and Industry, "Comprehensive energy statistics energy balance table (FY2021)," 2023.
7. Agency for Natural Resources and Energy, Ministry of Economy, Trade and Industry, "Outlook for energy supply and demand in FY2030 (Related Data)," 2021, https://www.enecho.meti.go.jp/committee/council/basic_policy_subcommittee/opinion/data/03.pdf
8. Ministry of the Environment, "Global warming countermeasures plan," 2021, https://www.env.go.jp/content/900440195.pdf
9. New Energy and Industrial Technology Development Organization (NEDO) Technology Strategy Center (TSC)/Global Technology Research Unit, TSC Foresight: Trends in Major Countries and Regions (the US, China ,the EU, the UK, Germany, Indonesia, India) Regarding Carbon Neutrality in the Run-up to COP27—From the Perspectives of Renewable and Fossil Energy," (in Japanese), 2022.
10. Heat Pump and Thermal Storage Technology Center of Japan, "Related policies and diffusion status of heat pumps in Europe and the United States," 2022.
11. United Nations Framework Convention on Climate Change (UNFCCC), "NDC registry," 2023, https://unfccc.int/NDCREG
12. United Nations Framework Convention on Climate Change (UNFCCC), "Long-term strategies portal," 2023, https://unfccc.int/process/the-paris-agreement/long-term-strategies
13. International Energy Agency (IEA), "The future of cooling in Southeast Asia," p. 11, 2019, 2023, https://iea.blob.core.windows.net/assets/dcadf8ee-c43d-400e-9112-533516662e3e/The_Future_of_Cooling_in_Southeast_Asia.pdf
14. Ministry of Economy, Trade and Industry, Ozone Layer Protection Promotion Office, etc., "Summary of the Montreal Protocol and Kigali Amendment," p. 1, 2017, https://www.env.go.jp/content/900530414.pdf
15. Ministry of the Environment and Ministry of Economy, Trade and Industry, "Overview of trends surrounding CFCs and the revised Fluorocarbons Emissions Control Law," p. 9, 2019, https://www.env.go.jp/earth/furon/files/briefing_2019_kaitaisanpai_0.pdf
16. Ministry of Economy, Trade and Industry, "Regulations for designated product manufacturers, etc.," 2023, https://www.meti.go.jp/policy/chemical_management/ozone/furon_seihin_seizo.html
17. Japan Refrigeration and Air Conditioning Industry Association, "About JRA GL-20," 2023, https://www.jraia.or.jp/eco/lg-20/index.html
18. ECHA (European Chemicals Agency), "ECHA Annex XV restriction report proposal for a restriction," 2023, https://echa.europa.eu/documents/10162/f605d4b5-7c17-7414-8823-b49b9fd43aea
19. Ministry of the Environment, 2021 White Paper on Environment, Circular Society and Biodiversity, "Section 2: Transition to circular economy," "Figure 2-2-1 circular economy," 2021, https://www.env.go.jp/policy/hakusyo/r03/html/hj21010202.html
20. Nippon Keidanren (Japan Business Federation), "Proposal for the realisation of a circular economy," 2023, https://www.keidanren.or.jp/policy/2023/008_honbun.html
21. Ministry of the Environment, Central Environmental Council, Circular Society Subcommittee, "Results of the second inspection of the progress of the Fourth Basic Plan for the Establishment of a Circular Society (Circular Economy Roadmap): Towards a circular society in 2050," p. 106, 2022, https://www.env.go.jp/content/000071596.pdf

22. Japan External Trade Organisation (JETRO), Brussels Office, Overseas Research Department, "EU circular economy policies (Phase 1): What the EU is aiming for in the first policy package for 2022," 2022, https://www.jetro.go.jp/ext_images/_Reports/01/60d6edca66cfec17/20220030_01.pdf
23. Japan External Trade Organisation (JETRO), Brussels Office, Overseas Research Department, "EU circular economy policies (Phase 2): The second round of the 2022 policy package with a focus on the draft packaging and packaging waste regulation," 2023, https://www.jetro.go.jp/ext_images./_Reports/01/60d6edca66cfec17/20220030_02.pdf
24. Ministry of Economy, Trade and Industry, Study Group on Accelerating Digital Transformation, WG1 Overall Report, p. 26, 2020, https://www.meti.go.jp/shingikai/mono_info_service/digital_transformation_kasoku/pdf/20201228_4.pdf
25. DIC Corporation, "DIC and Hitachi Ltd. commence full-scale collaboration to realise next-generation plants in resin manufacturing," 2021, 2023, https://www.dic-global.com/ja/news/2021/ir/20211214135954.html
26. NEC Corporation, "Energy management system (xEMS)," 2021.
27. Global Battery Alliance, "Battery Passport Pilots," https://www.globalbattery.org/action-platforms-menu/pilot-test/
28. Battery Pass, "Battery passport content guidance," 2023, https://thebatterypass.eu/resources/
29. Ministry of Internal Affairs and Communications, "Beyond 5G promotion strategy (Summary)," p. 5, 2020, https://www.soumu.go.jp/main_content/000702111.pdf
30. Ministry of Foreign Affairs of Japan, "Sustainable development goals (SDGs) and Japan's efforts," p. 3, 2023, https://www.mofa.go.jp/mofaj/gaiko/oda/sdgs/pdf/SDGs_pamphlet.pdf
31. Ministry of Education, Culture, Sports, Science and Technology, "Grand design of higher education towards 2040 (Report)," "Society realized by Society 5.0," 2018, https://www.mext.go.jp/content/1413715_010.pdf

3 Measures to Solve the Challenges of Heat Pump Technology

OVERVIEW

Heat pump technology has high decarbonization potential and is adaptable to a wide range of heat-using social systems. As a result, its applications are expanding into air conditioning, water heating, and industrial technology, and there are high expectations for the further promotion of heat pump technology.

On the other hand, the increased use of heat pump technology will increase the amount of electricity consumed by heat pumps, which will in turn emit a large amount of CO_2. Leakage of refrigerant could also accelerate global warming. They also use a lot of relatively expensive metals such as copper and aluminum, which also increase waste.

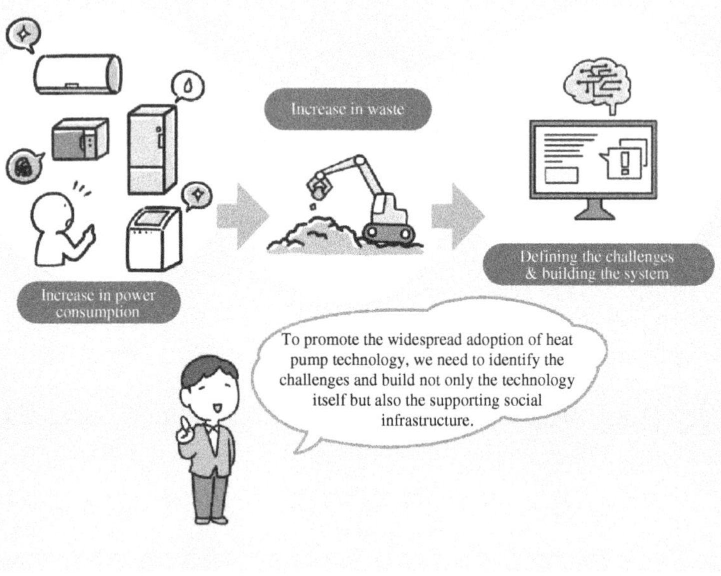

FIGURE 3.1 The whole of society should take action! "Solving the challenges of heat pump technology."

DOI: 10.1201/9781003666950-3

Thus, heat pump technology has many negative aspects. In the face of social demands to build a sustainable future, it is necessary to clarify the issues and develop the technology into the next phase of social systems.

This chapter explains the main challenges and requirements of heat pump technology at present and describes measures and initiatives to solve them (Figure 3.1).

3.1 CHALLENGES IN ACHIEVING DECARBONIZATION AND ENERGY SAVING

3.1.1 Challenges in Expanding Heat Pump Applications

The following three pillars are important for decarbonization through heat pumps

(1) Decarbonization through expansion of applications
(2) Energy saving through rational heat utilization
(3) Further energy saving through improved equipment performance

However, at present, (1) "expansion of applications" and (2) "rational heat utilization" have not progressed well. In addition, while it is important to improve the performance of equipment in realizing energy saving, there is a limit to just improving the performance of equipments, as has been the case in the past. It has been pointed out that the evaluation methodology for equipment performance, which serves as a driver for performance improvement, also deviates significantly from realistic equipment performance.

This section explains these issues, and describes measures and initiatives to resolve them.

3.1.2 Measures to Expand Applications for Cold Climates

The introduction of heat pump technology is by far the most effective way to reduce CO_2 emissions, as it allows combustion conversion through electrification. It is therefore important to further expand the use of heat pumps in areas such as heating, where combustion technology is mainly used. In particular, this section describes the situation in areas where the introduction of heat pumps is expected to have a positive effect.

3.1.2.1 Necessity of Low Outdoor Air Temperature Measures

Heat pumps that utilize air heat sources have not been widely used in cold regions, and it is particularly important to expand their application to these colder areas.

In cold regions, the outside air temperature is much lower than in temperate climate region, and less heat can be taken away from the outside air, which reduces not only the performance of the equipment but also its heating capacity. This is an inherent limitation of heat pumps, but this has led to the widespread use of electric heaters and combustion equipment that are largely unaffected by outdoor air temperatures and can provide a stable heat supply even at low outdoor air temperatures. Heat pumps for cold climates have therefore been developed that incorporate a two-phase injection, combining liquid and gas injection as described below.

In order to expand the use of heat pumps in cold regions, equipment that minimizes performance degradation and fully utilizes its heating capacity even in low ambient air temperatures must be developed.

3.1.2.2 Improving the Performance of Heat Exchangers

In heat pumps that use air as the heat source, improved performance of heat exchangers is essential for the generation of warmth (heating and hot water supply) in cold regions.

When the outside temperature is low, as it is in cold regions, water vapor in the air condenses on the heat exchanger and forms frost, causing a frost layer to grow. This reduces airflow and heat transfer performance, and the performance of the heat exchanger is significantly reduced, so it must be defrosted regularly, as shown in Figure 3.2.

Heat exchangers need to minimize the defrosting time by making the fin surfaces frost-resistant (e.g. with a hydrophobic coating, microfabrication, etc.) and by optimizing the cycle control.

Furthermore, in order to prevent interruptions in heating during defrosting, it is also effective to use the thermal-storage function, which stores the exhaust heat from the compressor to maintain continuous heating.

3.1.2.3 Technology to Prevent Compressor Function Degradation

When the ambient air temperature drops, the discharge temperature of the compressor becomes hotter. If the discharge temperature becomes too high, the service life

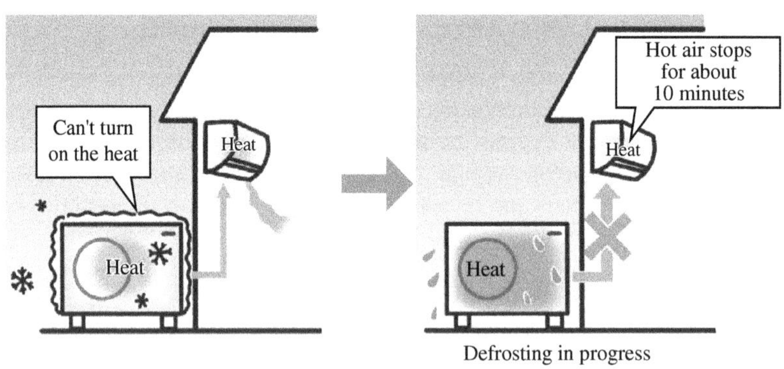

FIGURE 3.2 Defrosting function in cold climates. (From Panasonic Corporation and Panasonic Marketing Japan Co. Ltd, "Why air conditioner stops heating in winter" [1].)

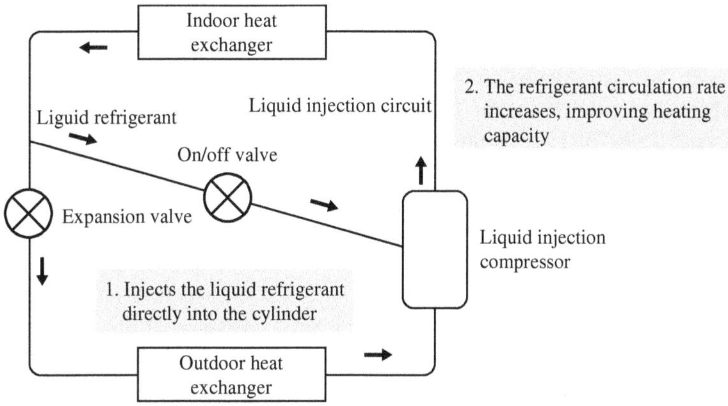

(a) Refrigeration cycle when the liquid injection circuit is operated during heating operation

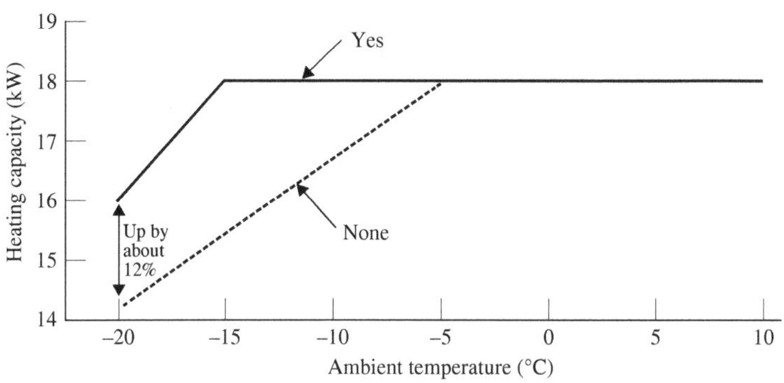

(b) Comparison of heating capacity with and without liquid injection

FIGURE 3.3 Liquid injection circuit and its effects. (Based on Mitsukuni Sano (Toshiba Carrier Corporation) et al., "Store and office air conditioners for cold regions that combine high heating performance and energy saving" [2].)

and reliability of the compressor are affected. Therefore, as an example of a system technology to reduce the discharge temperature, the introduction of a liquid injection system, as shown in Figure 3.3, is effective.

This is a technology in which part of the liquid refrigerant at the outlet of the condenser is sprayed into the compression chamber inside the compressor.

Liquid injection can lower the discharge temperature of the compressor and increase the refrigerant circulation by increasing the compressor speed, resulting in an increase in heating capacity.

Gas injection is another effective system technology that works to improve the coefficient of performance (COP) by lowering the discharge temperature.

This consists of a two-stage configuration as a means of pressure reduction by the expansion valve in the refrigeration cycle, with a gas–liquid separator in the middle to separate the gas refrigerant and liquid refrigerant. Direct only the gaseous refrigerant, at an intermediate pressure between the high and low sides, directly into the compressor cylinder during the mid-compression process. This increases the amount of refrigerant circulating in the condenser, thereby increasing the heating capacity. At the same time, the COP is improved due to reduced compression work.

3.1.3 Measures to Expand Industrial Applications through High Temperature Heat Generation

The applicable temperature range of heat pumps is limited to around 120°C on the high temperature side at the current level of practical application. In typical industrial applications, such as in factories, the utilization of steam is commonly at around 180°C (with a vapor pressure of around 10 kgf/cm^2), so heat pumps would need to generate heat up to around 200°C for this wider application.

In order to achieve 200°C, it is important to ensure the reliability and service life of the compressor in response to its high discharge temperature. One of the proposed countermeasures is the use of oil-free turbochargers that use magnetic bearings that do not require lubricating oil, research and development of which is underway in a project of the New Energy and Industrial Technology Development Organization (NEDO) (Figure 3.4).

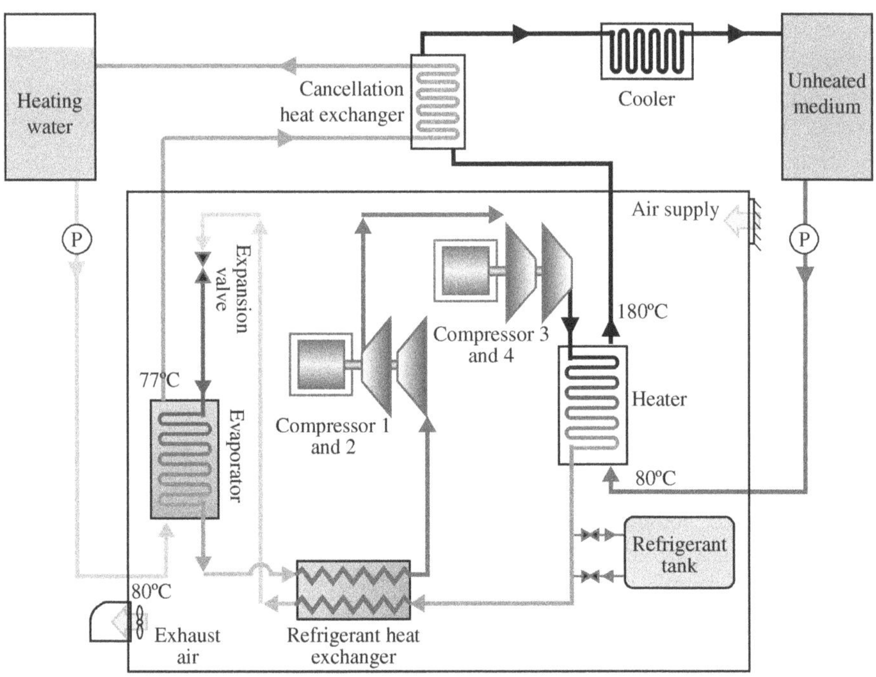

FIGURE 3.4 Example of 200°C-compatible technology: Oil-free turbo compressor. (Based on the New Energy and Industrial Technology Development Organization (NEDO), "Frontier to energy saving: Research and development of technology for innovative utilization of untapped heat energy" [3].)

Magnetic bearing technology is also undergoing innovation and prices are falling significantly. Until now, it has been too expensive to apply to heat pumps, but is expected to be adopted successively by expensive equipment such as industrial heat pumps in the future.

As the application temperature of heat pumps gets higher and higher, the insulation technology of the equipment will also become more important, as even the slightest heat dissipation will have a significant impact on its performance degradation. Although progress has been made in the development of vacuum insulation materials with extremely high insulation performance, there is still room for improvement, and further technological innovations are expected.

3.2 ENERGY-EFFICIENT INDUSTRIAL HEAT UTILIZATION TECHNOLOGIES

3.2.1 Measures to Improve Heat Transport Technology

Manufacturing, such as cleaning, sterilization, drying, distillation, and concentration processes involve heat treatment at different temperature ranges. As an industrial heat source for these processes, a steam conveying system is widely used in which a network of pipes draws steam from a centrally installed steam boiler to the required point of use.

Heat transportation by steam does not require pumps or blowers and has the advantage of being relatively easy to distribute. On the other hand, it also suffers from very large heat losses during transportation, along with exhaust and drain losses when the transportation distance is long.

In factories, there are many processes where the energy is used as hot water rather than steam on the end-use side, and it is often used for circulating heating applications to maintain a constant temperature in a hot water storage tank at a hot water temperature of 90°C or lower.

In such cases, air-source heat pumps for heating can be installed on-site near the point of heat utilization to prevent losses during heat transfer and increase energy efficiency (Figure 3.5). This is what's known as a decentralized heat pump system.

3.2.2 Use of Untapped Energy

Heat pumps can utilize various untapped energy sources as heat sources, such as geothermal heat, heat from rivers, waste heat from factories, the heat from sewage water used for daily life, as well as air heat from the atmosphere. This can lead to significant energy savings. In processes where these can be utilized, their effective use should be actively promoted.

Since heat pumps are also easy to install on-site, the potential for on-site use of these untapped energies can be greatly expanded if they are considered for utilization (Figure 3.6).

3.2.3 Simultaneous Use of Hot and Cold Energy Utilization Technology

In the industrial sector, cold and heat utilization processes are often combined. In the heating process, heating is mainly carried out with steam delivered from a steam boiler through steam pipes.

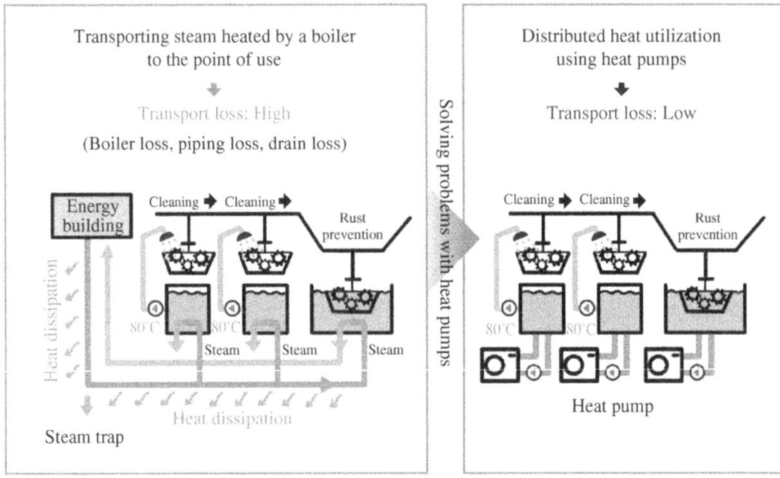

FIGURE 3.5 Distributed supply by heat pumps located in close proximity to the point of use. (Based on Japan Electro-heat Center/Industrial heat pumps.com, "Reducing steam losses" [4].)

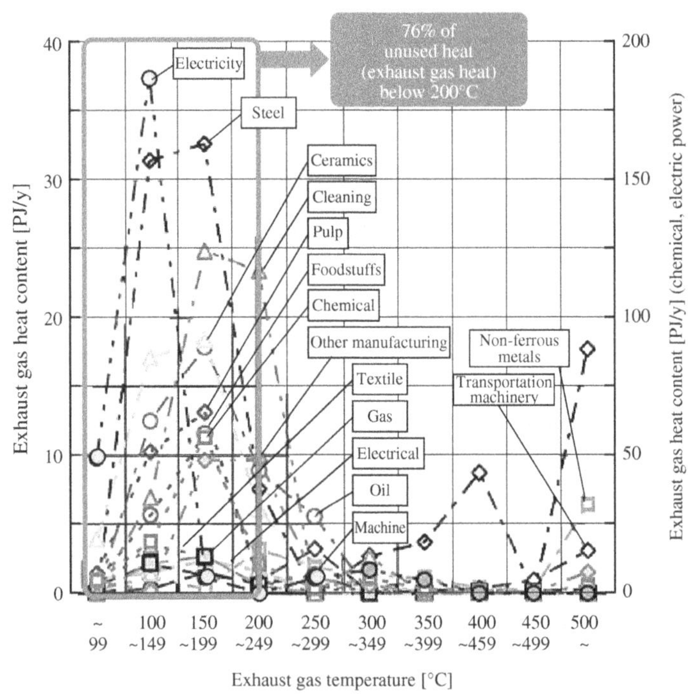

Note: Chemicals and electricity are on the right axis; other industries are on the left axis.

FIGURE 3.6 National estimates of unused heat (fuel gas heat) by industry and temperature range. (Based on the New Energy and Industrial Technology Development Organization (NEDO) news release, "Report on survey of exhaust heat in plant facilities in 15 industries" [5].)

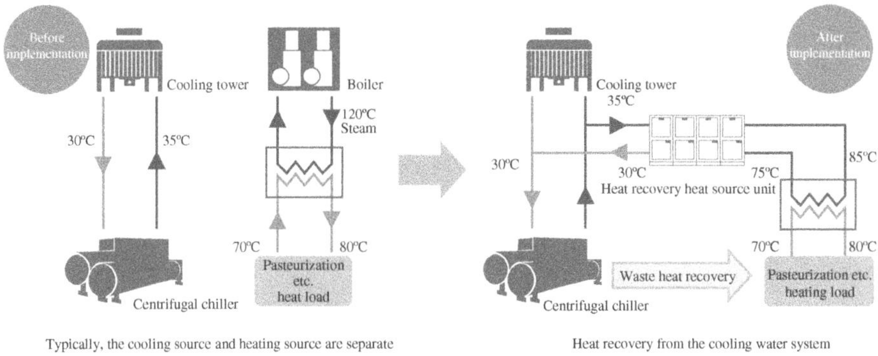

FIGURE 3.7 Simultaneous use of hot and cold energy utilization of heat pumps. (Based on Toshiba Carrier Corporation, General Catalogue of Heat Source Systems [6].)

In the cooling process, the heat is cooled by a chiller, etc., and the waste heat emitted during the cooling process is not utilized effectively but is simply released to the outside.

In such cases, efficiency can be improved by recovering the condensation waste heat discharged from the chiller and simultaneously utilizing it in the heating process (Figure 3.7). This is the known as the simultaneous use of hot and cold heat utilization in heat pumps.

The simultaneous use of hot and cold heat utilization in heat pumps is possible not only in industrial applications but also in consumer applications. Heat pumps and refrigeration units provide two such opportunities. Heat pump water heaters dissipate cold heat to the outside. Moreover, the warm exhaust heat from refrigerators and cooling systems is released directly into the open air without being used effectively, even though heating, such as in water heating, is also taking place. There is a need to eliminate such wastage of cold and heat energy and to promote a more efficient system going forward that uses all energy outputs.

For this reason, it is important from the perspective of effective energy utilization and CO_2 emission reductions to promote mutual utilization of simultaneous use of hot and cold and even integration of multiple devices. Note that, in this case, the required cold and hot energy usage times are not simultaneous, but the time gap can be absorbed by thermal storage.

3.2.4 Heat Dissipation in Refrigeration and Freezing Showcases

In many cases where refrigeration and freezing showcases are used in shops, during the summer months, the cold air leaks from the units, such that they cool the room as well as their contents, which is less efficient than air conditioning (Figure 3.8). There are also many integrated type showcases in use that dissipate exhaust heat directly into the room.

However, by adding an external exhaust function to these appliances, technology has been developed to achieve significant energy-saving effects and reduce the external load of air conditioning in summer.

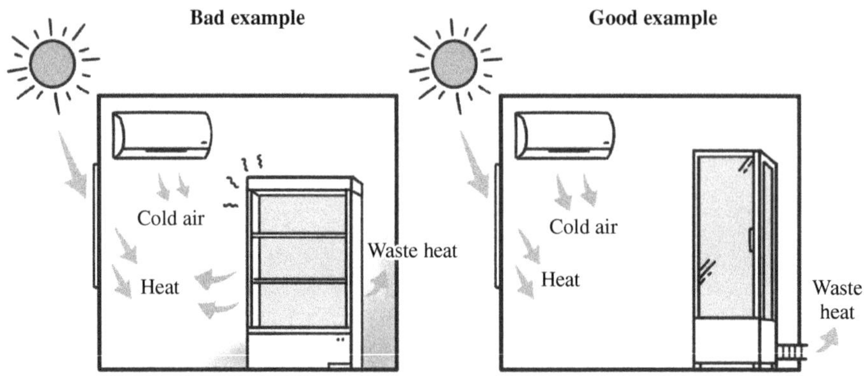

FIGURE 3.8 Example of showcase use.

As described above, the temperature controllability of heat pump technology can be used to promote the decentralized use of heat or, conversely, the effective centralized use of heat, such as simultaneous use of hot and cold.

In this case, it is essential to realize the coupled control of various equipment by using digital transformation (DX), etc. (see Section 3.6).

3.3 PERFORMANCE EVALUATION METHODOLOGY TO IMPROVE PERFORMANCE

In order to improve performance of heat pump technology, it is important not only to improve the performance of the equipment itself, but also to develop an advanced performance evaluation methodology that can drive improvements forward. This section describes these issues, needs, and measures.

3.3.1 The Need to Improve Heat Pump Equipment Performance

3.3.1.1 Equipment Performance Improvement Measures and Their Limitations

While heat pump equipments allow combustion conversion without CO_2 emissions through electrification, their widespread use will also increase indirect CO_2 emissions through electricity consumption.

For this reason, further performance improvements of the equipment itself are still essential to achieve carbon neutrality.

Various measures have been taken for component devices to improve the efficiency of heat pump equipment. For example, the following performance improvements have been promoted (Table 3.1).

An example of a heat exchanger with smaller diameter heat transfer tubes is the microchannel heat exchanger shown in Figure 3.9. Compared with conventional cross-fin heat exchangers, microchannel heat exchangers are not only lighter in weight due to the sole use of aluminum, and have higher performance and smaller

TABLE 3.1
Improvements for component Performance

Component	Improvements for Performance
Heat exchangers	Improved heat transfer performance, reduced refrigerant deflection, reduced diameter of heat transfer tubes to achieve higher density
Compressors	Reduced losses (reduction of refrigerant leakage loss in the device and higher efficiency of the built-in motor)
Blowers	Improved efficiency of motors

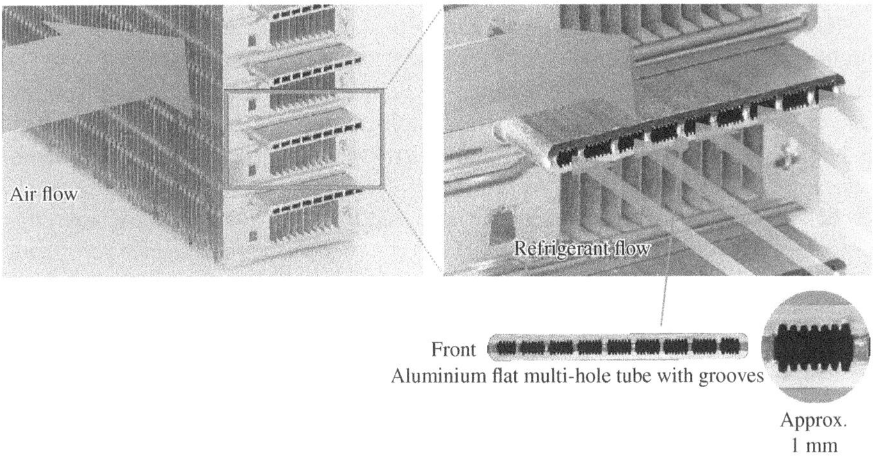

FIGURE 3.9 Microchannel heat exchanger. (Based on Daikin Industries Ltd, "The Challenge of Becoming Carbon Neutral" [7].)

size due to the finer flow paths in the tubes, but also greatly reduce the amount of refrigerant held, which is a substance contributing to global warming.

In this way, improvements in the performance of individual components have been promoted while improving the performance of the equipment as a whole. On the other hand, while the performance of components has improved significantly, the current situation is that for many components, the limits to further significant performance improvements are becoming apparent.

There is no doubt that component compactness remains important, but there is a common understanding that the improvement in components performance alone is not enough to achieve a significant improvement in overall performance that would contribute to the realization of carbon neutrality [8].

3.3.1.2 Improvements in Control Systems

The system technology, including operational control, has great potential to contribute to energy saving, so there is a need for its advancement.

The biggest innovation to system technology is that of inverter control. In conventional constant-speed equipment, the required capacity is adjusted by repeatedly starting and stopping the equipment in response to the required air conditioning and other loads. However, if such starts and stops are repeated, the performance of the equipment is greatly reduced. This is comparable to how driving a car through the city, which requires repeated starts and stops, is overwhelmingly less fuel-efficient than highway driving at a more constant speed.

On the other hand, inverters can be operated by adjusting the speed of the compressor. By operating at a speed commensurate with the load required, such as for air conditioning, continuous operation has been achieved without the equipment repeatedly starting and stopping. This has overwhelmingly improved the performance of such equipment.

Recently, however, the air conditioning load has been reduced, partly due to improved building insulation. When the load drops significantly, the equipment can no longer be continuously capacity-controlled by the inverter, resulting in intermittent operation and a significant drop in equipment performance (Figure 3.10). This is being addressed through new system control technologies that are less prone to intermittent operation.

At the same time, compressors are also needed that can expand the operating range with inverter and hardware technology to prevent unbalanced refrigerant flow in the heat exchanger that occurs at low loads. Such hardware technology also

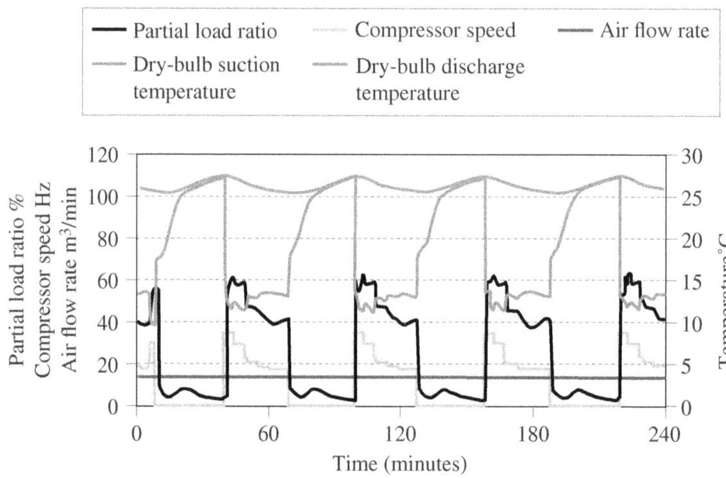

Test unit: Mitsubishi Heavy Industries Thermal Systems, Ltd. – 4-horsepower packaged air conditioner
Test conditions: Cooling mode, outdoor air temperature 29°C, load factor 25%, volume 147 m³
Test facility: Indoor Unit Room 2, Waseda University

FIGURE 3.10 Example of inverter operation at low load. EUT (Equipment Under Test): Mitsubishi Heavy Industries Thermal Systems Ltd, 4 hp package air conditioner. Test conditions: Cooling, ambient air temperature 29°C, load factor 25%, volume 147 m³. Test facility: Indoor unit room 2 at Waseda University.

needs to be aimed at improving system efficiency from the perspective of improving control performance, rather than improving system efficiency by improving device performance, as in the past.

3.3.2 The Need for Performance Improvement through Innovation of Performance Evaluation Methodology

Performance evaluation methodologies are important to accelerate equipment performance improvements. This is because equipment design naturally evolves in ways that increase the performance derived from these evaluation methods. At the same time, the annual power consumption and electricity costs are estimated and be made public based on the equipment performance values, which can serve as a major selection guideline for consumers.

Performance evaluation indicators have evolved from the coefficient of performance (COP)[1] to the annual performance factor (APF)[2] to more accurately evaluate equipment performance. However, users have pointed out that there is still a significant discrepancy with the operational performance in the actual use environment.

For example, the degradation of performance under low load operation has often been discussed, but it has been expressed in a lump sum by something like a simple performance degradation coefficient, regardless of the equipment.

Even if the performance degradation at low loads is improved by system technology, the effect is not reflected in the APF and the incentive to improve it does not work at all. Therefore, there is a strong need to eliminate the discrepancy between the performance derived from the performance evaluation methodology and the actual operational performance in order to realize equipment efficiency improvements.

To solve this problem, it is important to have a standard that can evaluate even dynamic operating characteristics that better reflect actual conditions. For equipment that uses air as the heat transfer medium, it is not even easy to determine the steady-state operating performance of the equipment due to the distribution of air temperature and air volume. For this reason, the performance of the equipment is measured in a test chamber where it can be correctly measured even with temperature and air velocity distributions, as shown in Figure 3.11. Test chambers are so diverse that no two have exactly the same shape, as they are highly dependent on the installation space and other factors. Trying to determine the dynamic operating performance in such a test chamber is even more difficult, as the performance of the equipment also depends on the structure and heat capacity of the test chamber.

At Waseda University, the operating environment is calculated virtually using emulators and given as operating conditions for the equipment, as shown in Figure 3.12, thereby eliminating factors influenced by the testing room, and a new testing method is being developed that enables equipment performance to be determined under the same operating environment. The establishment of such a method that enables equipment performance to be determined independently of the test chamber will give impetus to the development of more sophisticated equipment performance evaluation methodologies that reflect actual operation.

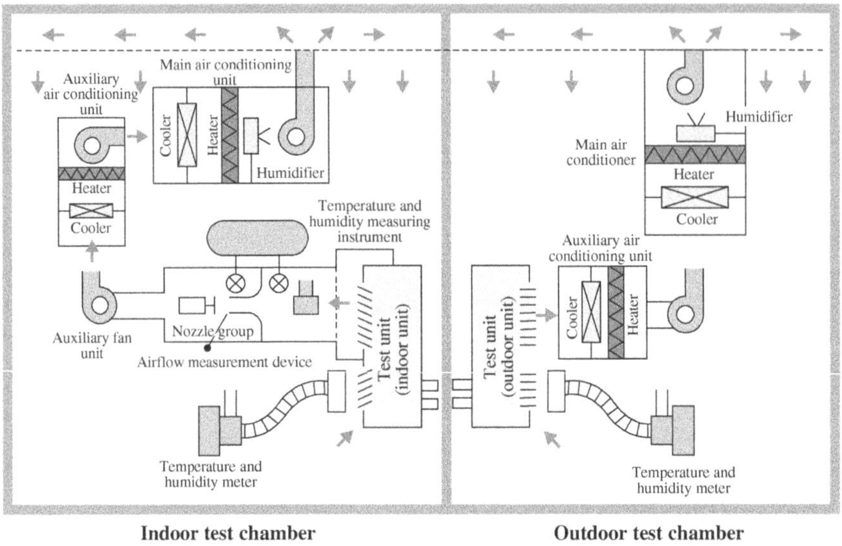

FIGURE 3.11 Example of air conditioner test equipment. (Based on Japan Air Conditioning and Refrigeration Testing Laboratory, "Introduction of test facilities" [9].)

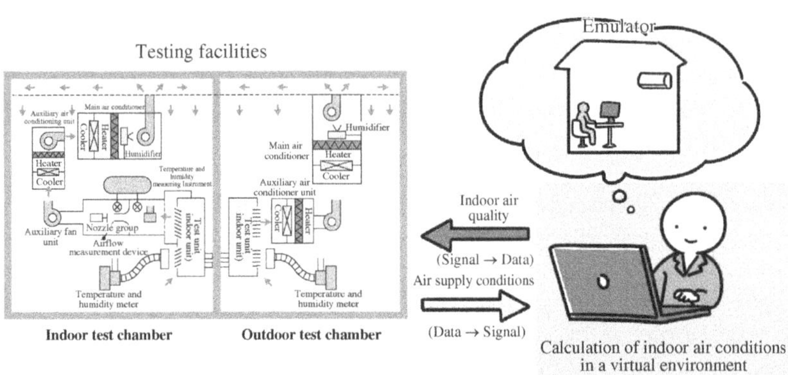

FIGURE 3.12 Example of test equipment using emulators. (Based on Saito et al., "Dynamic testing of air conditioners using emulators load testing equipment", Transactions of the Japanese Society of Refrigerating and Air Conditioning Engineers [10].)

3.4 NEXT-GENERATION REFRIGERANT CONVERSION

3.4.1 Regulation of Refrigerants Due to Environmental Issues

3.4.1.1 Why Refrigerant Conversion Is Necessary

As previously mentioned, refrigerant conversion from chlorofluorocarbon (CFC) and hydrochlorofluorocarbon (HCFC) refrigerants to hydrofluorocarbon (HFC) refrigerants as alternative refrigerants was originally initiated due to the depletion of the ozone layer. However, HFC refrigerants as alternatives have been found to be greenhouse gases (GHGs) with an extremely high global warming effect, and their drastic reduction is required.

These refrigerants have caused environmental problems that have had to be regulated, but conversely, the reason why they were introduced is because they are easy to handle and provide high equipment performance for a wide variety of temperature ranges. The development of heat pumps has been supported by the development of refrigerants with these excellent thermodynamic properties, most of which are now no longer available.

3.4.1.2 Low GWP Refrigerant Conversion

Future refrigerants will have to replace chlorine, the main causes of ozone layer depletion, with hydrogen, which is more flammable; to lower the global warming potential (GWP), the stability of the refrigerant itself will inevitably have to be decreased, which will present many challenges.

It goes without saying that we need to look for refrigerants with low GWP in order to achieve the Montreal Protocol, the Kigali Amendment, and carbon neutrality, but, on the other hand, we should not only focus on the development of low GWP refrigerants. This is because many refrigerants will be flammable in the future and, depending on the refrigerant, adverse human effects are also envisaged.

Therefore, it is essential to comprehensively consider equipment structures that can cope with different types of refrigerants, as well as changes in social structures, as follows

1. Refrigerants for which equipment performance can be guaranteed (including changes in performance during filling and over time)
2. Refrigerants that have no impact on humans or the environment (other than global warming), e.g. accumulation in the human body, toxicity, etc.
3. Social systems including equipment structure, safety standards, and legal systems that can guarantee safety of equipment
4. Social system including equipment structure, standards, and legal system to minimize refrigerant leakage
5. Social system including equipment structure, recovery system, technical standards, legal system, etc. to facilitate refrigerant recovery
6. Human resource development with a thorough knowledge of the characteristics of next-generation refrigerants, such as flammability and toxicity

In the next generation of low GWP refrigerant conversion, there will inevitably be trade-offs with various other issues, such as performance and safety, as already mentioned. Ultimately, the aim will be to convert to natural refrigerants, but the challenge will remain to deal with equipment that cannot be converted to natural refrigerants, including in terms of performance and safety.

At the present stage, as shown in Table 3.2, the adoption of natural refrigerants in refrigeration and freezing equipment is gaining ground. Large chillers in the air conditioning sector, such as centrifugal chillers, can be supported by HFO refrigerants. Integrated type equipment in which the dilatation type is used will

TABLE 3.2
List of Refrigerant Conversion Forecasts by Equipment

Type of equipment	Current main refrigerants	Candidate refrigerants in transition	Refrigerant candidates with GWP below 10 other than natural refrigerants	Final refrigerant candidate if only natural refrigerants are allowed
Residential air conditioning	R32	HFO mixed refrigerant (e.g. R454C)	HFO mixed refrigerant (under development)	A3 refrigerant (e.g. R290)
Commercial packaged air conditioners	R32 (partly R410A)	HFO mixed refrigerant (e.g. R454C)	HFO mixed refrigerant (under development)	Conversion to chiller with A3 refrigerant (e.g. R290) (*)
Multi-air conditioner for buildings	R410A	R32, HFO mixed refrigerant (e.g. R454B)	HFO mixed refrigerant (under development)	CO_2, conversion to chiller with A3 refrigerant (e.g. R290) (*)
Built-in refrigeration and freezing equipment	R134a, R404A, R410A (partly CO_2)	HFO mixed refrigerant (e.g. R454C)	HFO mixed refrigerant (under development)	CO_2, A3 refrigerant (e.g. R290)
Condensing units				
Refrigeration and freezing units	R404A, R410A	HFO mixed refrigerant (e.g. R454C)	HFO mixed refrigerant (under development)	CO_2, A3 refrigerant (e.g. R290)
Medium-sized condensing units	R404A, R410A (partly CO_2)	Possible A2L HFO mixed refrigerant conversion, but with challenges.	– HFO mixed refrigerant (under development)	CO_2, separate type showcases to be switched to built-in type and conversion to chiller with A3 refrigerant (e.g. R290) (*)
Large condensing units	R404A, R134a, R407C	– Possible A2L HFO mixed refrigerant conversion, but with challenges	HFO mixed refrigerant (under development)	A3 refrigerant (e.g. R290) chiller conversion (*)
Chiller				
Scroll chiller	R410A	R32, HFO mixed refrigerant (e.g. R454C)	HFO mixed refrigerant (under development)	A3 refrigerant (e.g. R290)
Screw chillers	R404A, R134a	HFO mixed refrigerant (e.g. R454C)	HFO mixed refrigerant (under development)	A3 refrigerant (e.g. R290)

(Continued)

TABLE 3.2 ***(Continued)***
List of Refrigerant Conversion Forecasts by Equipment

Type of equipment	Current main refrigerants	Candidate refrigerants in transition	Refrigerant candidates with GWP below 10 other than natural refrigerants	Final refrigerant candidate if only natural refrigerants are allowed
Centrifugal chiller	Low pressure refrigerant: R123, R245fa High pressure refrigerants: R134a, R513A	–	Low pressure refrigerant: R514A, R1233zd(E), R1224yd(Z) High pressure refrigerants: R1234yf, R1234ze(E)	–
Vehicle air conditioner	R134a	–	R1234yf	CO_2, R290
Transport refrigeration and freezing units	R404A, R134a	R452a	–	–
High temperature heat pumps	Below 100°C: CO_2, Above 100°C: R1234ze(E), R1233zd(E), R1336mzz(E)	–	Below 100°C:–, Above 100°C: R1234ze(E), R1233zd(E), R1336mzz(E)	Below 100°C: CO_2 Above 100°C: A3 refrigerant (e.g. hydrocarbons)

Note: When converting to chillers, it should be noted that there may be negative effects such as increased equipment installation costs and increased power consumption.

increasingly adopt HFO refrigerants and natural refrigerants, as refrigerant management within the equipment is easier.

On the other hand, hydrofluoroolefin (HFO) refrigerants are candidates for residential air conditioners, commercial packaged air conditioners, and multi-air conditioners for buildings, but a single refrigerant (e.g. HFO1123 or R1132(E)) causes disproportionation reactions[3] [11], so mixed refrigerants must be used, and their development is proceeding at a rapid pace.

However, the performance is expected to be inferior to that of conventional HFC-based refrigerants and, to avoid this, the piping diameter and the size of components, including heat exchangers, must be increased. A search is underway for a refrigerant that will maximally increase equipment performance while minimizing the size increase of the equipment, while also suppressing disproportionation reactions. With regard to compact machines, there has been a concerted move toward the introduction of hydrocarbons.

As regulations on chemicals that affect human health are being tightened internationally, some HFO refrigerants have been pointed out as being subject to

Perfluoroalkyl and Polyfluoroalkyl Substances (PFAS) regulations. If they become subject to the regulations, their use may be banned, so it is necessary to promote refrigerant conversion while keeping an eye on the trends in these regulations.

On the other hand, if refrigerants such as hydrocarbons and ammonia, which are natural refrigerants but flammable and toxic gases, are to be used, the relevant standards, such as the High Pressure Gas Safety Act, need to be fully considered.

3.4.2 Measures Required for Refrigerant Conversion

Considering that future refrigerants will be flammable, Japan, which is prone to natural disasters, must first develop technologies that can guarantee safety. At the same time, safety management will be necessary, and human resource development, including for installers, is an urgent task. Cooperation involving academia and other fields is necessary, including the enhancement of certification systems.

Prevention of refrigerant leakage and recovery/recycling are also important measures. On the equipment side, a connection system for piping that minimizes leakage and a mechanism that enables easy recovery of the refrigerant contained in the equipment are needed. Innovations in piping, such as easy-to-connect and inexpensive piping, are also necessary, and a bold shift from direct expansion type to inter-expansion type should also be considered. However, in some cases—for example, the conversion of air conditioning to water piping—will require major building renovations, which cannot be handled by the equipment alone, but must be done in cooperation with building management.

If refrigerants with environmental impact and safety issues are to continue to be used, the visualization of information management on refrigerants is essential, which is discussed in Section 3.6.

A certain period of time will be required for technological development, and it will be necessary to take sufficient time to anticipate regulatory trends and other factors. To accelerate the transition period for refrigerants, it is the accompanying measures that should be promoted as soon as possible.

3.5 RESOURCE CIRCULATION THROUGH THE CIRCULAR ECONOMY

3.5.1 Policies to Promote Resource Circulation

With policy support, the transition to a circular economy is expected to accelerate further in the future. Heat pumps are no exception and this needs to be achieved.

The prerequisite for this is the recovery of components, materials, and refrigerants from end-of-life equipment alongside their re-use in equipment manufacturing.

In heat pump technology, heat utilization is vital because of the importance of heat transfer, which means that materials with good heat transfer performance must be used. For this reason, relatively expensive materials such as copper and aluminum are often used. On the other hand, the number of constituent components is not very large. For this reason, the advancement of resource circulation is easy to aim for and has great merit.

TABLE 3.3
Current Status of Resource Circulation in Heat Pump Technology

Resource type	Main areas of use in heat pump technology	Current status of resource circulation
Iron	Outdoor unit cabinets, etc.	Waste sorted and recycled by waste disposal companies, etc. However, there is no system to ascertain the actual status, and the extent of recycling is unknown
Non-ferrous metals (copper, aluminum)	Heat exchangers, internal piping, internal/external piping, etc.	Waste sorted and recycled by waste disposal companies, etc. However, the extent of recycling is unknown as there is no system to ascertain the actual situation Copper, in particular, has a large environmental impact from mining and smelting
Plastics	Indoor unit cabinets, etc.	Dismantling and recovery of plastic parts is considered to be limited
		Use of recycled plastics and bioplastics in heat pump technology is limited
Fluorine compounds	Refrigerants (e.g. HFC)	The production of fluorite, a raw material, is unevenly distributed in some countries outside Japan, and there is a risk of resource constraints

According to the circular economy vision[4] published by the Ministry of Economy, Trade and Industry in May 2020, the current status of resources used in heat pump technology is shown in Table 3.3.

In heat pump technology, there are two types of product technology: those that require on-site installation work and those that do not (integrated type). The supply chain for both types is shown in Figure 3.13.

For this reason, for large equipment (e.g. centrifugal chillers), which are often supported by the manufacturer from installation to disposal, the circulation of equipment and refrigerant is also visible to the manufacturer.

For many other types of equipment, on the other hand, ownership transfers with the sale of the equipment, in the flow from supply of the product by the manufacturer to disposal of the equipment. The decision-making body changes throughout the equipment's life cycle, such as the dealer at the time of sale and the user at the time of use, and these are not interconnected, so the movement of equipment and refrigerants is only partially visible. Against the background of this current situation in the supply chain, resource circulation is not sufficiently advanced.

Residential air conditioners and other home appliances are positioned as "designated reuse promotion products"[5] and "designated resource-saving products"[6] under the Law for the Promotion of Effective Utilization of Resources, and the recycling rate of residential air conditioners is high at over 92% based on the Home Appliance Recycling Law. Although at first glance it appears that the majority of products are recycled, the reality is that only about 30% of equipment is ultimately recycled, due to the fragmentation of information between product sales and product use.

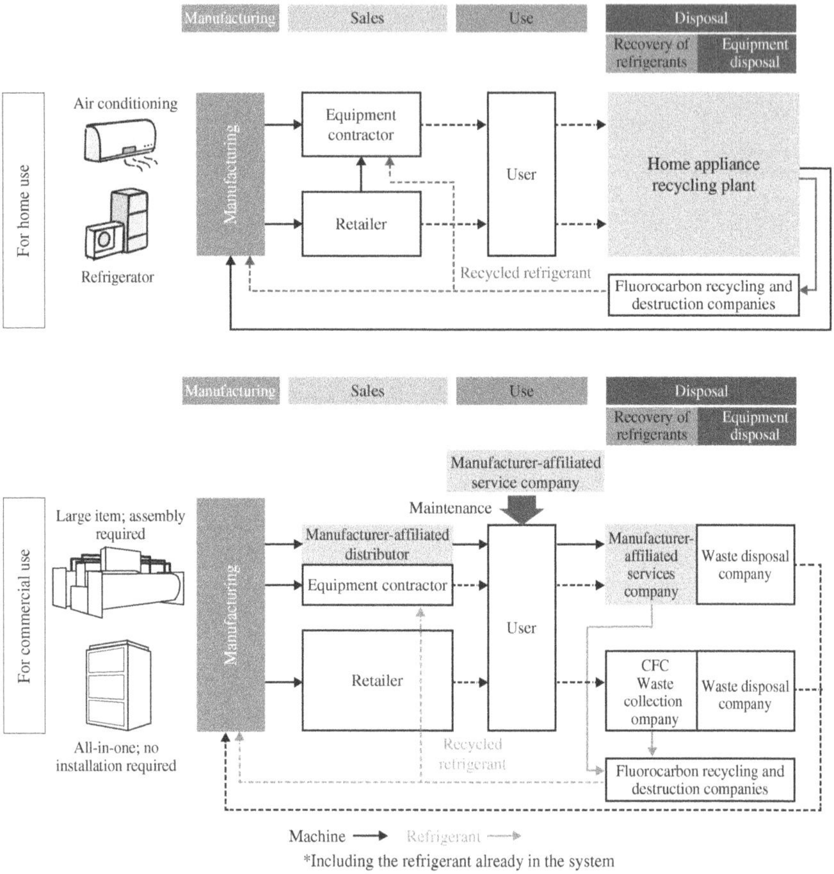

FIGURE 3.13 Current supply chain of heat pump technology. *Note:* Solid lines indicate flows visible to the manufacturer; dotted lines are invisible. Note that, in reality, there may also be distributors, users, etc., in layers, with invisible flows occurring within each of them.

3.5.2 Recycling Status of Commercial Equipment

3.5.2.1 Recycling Rates Lower than for Residential Use

For commercial air conditioning and refrigeration and freezing equipment, there is no such thing as a recycling law and they are disposed of as industrial waste, as shown in Figure 3.14. Of course, they are also broken down into iron, copper, and aluminum for reuse, but the actual situation is not known. In any case, the construction of equipment structures and social systems that enable equipment recovery is required for the many residential and commercial unit sold.

The recovery of air conditioning and condensing units (e.g. separate type showcases) for commercial use, where recovery rates are low, should be promoted first. However, at present, a system is not in place to ensure that these appliances are placed on the collection route, despite the tightening of regulations due to previous

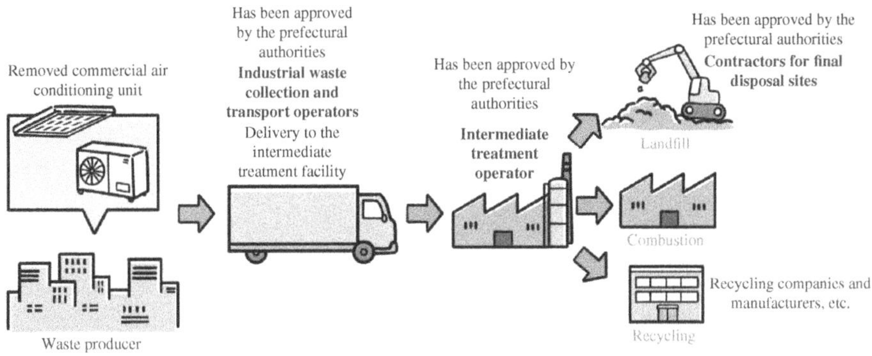

FIGURE 3.14 Disposal of industrial waste. (Based on Eco Plan Inc., "Commercial air conditioners waste flow" [12].)

legal changes. A system to ensure that the flow of equipment can be reliably identified and appropriate treatment ensured needs to be put in place as soon as possible.

As well as a system to ensure that equipment can be properly collected, consideration should also be given to the ease with which equipment can be dismantled and re-used. As already mentioned, heat pump technology has a relatively small number of components and the reuse of them can be easily promoted. However, as usual, heat exchangers and compressors are focused only on compactness and performance improvement. Fundamental changes in equipment design are needed to facilitate the realization of a circular economy.

It is reasonable to proceed not only with the equipment, but also with the recovery of refrigerants at the same time.

If the two are pursued separately, neither equipment nor refrigerant recovery rates will improve unless they are legally binding.

3.5.2.2 Recovery of Fluorocarbons

In addition to the fact that the production and consumption of fluorocarbons used as refrigerants has been reduced in the wake of the Kigali Amendment, but their raw material, fluorspar, is also a resource that is highly concentrated in specific regions of the world. If the fluorocarbons in equipment on the market in the future leak at the current rate and cannot be recovered sufficiently at the time of disposal, the amount recovered cannot be recycled and the direction of refrigerant conversion will be narrowed.

Simply strengthening regulations and initiatives in each cross-section of production, use, and disposal is not enough to fully understand the overall flow and the problems that lie in each individual case, and there are limits to effective countermeasures. Here, too, it is important to ensure traceability of the equipment itself and refrigerants using DX (see Section 3.6).

Furthermore, the current commercial distribution system is not always optimal for realizing a circular economy of resources. Nor is it without inconveniences; so, for example, in residential air conditioning and built-in refrigeration and freezing equipment, a move from the sell-out model to new business models such as leasing and subscription (where traceability of equipment can be easily ensured) should be explored.

3.6 CHALLENGES IN PROMOTING DX AND EXPECTATIONS FOR THE FUTURE

3.6.1 Lack of Digital Transformation in Heat Pump Technology

The various challenges and needs mentioned above make DX essential in finding a solution. This section therefore summarizes the challenges and needs of DX and measures to solve them.

In heat pump technology, there are many heat and fluid phenomena that occur inside the equipment, the detailed physical quantities of which cannot be measured inexpensively and, as a result, it is difficult to digitalize information on energy-related heat quantities and efficiency. For this reason, heat pump technology can be regarded as an analogue technology that is difficult to digitally transform.

This situation also means that, in practice, in the overwhelming majority of cases, it is not well understood how the equipment is being operated, which makes it difficult to discover how to improve the performance of the equipment as a system technology.

3.6.2 Aiming at Energy Saving through the Use of EMS

With regard to equipment control, initially analogue control centered on a simple, whole unit, on–off control. However, digitalization is gradually progressing, including the introduction of inverter control, and capacity control is becoming easier.

Furthermore, it is important to control not only individual equipments but also multiple equipments externally using an energy management system (EMS), as shown in Figure 3.15, in order to save energy.

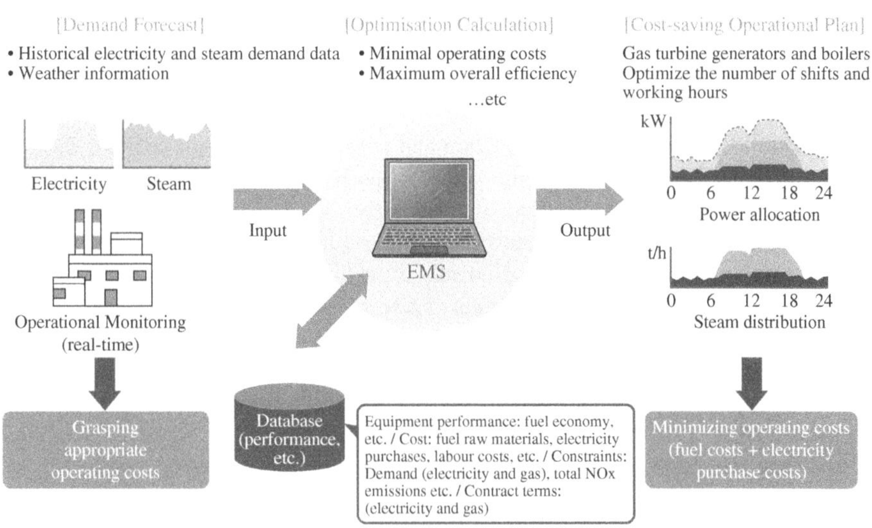

FIGURE 3.15 Image of energy saving with EMS. (Based on Kawasaki Heavy Industries Ltd, "Reduction of operating costs by optimising the energy management system (EMS) equipment configuration" [13].)

Currently, external equipment control is mostly restricted due, to a large extent, to equipment protection controls to protect rotating machinery. However, several manufacturers have started collecting real-time equipment operating data through their own Internet of Things (IoT) technology.

Services have also started to collect data through the internet, analyze operating information and feed back the results to the user. The level of achievement of the technology is unknown, as the details are rarely revealed to the outside world and there is no way to verify its effectiveness, but if it becomes possible to collect energy and efficiency information from equipment, a more advanced EMS can be built. In the future, rational heat utilization by EMSs will also progress.

There are also high expectations for equipment simulation technology. With regard to heat pump technology, which involves heat and fluid phenomena, its simulation is quite challenging. On the other hand, there has been progress in numerical fluid analysis, and the field is being prepared for a further leap forward in technology. As simulation technology becomes more sophisticated, it can be used for a wide range of applications, including prediction of equipment operating performance, fault diagnosis, optimum design, and control. Waseda University has already achieved transient analysis, including component separation of non-azeotropic mixed refrigerants.

The use of such simulation technology will contribute to the development of digital twins etc., and if verification of the effectiveness of EMS is also advanced, it will contribute to the development of advanced EMS.

3.6.3 Digitalization of Production Management

Furthermore, for leakage control and recovery of refrigerants, it is essential to have corroboration in the form of data, from the production of refrigerants to their recovery. Of course, fragmentary information is beginning to be collected, but this does not facilitate progress in refrigerant management: it is expected that DX will enable the establishment of technologies and systems that can digitally and effectively collect, store, and manage information on refrigerants from production to recovery (e.g. filling volume, recovery volume, leakage volume, regeneration volume) in a consistent manner.

Furthermore, in heat pump technology, which makes extensive use of valuable resources, it is also important to promote the recovery and re-use of the equipment itself. It is expected that the digitalization, collection, accumulation, and management of product information on equipment and information related to recovery and recycling will facilitate the recovery and re-use of equipment. Methods for collecting and managing this information have already been proposed in Europe and elsewhere by the digital product passport (DPP) and other such schemes.

3.7 GLOBAL HEAT PUMP TECHNOLOGY TRENDS

3.7.1 Penetration Rate of Inverter Air Conditioning

In order to achieve carbon neutrality, the first priority in the future will be the proliferation of inverter machines as high-performance machines. Japan was the first country in the world to promote inverter-equipped residential and commercial

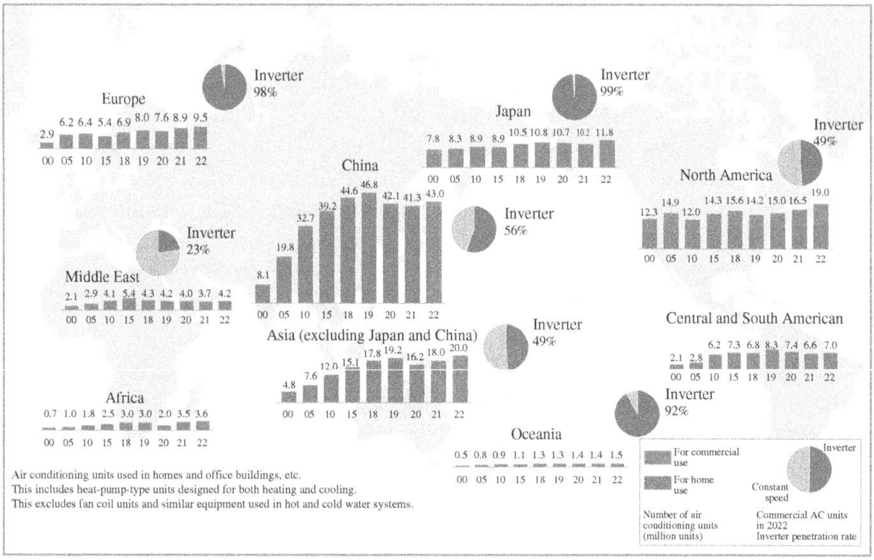

FIGURE 3.16 Global demand for air conditioner and inverter air conditioner conversion rate. *Note:* 2000–22/Estimate: 117.77 million units (2022). (Based on Japan Refrigeration and Air Conditioning Industry Association, "Estimated global air conditioner demand" [14].)

air conditioners, and the recent market size has remained flat at around 10 million units, whereas the global demand has expanded to over 100 million (Figure 3.16).

As shown in this demand trend, Japan has widely adopted air conditioning and has generally completed the shift to the inverter type. On the other hand, the situation varies in regions where air conditioning is not widespread (e.g. Africa) and regions where air conditioning is widespread, but inverter air conditioning equipment is not (e.g. North America).

3.7.2 Air Conditioning and Water Heater Utilization by Region

In order to consider the challenges of refrigeration and air conditioning equipment as a whole, it is important to capture the global market.

However, different regions use different types of equipment and different ways of using them due to their natural environment and cultural background. The differences are particularly significant in air conditioning and water heating, which differ significantly between Japan, Europe, and North America. Table 3.4 shows the characteristics of air conditioning and water heaters by region.

Air conditioning systems can be broadly divided into ducted systems that send heated air throughout the building as used in North America (central method whole-house air conditioning) and individual air conditioning systems without ducts in other regions.

There are also differences between regions, particularly in heating systems, reflecting such factors as the magnitude of cooling and heating demand.

TABLE 3.4
Characteristics of Air Conditioners and Water Heaters by Region

Region	Air conditioning		Hot water supply
	Cooling	Heating	
Japan (temperate climate region)	Ductless and individual air conditioning		Gas water heaters (instantaneous type) and heat pumps (hot water storage type: capacity 370–560 L level)*2
Japan (cold regions)		Combustion type (kerosene/gas)	
Europe	Ductless type and individual air conditioning	Hot water circulation type (combustion type, electric heater)*1	Electric heaters and gas combustors (hot water storage type: 200 L capacity level)
North America	Ductless, individual air conditioning, window air conditioning	Ductless type, individual air conditioning, window-air conditioning, gas combustion, electric heater, hot water fan coil	Gas-fired, electric heaters (hot water storage type: 200 L capacity level)
Tropical Asia (India, Southeast Asia, etc.)	Ductless, individual air conditioning	Ductless, individual air conditioning	Electric heaters (hot water storage type: capacity 25–200 level)
China (China)	Ductless, individual air conditioning		Electric water heaters (hot water storage type; capacity (100–200 L level)*3
		Hot water circulation type	

Notes: *1 Heat pump hot water heating systems are also increasingly being introduced.
*2 Heat pumps are also increasingly being installed in houses.
*3 Introduction of heat pumps is also progressing.

Air conditioning in Europe was based on the use of heating only, and building equipment has been designed around heating. Heating has often been of the hot water circulation type, with combustion equipment and electric heaters predominantly, although heat pump hot water heating systems (air to water)[7] are also being introduced. In the case of North America, gas combustion and electric heaters are also used for heating.

On the other hand, when looking at water heating, heat pumps are used only in Japan and China and are not yet widespread worldwide. Combustion equipment and electric heaters are the mainstays.

Thus, the developed markets of Europe and North America are undergoing distinctive market formations. In addition, the rapidly expanding markets of India, tropical Southeast Asian countries and China are close to the Japanese system.

Table 3.5 shows examples of typical air conditioning equipment (see Figure 3.17 for photographs).

3.7.3 Refrigerant and Energy-Saving Compliance Status by Region

The refrigerants used in heat pump technology are mainly fluorocarbons, but different types of equipment use different refrigerants. Recently, refrigerant conversions

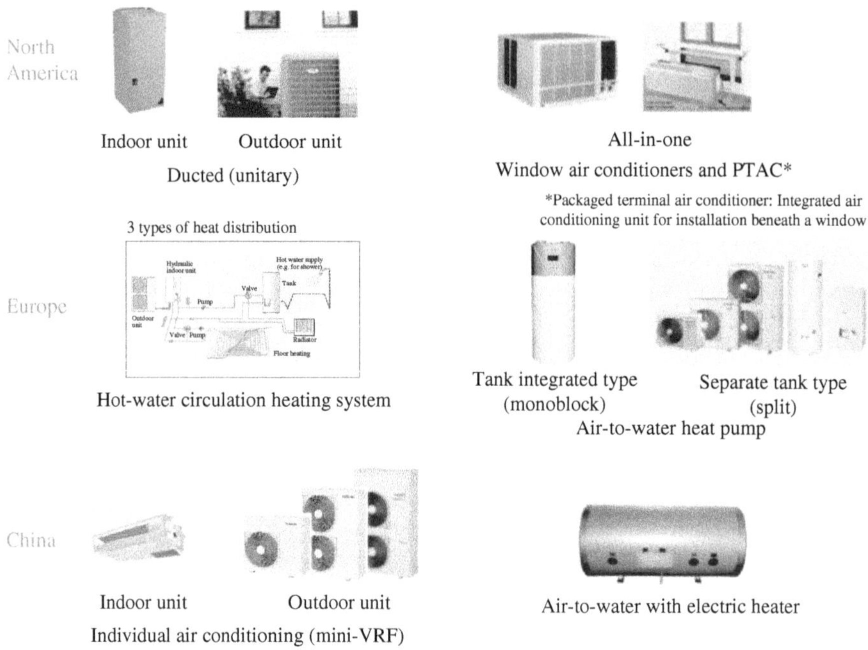

FIGURE 3.17 Characteristic air conditioners and water heaters by region.

TABLE 3.5
Examples of Air Conditioning Equipment Used in Various Regions

Country/region	Manufacturer	Product name
North America	Carrier	Vertical AHU 40VMV [15]
		PRODUCTS, INFINITY SYSTEM [16]
		Window Room Air Conditioner R-22 Refrigerant [17]
	AJ Madison	Packaged Terminal Air Conditioners (PTACs) [18]
Europe	FUJITSU GENERAL in Europe	SPLIT TYPE High Power Series, System Configuration (ATW SYSTEM) [19]
	MICOE CORPORATION	Low Energy Consumption Monoblock Indoor Heat Pump Water Heater [20]
	Toshiba Carrier Corporation	Air to Water Heat Pump System, ESTIA R410A series5 [21]
China	Dongshi Air Conditioning Sales (Shanghai) Co.	MiNi SMMS Series [22]
		Outdoor Unit [23]
	Focus Technology Co.	Made-in-China 'Horizontal Hot Water Tank [24]'

have been progressing in line with the trend toward refrigerant regulation, but in Europe, where the world's most stringent fluorocarbon regulations are in place, the use of natural refrigerants is progressing. In North America, HFO refrigerants are being actively developed and new ones are being adopted. In developing countries such as tropical Southeast Asian countries and China, refrigerant conversion

TABLE 3.6
Comparison of Efficiency Regulations for Air Conditioner by Country (as of March 2020)

Country	Metric	MEPS	Equipment Capacity	Normalized to Japan APF
Australia/New Zealand	EER/AEER	3.66	<4 kW	N/A
China	SEER	4.30	<4.5 kW	4.69
EU (GWP>150)		4.60	<6 kW	4.2
EU (GWP<150)		4.14	<6 kW	3.84
Japan	APF	5.50	4-5 kW	5.50
Korea	EER	3.50	<4 kW	N/A
USA	SEER	3.81	<19 kW	4.28
USA	SEER2	3.93	<13 kW	N/A

Note: MEPS regulations of various countries are standardized and compared with the Japanese APF. Note that direct comparison is difficult due to differences in measurement conditions.
Source: Based on Cadeo Group LLC, "Domestic Air Conditioner Test Standards and Harmonisation Final Report" [26].

is progressing slowly, although residential air conditioners employing R290 are becoming more widespread in China [25].

Energy-saving initiatives are also being implemented in many countries, and minimum energy efficiency regulations (Minimum Energy Performance Standards (MEPS)) for products have been established. In Europe, Lot 10 of the Energy-related Products (ErP) Directive and the Ecodesign Regulation specify minimum energy efficiency regulations, while the USA also has a MEPS regulation. It should be noted, however, that the efficiency specified overseas differs in terms of the temperature conditions under which it is measured and other evaluation methods (Table 3.6).

As described above, the situation differs significantly from region to region around the globe in air conditioning equipment and small water heaters, but there are no significant differences in refrigeration equipment and industrial heat pump technology.

For example, with regard to air conditioning systems, the duct system is not a good system from the perspective of energy saving, because of the large heat losses during air transportation in ducts. However, it is not easy to change the equipment or system, as the installation method is strongly related to the building construction. Therefore, it is necessary to introduce appropriate equipment in a way that fits each region. In Southeast Asia and India, where air conditioning systems are expected to become more widespread in the future, the proliferation of high-performance equipment, mainly Japanese inverters, should be promoted as soon as possible, as these regions do not have their own specific air conditioning cultures.

With regard to refrigerant conversion, it is expected that all countries will eventually align their approaches, albeit at different speeds.

Having discussed the situation surrounding the equipment above, there is a huge difference users between Japan and the rest of the world in the public awareness of air conditioning in terms of set temperatures. Japan adheres to set temperatures relatively well. For example, air conditioning is often set at around 26–27°C, whereas in the rest of the world, it is common to find temperatures set below 20°C for cooling.

Simply changing this set temperature can have an energy-saving effect of up to several tens of percent—an important fact that should be shared with the world.

NOTES

1. Coefficient of performance (COP): An energy efficiency indicator that expresses energy consumption efficiency. It evaluates performance at the rated point.
2. Annual performance factor (APF): An energy saving indicator that expresses year-round energy efficiency, which was defined as a reference value for energy saving indicators from COP by the Law Concerning the Rational Use of Energy (Energy Conservation Law), which was revised in September 2006. Indicates operating efficiency after 1 year of operation under a defined operating environment based on Japanese Industrial Standards (JIS) standards.
3. Disproportionation reaction: A chemical reaction in which two or more molecules react with each other and break down into two or more different types of products, depending on the energy provided externally.
4. Circular economy vision: Officially known as Circular Economy Vision 2020, a policy aimed at, among other things, shifting to a more circular business model.
5. Designated reuse promotion products: Refers to products that are required to promote the use of recycled resources or recycled parts.
6. Designated resource-saving products: Products that are required to rationalize the use of raw materials and other materials, promote long-term use, and reduce the generation of other used goods.
7. Hot water heating system: A system in which hot water is circulated for heating and hot water supply. Combustors and electric heaters used to be the norm, but heat pump systems are becoming more widespread. Some systems can also be used for cooling.

REFERENCES

1. Panasonic Corporation and Panasonic Marketing Japan Co. Ltd., Cold climate air conditioner special website, "Why air conditioner stops heating in winter," 2023) https://panasonic.jp/aircon/kanreichi/enecharge.html
2. Mitsukuni Sano (Toshiba Carrier Corporation) et al., Toshiba Review, "Store and office air conditioners for cold regions that combine high heating performance and energy saving," vol. 75, No. 4, p.76, 2020, https://www.global.toshiba/content/dam/toshiba/migration/corp/techReviewAssets/tech/review/2020/04/75_04pdf/b07.pdf
3. New Energy and Industrial Technology Development Organization (NEDO), "Frontier to energy saving: Research and development of innovative utilization technology for untapped heat energy," p.8 (January 2023), https://www.nedo.go.jp/content/100927351.pdf
4. Japan Electro-Heat Center/Industrial Heat Pumps.com, "Reducing steam losses," 2023, https://sangyo-hp.jeh-center.org/heatpump_steam_loss.html
5. New Energy and Industrial Technology Development (NEDO), News release, "Report on survey of exhaust heat in factory facilities in 15 industries," 2019, https://www.nedo.go.jp/news/press/AA5_101074.html
6. Toshiba Carrier Corporation, General Catalogue of Heat Source Systems (July 2022), p.178.

7. Daikin Industries Ltd, "The challenge of becoming carbon neutral," 2023, https://www.daikin.co.jp/tic/technology/carbon
8. Japan Refrigeration and Air Conditioning Industry Association, The Nikkan Kogyo Shimbun Ltd, "Practical performance and potential of 'heat pumps'", "Chapter 4: Efficiency limits and future prospects of practical equipment (subject: air conditioning)", pp.77–79, 2010.
9. Japan Air Conditioning and Refrigeration Testing Laboratory, "Introduction of test facilities," 2023, https://www.jatl.or.jp/testing-facilities.html
10. Kiyoshi Saito (Research Institute for Sustainable Environmental Energy Society Co-Creation, Waseda University) et al., Collection paper of the Japanese Society of Refrigerating and Air Conditioning Engineers, "Dynamic testing of air conditioners using emulator load test equipment", Fig. 8, p.1, 2023, https://www.jstage.jst.go.jp/article/tjsrae/advpub/0/advpub_22-17_OA/_pdf/-char/ja
11. New Energy and Industrial Technology Development Organization (NEDO), "Successful suppression of autolysis reactions of low GWP (global warming potential) refrigerants with potential application to residential and commercial air conditioners", 2022, https://www.nedo.go.jp/news/press/AA5_101539.html
12. ECO-PLAN Co., Ltd., "Infiltration of the industrial waste treatment site! What happens after disposal of removed commercial air conditioners? SCOPE 3-5 reduction tips, 'Commercial air-conditioners the flow of waste'," 2023, https://www.ecology-plan.co.jp/information/34710/
13. Kawasaki Heavy Industries, Ltd, "Energy management system (EMS): Reducing operating costs by optimising equipment configuration," 2023, https://www.khi.co.jp/energy/management/energy_management_system.html
14. Japan Refrigeration and Air Conditioning Industry Association, "Global air conditioner demand estimation," (2023), https://www.jraia.or.jp/statistic/2306_aircon.pdf
15. Carrier, "Vertical AHU 40VMV," 2023, https://www.carrier.com/commercial/en/us/products/variable-refrigerant-flow/carrier-vrf-products/carrier-vrf-indoor-units/40vmv/
16. Carrier, "Products, infinity system," 2023, https://www.carrier.com/residential/en/us/products/infinity-system/total-comfort/
17. Carrier, "Window room air conditioner R-22 refrigerant," 2023, https://www.carrier.com/commercial/en/sa/products/residential/window-room-air-conditioners/51h/
18. AJ Madison, "Packaged terminal air conditioners (PTAC)," 2023, https://www.ajmadison.com/guides/air_conditioner/ptac/
19. FUJITSU GENERAL in Europe, "SPLIT TYPE High Power Series, System Configuration (ATW SYSTEM)," 2023, https://www.fujitsu-general.com/g-eu/products/atw/split-highpower/system.html
20. MICOE CORPORATION, "Low Energy Consumption Monoblock Indoor Heat Pump Water Heater," 2023.
21. Toshiba Carrier Corporation, "Air to Water Heat Pump System, ESTIA R410A ESTIA series5," 2023. https://www.toshiba-carrier.co.jp/global/products/heating/estia-r410a/index.htm
22. 东芝开利空调销售 (Shanghai) Co., Ltd., "MiNi SMMS series", 2023.
23. 东芝开利空调销售(上海)有限公司, "Outdoor unit," 2023, https://www.toshiba-air-conditioning.com.cn/dongzhi/product/list?category_id=60138d80c889453c8d6b6da6
24. Focus Technology Co Ltd, Made-in-China "Horizontal hot water tank," 2023
25. Acceleratejapan, "R290 RAC sales in China exceed 270,000 units," 2023, https://acceleratejapan.com/chinese_rac_makersandrsquo_domestic_r290_sales_exceed_270_000_units/
26. Cadeo Group LLC, "Domestic air conditioner test standards and harmonization final report," p.25, 2020, https://www.iea-4e.org/wp-content/uploads/2020/12/AC_Test_Methods_Report_Final_V2_incl_JP_KO.pdf

4 Social Implementation of Next-Generation Heat Pump Technology

OVERVIEW

Heat pump technology has been used for a wide range of applications and has pervaded society thanks to its excellent properties. However, social conditions and trends surrounding heat pump technology are forcing changes.

The response to this is to aim for overall optimization of heat pump technology, including the complex value chain surrounding it. A grand design for overall optimization needs to be carried out as soon as possible, and backcasting, including rule-making, is needed to resolve issues.

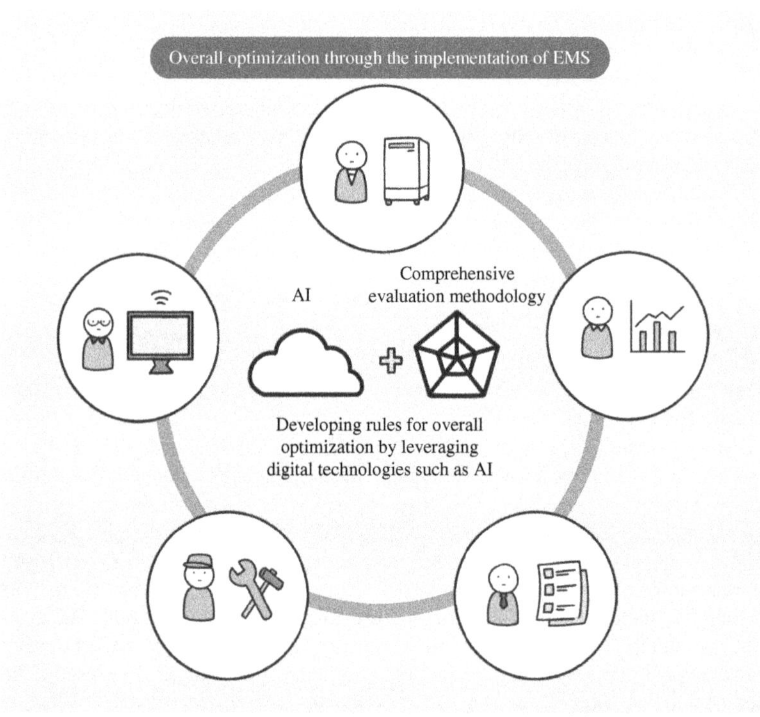

FIGURE 4.1 Overall optimization of next-generation heat pump technology.

DOI: 10.1201/9781003666950-4

This chapter therefore provides an example of the limitations that can arise when focusing only on individual events when proceeding with solutions. It then presents an example of how next-generation heat pump technology can be achieved through initiatives aimed at overall optimization (Figure 4.1).

4.1 LIMITATIONS OF INDIVIDUAL PROBLEM-SOLVING

4.1.1 Necessity of Introducing EMS across Manufacturers

It has already been mentioned in previous chapters that the technology for energy management systems (EMS) for coupling and optimizing the operation of multiple equipments in order to achieve rational heat utilization technology is also making steady progress. However, information related to heat energy, which is essential for heat pump technology, is currently often not disclosed. There is no so-called visualization.

Of course, even under such circumstances, various types of coupled control are expected to progress if the equipment is from the same manufacturer, but this will only result in partial energy optimization. Also, the effects of the EMS on the user side remain largely unknown, and the momentum for its introduction is not likely to increase.

However, it is important to promote coupled control even between equipment from different manufacturers and to promote overall optimization of energy use throughout the process. If the rate of disclosure and dissemination of energy-related information remains stagnant, this is likely to become a bottleneck in the advancement of EMS (Figure 4.2).

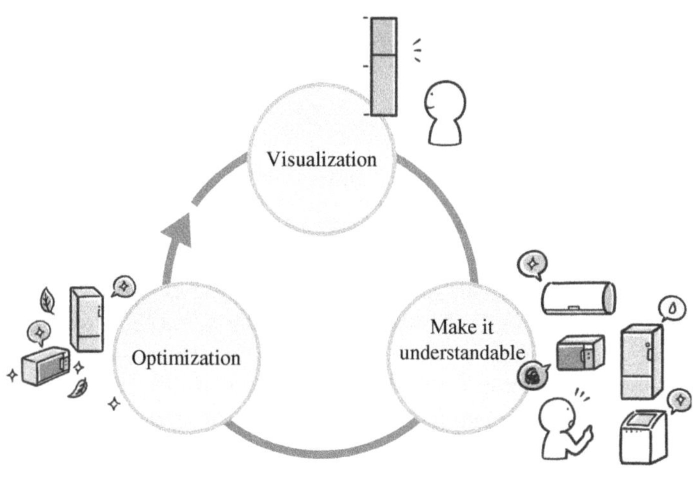

FIGURE 4.2 An environmental management system.

4.1.2 Challenges in the Practical Application of Low GWP Refrigerants

The low global warming potential (GWP) refrigerants that will be put to practical use in the future are flammable and toxic, so safety measures will naturally be required. However, this may also reduce equipment performance, so even if the GWP is lowered, if equipment performance is significantly reduced, energy saving will not be achieved.

Refrigerant conversion is also required by law and will certainly progress. However, if, for example, in air conditioning, standards are left in place that do not take into account performance improvements through system technology or the effect of reduced air conditioning load due to building insulation, improved equipment performance will have to be met by larger devices (equipment). If this continues, equipment will become isolated from global standards, hindering the global deployment of high-performance technology.

The Japanese residential air conditioner indoor units are already abnormally large. It is a well-known fact that heat pump water heaters are rarely sold overseas because the equipment has been optimized for hot water usage tailored to domestic conditions in Japan.

4.1.3 The Need for Linkage with Refrigerant-Related Technologies

The importance of refrigerant and equipment distribution information management has been discussed above, and manufacturers will be hesitant to introduce low GWP refrigerants if rules for safety are not developed at the same time. The inter-expansion requires major changes on the construction side, including changing from refrigerant piping to water piping. Manufacturers alone cannot push ahead with this.

No matter how much effort is put into improving equipment performance and refrigerant conversion, if resource circulation does not proceed and equipment prices soar, low-end products with poor performance will appear on the market, which may delay the reduction of CO_2 emissions. Such concerns have already begun to emerge in South-East Asia and elsewhere. The flood of low-end products with poor performance will, in the long run, slow down investment in environmental technology.

4.1.4 The Need to Build Social Systems for Social Implementation

Heat pump technology is limited to focusing only on the issues of individual, compartmentalized events, but issues such as decarbonization, energy saving, refrigerant conversion, and resource circulation must all be resolved as a set.

Furthermore, even if we look only at mechanical technology, we cannot expect significant advances in heat pump technology. If we do not go beyond the technology and consider the construction of a supply chain that can safely and rationally proceed with refrigerant recovery and equipment recovery, as well as human resource development, we will not be able to deliver safe and reasonably priced products to consumers.

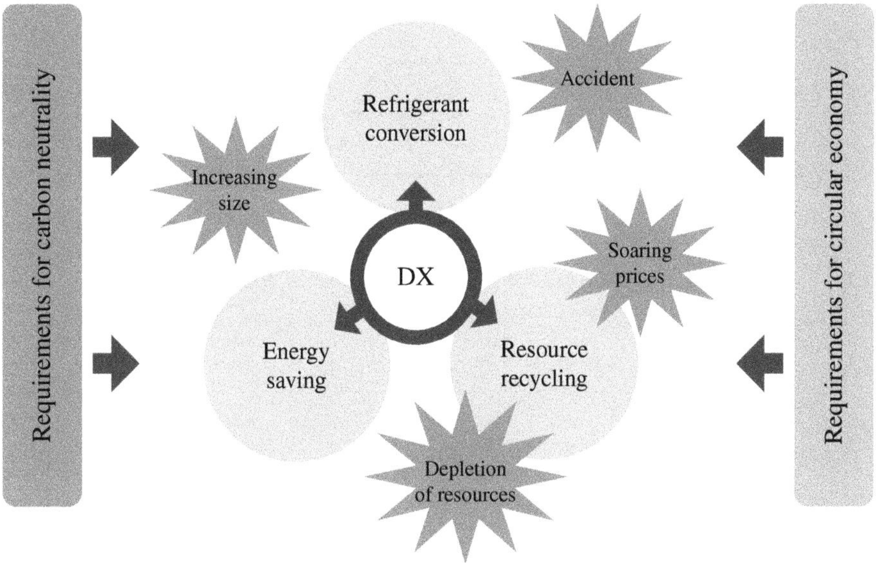

FIGURE 4.3 Current status and interdependence of individual issues related to heat pump technology.

In order to make it a sustainable next-generation heat pump technology, it is necessary to go beyond mere mechanical technology and build social systems based on the needs, demands, and usage of the society around it (Figure 4.3).

4.2 ESTABLISHING SYSTEMS FOR OVERALL OPTIMIZATION

4.2.1 The Necessity for Centralization of Systems

So far, we have looked at the events that heat pumps need to deal with and the direction in which each of these issues should be resolved. However, as the individual events are not independent but are strongly interrelated, simply responding to each one individually is short-sighted and limiting. For this reason, the basic idea of overall optimization is to try to solve challenges by taking a cross-sectional view of the individual events, while trying to find the optimum solution for the technology and social system as a whole.

Digital transformation (DX), such as Internet of Things (IoT), simulation, digital twin, and next-generation communication technologies, are particularly important foundations for the means to realize overall optimization. This is because they enable the visualization of diverse information, such as information on equipment operation, equipment life cycle, and refrigerant management, and the accumulation of that information, so that no one aspect is overshadowed and measures can be taken as a whole.

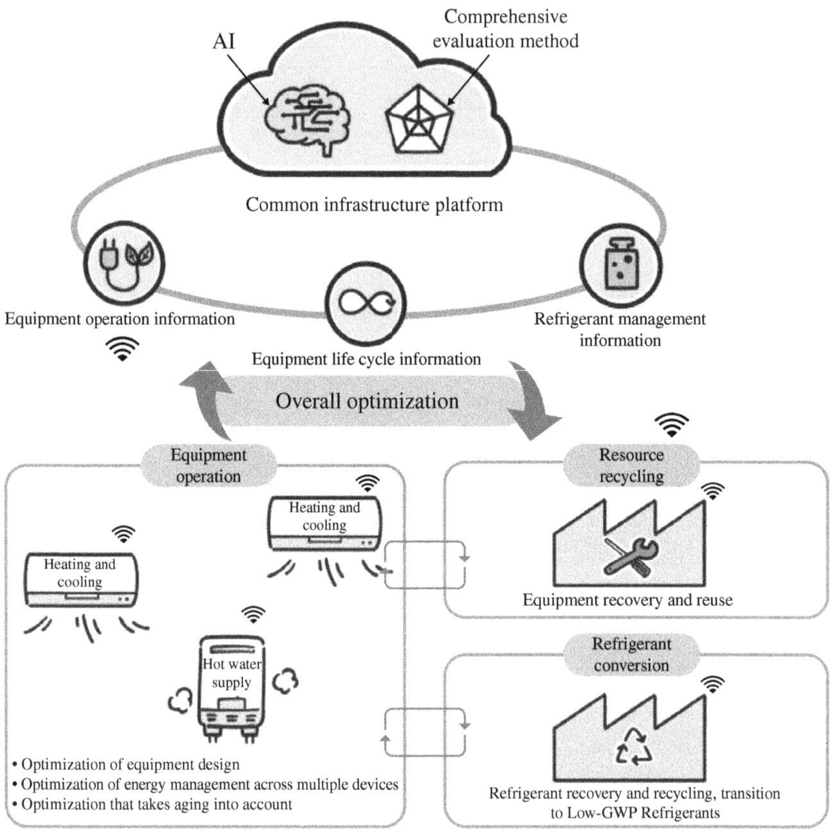

FIGURE 4.4 Image of next-generation heat pump technology that has achieved overall optimization.

The systemization of equipment that has been integrated and optimized with digital data in various aspects through such digitalization and networking, and the common platform that surrounds it, is what next-generation heat pump technology is aiming for (Figure 4.4).

If the aim is to achieve overall optimization through the resolution of individual issues while the ultimate goal is unclear, the rules for individual issues and the organization for dealing with them will also be systematized.

Furthermore, it is easy to imagine that multiple information management systems corresponding to each individual event will have been constructed, which will be extremely difficult to modify by the time overall optimization is achieved. An example is the current attempt to centralize management of administrative information by linking it to the My Number system, which is extremely difficult due to the multiplicity of information in different formats.

Therefore, the most important thing is to establish a grand design for overall optimization toward the realization of next-generation heat pump technology as quickly as possible. The key elements for the realization of the next generation of heat pumps are

1. Realization of centralized management of all information through DX
2. Establishment of a comprehensive evaluation methodology

4.2.2 The Construction of an Information Platform

An important part of promoting overall optimization in next-generation heat pump technology is the creation of a platform that serves as a common platform for various information systems. This is because if the rules for data exchange and mutual control of equipment are not properly established, mutual cooperation between different events, manufacturers, etc. will never progress.

For example, as shown in Figure 4.5, the digital product passport, which will be mandatory in Europe, will digitalize information such as the product's origin, materials used, recyclability, and dismantling method. The aim is to share this information with actors involved in the product value chain.

It is important to carry out the basic design of such a common platform, but Europe is very good in this area, while Japan is inevitably lagging behind. This is considered to be the biggest factor slowing down technological innovation in Japan. Urgent action is needed.

Centralization of all information should include a wider range of dynamic data than the concept of a digital product passport, which accumulates static information.

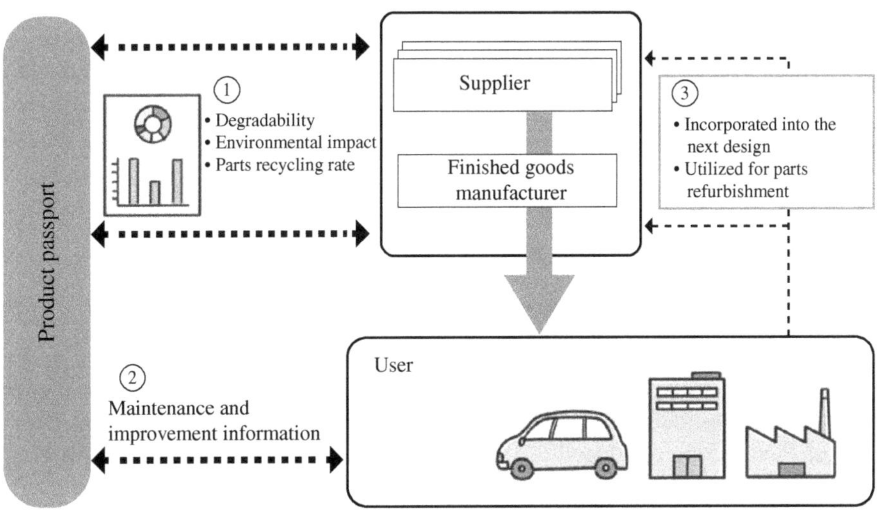

FIGURE 4.5 Mechanism of the digital product passport. (From Hitachi Research Institute, "1. What is product passport" [1].)

Centralization of information is already underway, including not only static information on equipment such as battery passports but also dynamic information, including real-time equipment performance.

For batteries, however, the amount of information required is far less, and the technology is also electrical, making such information integration relatively easy. For heat utilization equipment, the amount of information required is much greater than that for electrical equipment, making the integration of information overwhelmingly difficult. It is precisely because it is so difficult that, if realized, it will be a technology in which Japan will have an advantage.

4.2.3 Aim for a Centralized Management System for All Information

In the future, it is expected that not only systems for managing the equipment itself, but also systems for managing the operating status of the equipment as well as systems for managing refrigerant will be built for different purposes. However, before each of these systems is constructed separately and interoperability becomes difficult, it is necessary to present a concept for the centralization of the whole system. If they are eventually integrated as a whole and centralized management of all information is achieved, overall optimization on a wider scale will be possible.

If this is achieved, it will have a variety of effects. For example, if the system can be used where information is currently divided between design, construction, and operation entities, it will be possible to feed back actual operational conditions to design and construction for improvement. In addition, by combining information on *usage* obtained from connected equipment along with fault information, and analyzing it with big data, it may be possible to estimate the degree of wear and tear of products and components, and promote resource circulation. In the future, fault diagnosis will also be possible.

The protection of users' personal information and other confidential information is also important, but this can be handled using information protection technologies such as blockchain technology and quantum cryptographic keys, which are expected to be utilized in the future.

4.3 CONSTRUCTION OF A COMPREHENSIVE EVALUATION METHODOLOGY FOR HEAT PUMP TECHNOLOGY

4.3.1 What Is Life Cycle Assessment?

When information can be centralized management, it will be used as a basis for overall optimization. This will require the construction of evaluation methodologies and indicators to integrate and evaluate different values. Whether or not an effective method can be developed is a key point in guiding the process toward overall optimization.

For example, life cycle assessment (LCA) is an evaluation methodology that identifies the environmental impact of individual products at each stage throughout their life cycle, from production, transportation, sales, use, disposal, and re-use, and comprehensively evaluates measures for improvement. When impact categories with

TABLE 4.1
Examples of Impact Category Weighting in LCA

Impact category	Weighting 1	Weighting 2	Weighting 3	Weighting 4	Weighting 5
Greenhouse effect	0.5	1	0.59	1	–
Shadow on ozone layer	0.5	1	3.0	–	–
Acidification	1	1	2.9	–	–
Eutrophication of terrestrial systems	0.5	1	1.8	–	–
Eutrophication of aquatic systems	0.5	1	1.8	–	–
Toxicity to humans	0.33	1	0.66	–	0.33
Toxicity to water system ecology	0.17	1	0.34	–	0.17
Toxicity to terrestrial system ecology	0.17	1	0.34	–	0.17
Photochemical oxidants	0.33	1	0.66	–	.33
Use of non-biological resources	1	1	–	–	–
Biodiversity	0.5	1	–	–	–
Life support systems	0.5	1	–	–	–

Source: Based on Ministry of the Environment/Mitsubishi Research Institute, "Effective SEA and Case Analysis" [2].

different implications[1] are evaluated collectively, it is not possible to simply compare each impact category. However, different impact categories are weighted and an integration index is calculated (Table 4.1).

4.3.2 Life Cycle Indicators Considering Global Warming Impact

In addition to energy-saving performance, from the perspective of global warming countermeasures, life cycle assessment is also carried out using indicators such as product life climatic load (LCCP)[2], LCA and life cycle CO_2 ($LCCO_2$)[3] (Figure 4.6).

These assess how much air conditioners impacts global warming in its overall LCA (resource extraction, manufacturing, transportation, use, disposal, recycling, etc.).

Although these values should be minimized as a measure against global warming, the actual leakage rate of refrigerants and the amount of CO_2 emissions per unit of electricity consumed differ from region to region and from country to country, so the evaluation needs to take into account the characteristics of each geographic locale.

Such an approach involves a large number of impact categories and therefore requires a huge amount of data and a lot of weighting. If a high degree of perfection is sought from the outset, nothing will ever move forward. The key point is how simple and impactful methods and indicators can be constructed as a guide to lead the technology toward overall optimization (Figure 4.7).

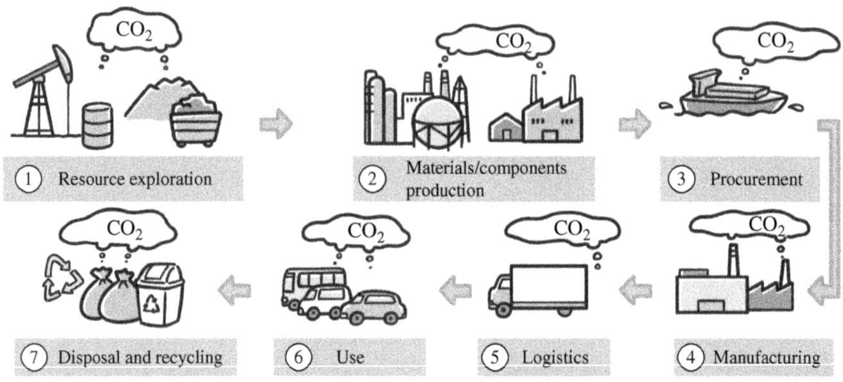

FIGURE 4.6 The life cycle assessment. (From Mitsui & Co, Ltd. LCA Plus decarbonization platform, "Towards a clean society created through decarbonization" [3].)

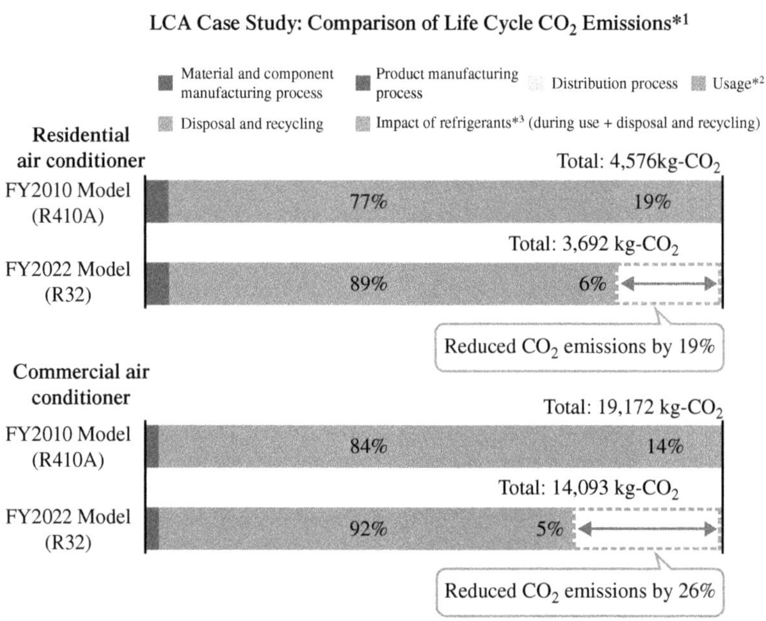

FIGURE 4.7 Example of an LCA. *1Calculations based on Daikin standards for residential air conditioners in the 2.8 kW class and for commercial air conditioners in the 14 kW class. *2Periodic energy consumption: Japanese Industrial Standards (JIS) for residential use and the standards of the Japan Refrigeration and Air Conditioning Industry Association (JRAIA) for commercial use. *3Refrigerant impact is calculated from the global warming potential per unit weight, taking into account the average leakage rate during use and during disposal/recycling. *Notes:* (From the Daikin Group Sustainability Report 2023 [4], Daikin Industries Ltd.)

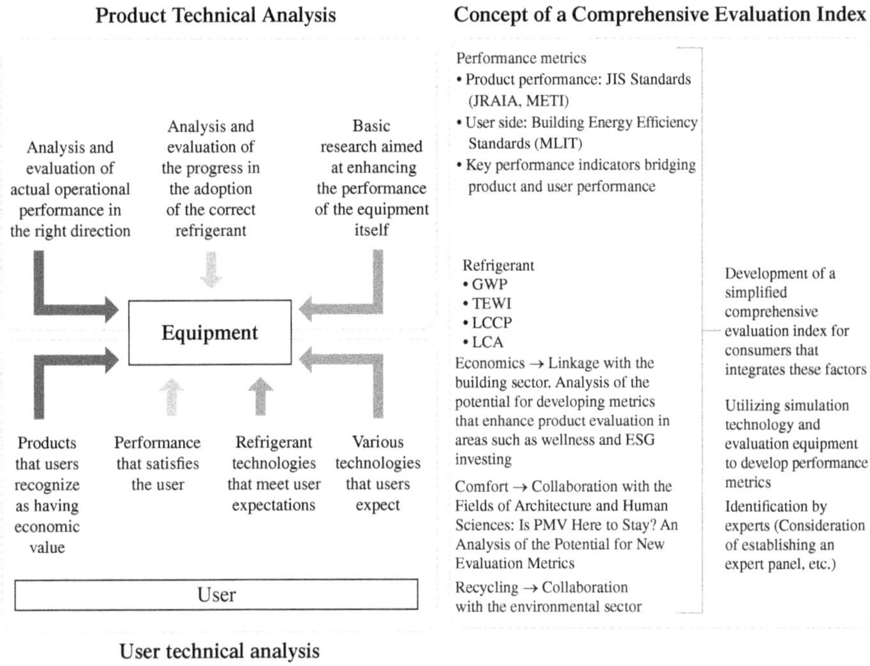

FIGURE 4.8 Image of overall evaluation index. (From Consortium for Strategic Research on Next-generation Heat Pump Technology, Open Innovation Strategy Research Organisation, Waseda University, "User technology analysis" [5].)

4.3.3 Construction of an Overall Evaluation Index for Heat Pumps

Meanwhile, the Consortium for Strategic Research on Next-Generation Heat Pump Technology at Waseda University is developing an overall evaluation index for heat pumps, as shown in Figure 4.8. The aim here is to construct an index that includes not only inherent properties (energy efficiency, GWP of refrigerant, safety, resource consumption, etc.) but also evaluation factors that are important for the social system in which the equipment is used. For example, in the case of air conditioning, comfort and health are evaluated not only in terms of temperature but also humidity and air flow.

The aim is to introduce the system as early as possible as a relatively simple one. Each category is weighted and evaluated, and the evaluation results are derived on a 100-point scale.

The Consortium for Strategic Research on Next-Generation Heat Pump Technology, Waseda University, is currently studying the following overall evaluation index for residential air conditioners.

4.3.3.1 Purpose of the Overall Evaluation Index

The objectives of the overall evaluation index are as follows in (1) and (2). As it is essential to meet user needs when marketing a product, the objectives indicated in (1) have first priority, with those in (2) as a secondary objective.

(1) To formulate indicators that enable a comprehensive judgment of the quality of individual heat pumps from the point of view of the user of heat pump technology.
(2) To formulate indicators that contribute to the company's technology development strategy in relation to heat pump technology, suggesting a direction regarding its development, which allows positioning of the product technology.

4.3.3.2 Procedure for Implementing the Overall Evaluation

The procedure for conducting a comprehensive evaluation is presented below. The example of residential air conditioners is given as a reference. In the case of heat pumps for commercial, industrial, and other applications, the evaluation procedure can be applied universally, although the evaluation indicators may differ from those for residential air conditioners (Figure 4.9).

4.3.3.3 Evaluation Items and Evaluation Indicators

The evaluation items and indicators adopted are shown in Table 4.2.

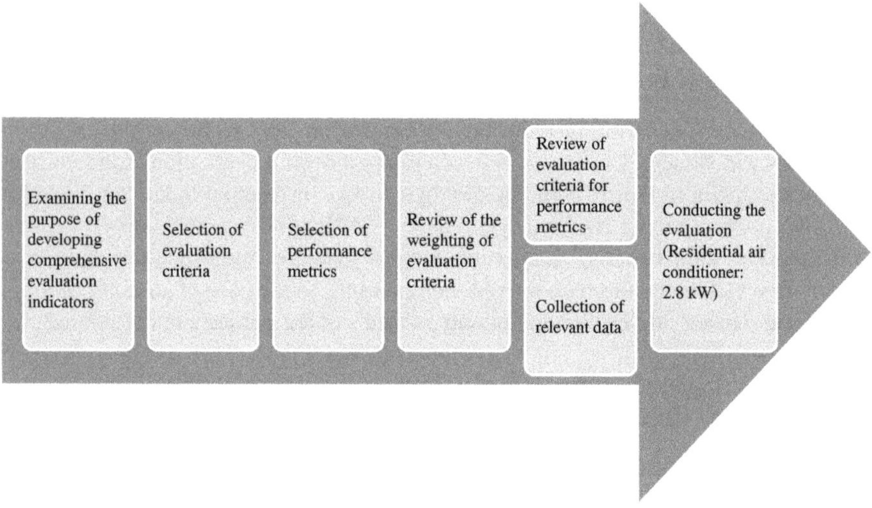

FIGURE 4.9 Evaluation implementation procedure for the overall evaluation index under consideration.

TABLE 4.2
Evaluation Items and Evaluation Indices for the Overall Evaluation Index under Consideration

Evaluation item	Evaluation indicators	
Energy efficiency	Annual performance factor (APF)	
Safety	Flammability and toxicity of refrigerants	
Environmental issues	Global warming	GWP of refrigerant × amount of refrigerant contained
	Waste	Whether the product is recyclable or not, recommodification rate, mass (indoor unit + outdoor unit)
Comfort	Energy-saving function/comfort control, AI function, operating noise (indoor unit + outdoor unit)	
Health	Ventilation function, humidification function, sterilization and disinfection function	
IoT connectivity	Remote control/monitoring function, linkage with other products	

4.3.3.4 Points for Overall Evaluation Index

Normally, for residential air conditioners, for example, users look at energy efficiency, such as the annual performance factor (APF), and healthiness, such as ventilation function, in a discrete way and make an overall decision on which air conditioner is best.

If the health-related performance of the air conditioner is improved, it will require more power, which means that its energy efficiency will be inferior, and the user considers which is more important to them when making a purchase after judging such advantages and disadvantages.

The overall evaluation index is a method for making such judgments using a clearly defined score, and enables users to carry out evaluations in a clear manner, whereas previously they had to make judgments more ambiguously. For air conditioning manufacturers, it also brings the benefit of being able to decide on the direction of technological development based on the evaluation presented in such a clear form.

4.4 OVERALL OPTIMIZATION OF THE VALUE CHAIN

4.4.1 Optimization of the Value Chain through the Establishment of a Comprehensive Evaluation Methodology

If a comprehensive evaluation methodology can be established alongside a centralized information management system, it will also bring about technological innovation, changes in social systems, and a change in the orientation of purchasing for users. This will enable a holistic approach to the supply chain, including resource circulation and refrigerant conversion, and facilitate overall optimization of the entire value chain. As mentioned above, if we focus only on individual events to solve the problems of heat pump technology, it will take longer to realize the ultimate goal of next-generation heat pump technology and we may reach the limit of change earlier.

4.4.2 Establishment of a National Framework

Such overall optimization is extremely difficult to achieve by manufacturers alone and requires a nationwide response, including from industry associations, governments, and academic institutions. Due to the inherent difficulty involved, the regions and countries that are quick to respond will play a leading role in the global heat pump market in the future. Although system integration has long been a challenge for Japan, its achievement is assured when the nation works together as one. It is no exaggeration to say that Japan's heat pump technology, which is currently the most widely accepted in the world, is facing a critical moment to see whether it can continue to lead the world.

Figure 4.10 illustrates the overall optimization of the value chain with the next-generation heat pump technology proposed in this publication.

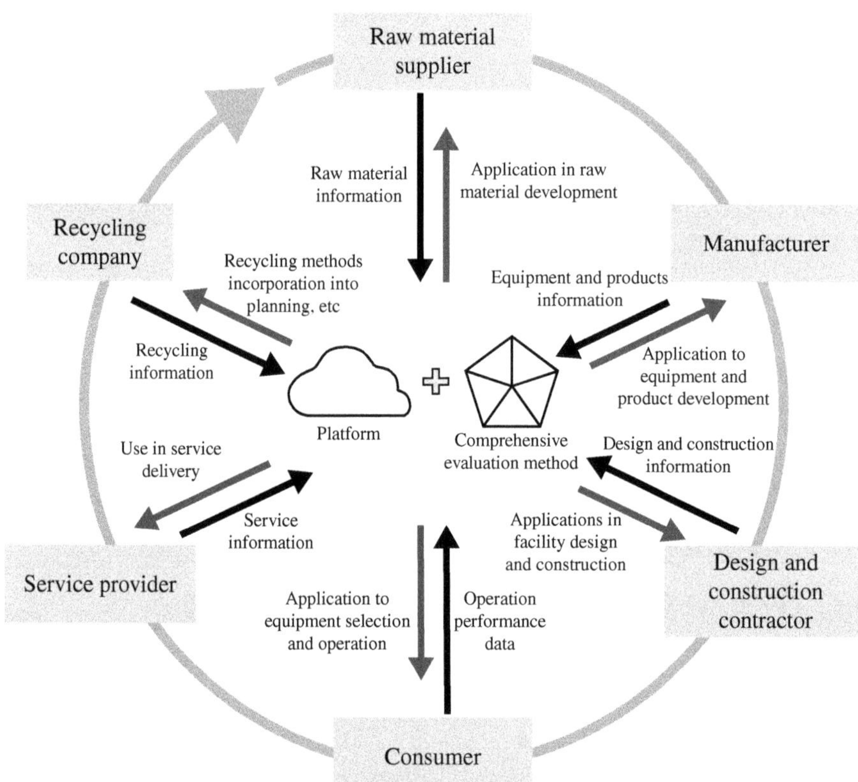

FIGURE 4.10 Optimization throughout the value chain with next-generation heat pump technology.

If overall optimization can be achieved, the entire value chain can be developed globally as a business. A good example of this is the global development of the Shinkansen bullet train as a set of operating systems.

4.5 EFFECTS OF OVERALL OPTIMIZATION

This section provides some examples of what is possible and what effects can be achieved if overall optimization is realized. The important thing is that partial optimization is acceptable in the middle of the transformation process, and that efforts should always be made to envision the next-generation heat pump technology and work toward overall optimization.

4.5.1 Overall Optimization for Energy Saving

4.5.1.1 Use in Processes where Simultaneous Cold and Hot Use Is Required

At present, attention is inevitably focused only on improving the performance of individual pieces of equipment. Due in part to the lack of *visualization* of actual equipment operating conditions and performance, users are forced to select equipment based on fragmentary information such as standardized performance. For example, cooling and heating may be required simultaneously in factories and low-temperature distribution centers. In addition, in large buildings and city centers where heat demand is concentrated, multiple temperature ranges of heat demand such as hot water, air conditioning, and refrigeration are generated in close proximity.

In such cases, it would be inefficient to respond to each individual heat demand with separate equipment, but more efficiency can be achieved by using, for example, heat pump technology with simultaneous use of hot and cold (Figure 4.11).

4.5.1.2 System Design Using Digital

To realize the process shown in Figure 4.11, it is necessary to design and manage the entire heat supply system as a whole, rather than on an individual equipment basis.

Even when equipment simply interacts with each other, it is necessary to optimize the configuration and layout of both pieces of equipment, rather than designing

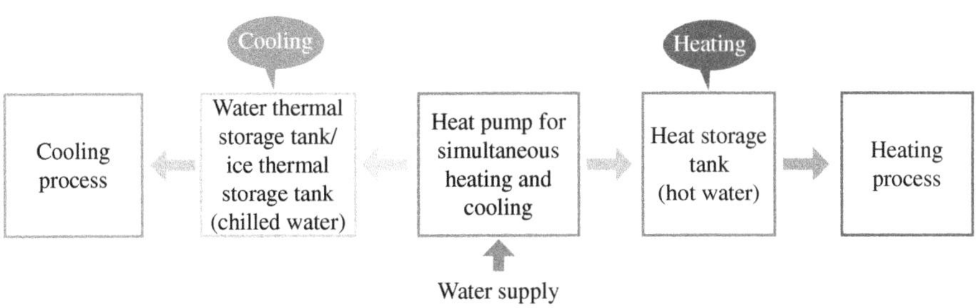

FIGURE 4.11 Image of heat supply using heat pumps with simultaneous use of hot and cold (thermal storage tank use). (Based on Japan Electro-Heat Center, "Simultaneous cold and hot supply system" [6].)

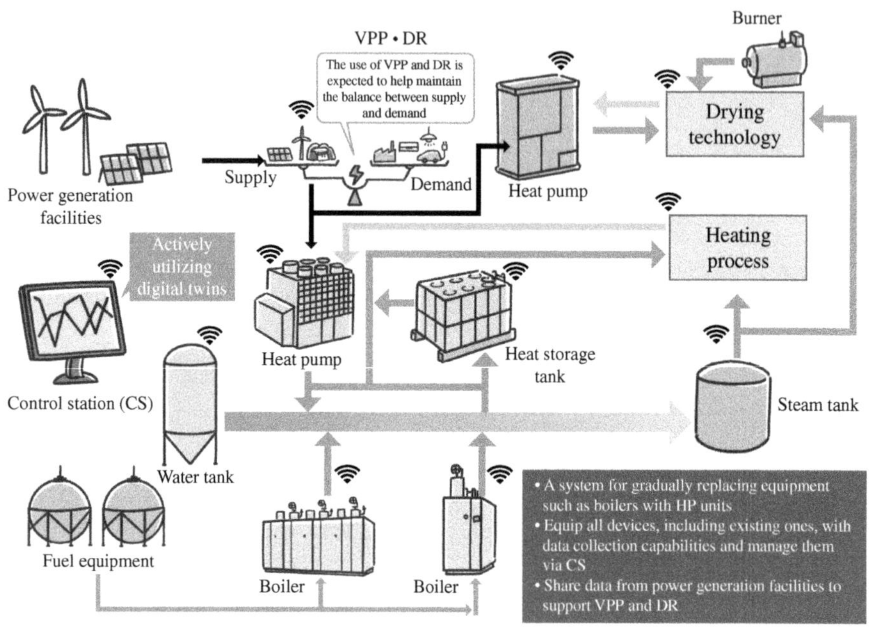

FIGURE 4.12 Image of comprehensive optimization of energy supply and demand, including heat, by EMS. (From Compiled based on the Council on Competitiveness-Nippon(COCN) "[Construction and social implementation of an Ambient Energy Platform]—Towards early realisation of an integrated EMS including heat" [7].)

and installing air conditioning and refrigeration equipment on a stand-alone basis. Alongside this, operations can be optimized by using information platforms to understand the operating status of equipment from different manufacturers, and by linking them through EMS and other means.

Even though it is known that such equipment integration and coupled operation are highly effective, it is hardly quantifiable how effective they are when the equipment is actually installed. Here, too, DX, including digital twin technology, will have a significant effect (Figure 4.12).

4.5.1.3 Predicting Running Costs Using Digital

Until now, the rationalization of heat has been considered for implementation based on running cost forecasts by ESCO operators. However, there were certain limitations in the accurate understanding of the energy use and reduction effects at the site, but if these effects can be predicted for performance and running costs based on the actual installation environment with a digital twin created within proper rules, this will have the effect of greatly reducing the hurdles to introducing equipment. Similarly, if the digital twin can not only predict the introduction of equipment, but also quantitatively verify the effects of introduction, it will give momentum to the consideration of new introductions, which will be effectively utilized by more advanced EMS.

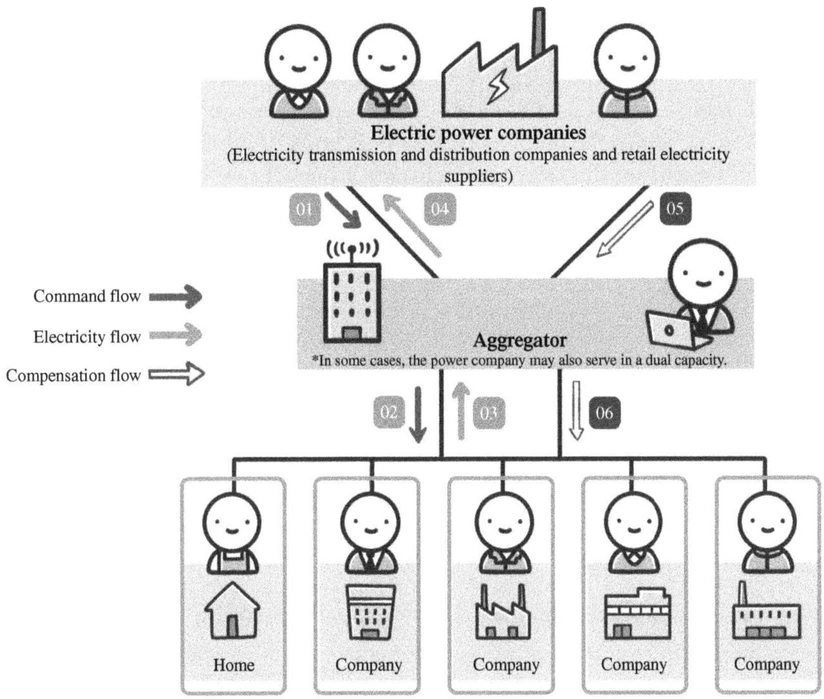

FIGURE 4.13 How demand response works. (From Nihon Business Publishing Co. Ltd, the Environmental Glossary "Demand Response" [8].)

Demand response (DR) is also essential for the mass introduction of renewable energy, and the industrial sector will also be required to introduce DR in the future (Figure 4.13). In such cases, factories that make extensive use of boilers will not be able to adjust their power and will not be able to meet the demand for power supply and demand adjustment. On the other hand, if heat pump technology with thermal storage is introduced, it will be possible to respond to DR by making effective use of thermal storage. If optimization is oriented in this way, even taking into account the nature of energy supply and demand in the next era, economic incentives such as contractual arrangements for DR-compatible consumers will be strengthened, and this will also provide incentives for the introduction of heat pump technology.

4.5.1.4 An EMS Platform to Manage Heat Energy

'Development and common platform for heat energy management systems' is being promoted by a consortium of companies led by Waseda University in the third phase of the Strategic Innovation Promotion Program (SIP). In heat utilization technology, there is a limit on significant performance improvement and decarbonization by building on conventional performance improvements of equipments and individual systems. Therefore, as an integrated systemization technology, an EMS platform will be built to link different industries and systems, including not only EMS software but also hardware such as components and sensors.

Using this platform, several integrated systems can be realized and advanced to social implementation, taking into account rapidly changing social demands, such as COVID-19 infections, and the introduction of renewable energy, the use of which will increase significantly in the future. At the same time, we will build an ecosystem that enables the prediction and verification of implementation effects by utilizing digital twin technology, which makes it possible to visualize the effects of EMS implementation that were previously unclear. The aim is to contribute to the early realization of a vision of a future society with a flexible and robust energy supply and demand system.

In the air conditioning sector, until now it has not been possible to determine under what conditions and at what performance air conditioning equipment is operated in actual operation. The APF, which is listed as catalogue performance, does not reflect improvements in control technology, which contribute significantly to improving equipment performance, and so there was no incentive to develop technology to improve actual operational performance. However, once energy-related information is shared, dynamic performance evaluation will become possible. If evaluations that take into account the actual operating conditions of the equipment are carried out, these studies, which are naturally effective, will also progress.

4.5.1.5 Performance Evaluation during Actual Operation Using Digital

As equipment continues to be used, its performance tends to deteriorate significantly due to aging and contamination, but the rate of decline is not constant from one piece of equipment to another. For this reason, equipment that performs better throughout its period of use, rather than as new equipment, is superior. It is desirable to be able to select such equipment or to have the option to choose equipment with countermeasures in place (Figure 4.14).

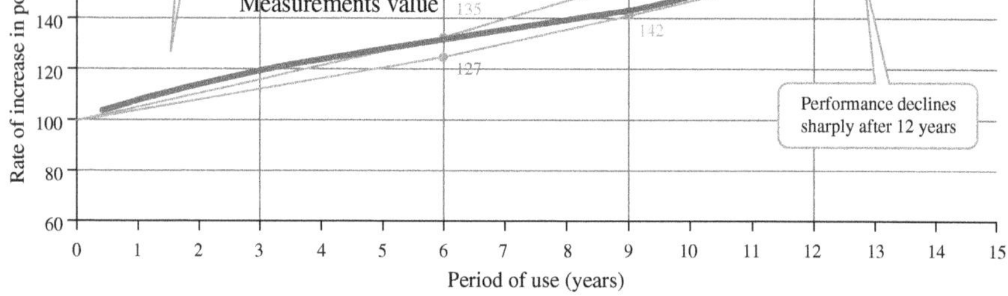

FIGURE 4.14 Duration of use and power consumption of air conditioners. *Note:* Calculated from actual measurements taken by 41 residential air conditioners with a cooling capacity of 2.2 kW (conducted in 2009). (From Oji Engineering Co. Ltd “Energy-saving measures for air conditioners” [9].)

If optimization in DX is advanced, it will be possible to utilize data from IoT-connected equipment to continuously monitor operating conditions in real time and trace actual performance in the market. Based on this data, simulations and comparisons with digital twins can be made to assess changes over time, failures, etc., and to improve overall optimization throughout the period of use. Naturally, the industrial sector is similarly evaluated, including the coupling of multiple equipments through EMS. This will open up new avenues for energy saving in the air conditioning sector, where further energy-saving measures are required.

4.5.2 Overall Optimization to Realize Early Refrigerant Conversion

4.5.2.1 System Design Based on Global Warming Measures

Linking with system information on refrigerants makes it possible to predict refrigerant leakage, for example, due to reduced equipment performance. Naturally, this is expected to lead to feedback regarding equipment development and technological innovation to further improve equipment performance. Centralized management of diverse information is very useful here.

Originally, energy-derived decarbonization and refrigerant conversion were both to be promoted as initiatives aimed at preventing global warming, but at present they are being considered almost as independent events. As mentioned above, conversion to low GWP refrigerants could result in lower equipment performance, which in turn could have a greater impact on global warming, or lead to equipment becoming larger and isolated from global standards making high-performance equipment unsaleable.

Greenhouse gas (GHG) emissions from heat pump technology can be broadly categorized into energy-related CO_2 emissions from energy consumption during use and leakage of fluorocarbons.

When using refrigerants with high GWP, the latter impact is at a non-negligible level (Figure 4.15). Therefore, when considering the environmental impact of heat pump technology, not only are energy savings with cost benefits important but also refrigerant measures. However, energy savings and refrigerant conversion often have a trade-off relationship and are difficult to reconcile. Both need to be taken into account when designing the optimum system.

4.5.2.2 Use of LCCP and $LCCO_2$ Evaluation Methodology

Although their use is still limited, in the future, evaluation methodologies such as LCCP and $LCCO_2$, which take GHG emissions into account at various stages of the life cycle, can be used to explore the balance between energy savings and refrigerant conversions for individual appliances. Furthermore, it is possible to optimize the overall optimization of society, including overall optimization, by combining multiple pieces of equipment and by preventing refrigerant leakage, and developing recovery and recycling systems. In this way, unreasonably large equipment, among other things, can be avoided and appropriate technological progress can be brought about.

4.5.2.3 Measures Against Refrigerant Leakage

Refrigerant measures are simultaneously required to reduce refrigerant production and consumption in accordance with the Kigali Amendment and to reduce refrigerant emissions in a move toward carbon neutrality. In both cases, preventing

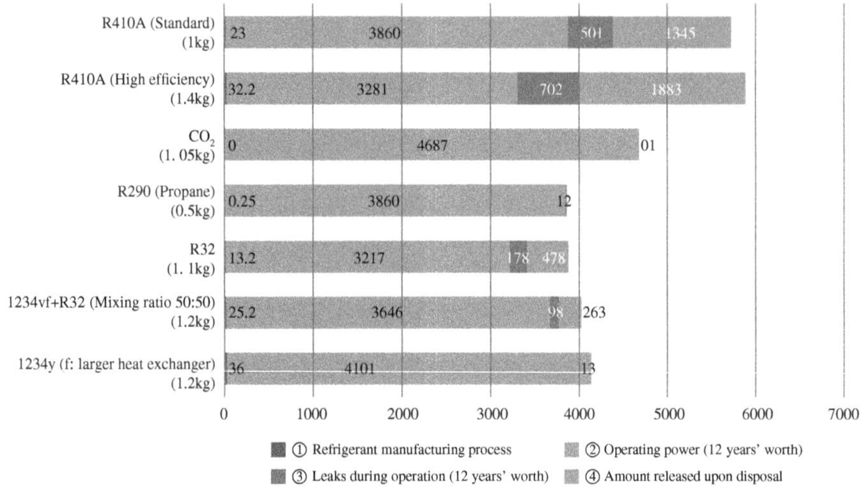

FIGURE 4.15 Overall GHG emissions by refrigerant type (4 kW residential air conditioners). *Note:* Calculation assumption: Residential air conditioners 4 kW (CO_2 emission factor 0.425 [CO_2 kg/kWh], 12-year service life, 9 hr/day operation time, 2%/year refrigerant leakage rate during operation, 30% refrigerant recovery rate at disposal). Horizontal axis: kgCO_2. (From the Ministry of the Environment's "JRAIA's efforts to prevent global warming and to use next-generation refrigerants" (2011) [10].)

refrigerant leakage and promoting recovery and regeneration will reduce the input of new refrigerants and also reduce emissions. Therefore, efficient refrigerant conversion will be promoted if the two measures—low GWP and recovery, and regeneration of existing refrigerants—are pursued at the same time.

In addition, as air conditioner is currently limited to the use of the mildly flammable refrigerant A2L, a flammable refrigerant, it is difficult to make progress in equipment safety measures, refrigerant leakage prevention technology, and refrigerant recovery technology. If flammable refrigerants are introduced in the future, it will be effective to simultaneously advance equipment safety measures as hardware and social systems as software. By highly evaluating these effects in the overall evaluation index, etc., it will also be possible to promote them.

We already know that safety measures are essential for the early introduction of next-generation refrigerants. Faster refrigerant conversion to next-generation refrigerants will be possible if safe refrigerant recovery technologies, effective leakage prevention measures, and social systems for recovery are developed.

4.5.3 Overall Optimization to Realize Resource Circulation

At present, energy saving and refrigerant conversion are being promoted individually, so little attention is being paid to the resultant increase in equipment size and cost. Forcing manufacturers to do too much will eventually reduce their profits, curb their environmental investment, and delay the development of more products in the future.

If the concept of overall optimization is promoted, for example, the system for equipment recovery and refrigerant recovery will be unified, and both will be recovered under a very simple management system. The development of equipment structures that enable simultaneous equipment and refrigerant recovery, as well as the creation of new commercial flows, can also be expected. This would also minimize the cost of next-generation refrigerant conversion to the consumer.

As the digital product passport system is developed, a system will also be established for each piece of equipment to manage not only information on the materials used in the equipment, but also comprehensive information on the use of refrigerants and the energy use of the equipment. There are many challenges that need to be solved, such as whether this information should be managed by the manufacturer or centrally by an organization. However, if optimization of resource use, optimization of energy use, and minimization of CO_2 emissions are achieved simultaneously, this will also make a significant contribution to the realization of a sustainable society.

NOTES

1. Impact category (impact area): Itemization of environmental issues of concern, such as global warming and depletion of the ozone layer.
2. Product life cycle climatic load: life cycle climate performance (LCCP), also called life cycle warming characteristics, is a comprehensive assessment of the impact of a single piece of equipment over its life cycle, not only in terms of GWP, but also in terms of the impact of refrigerant and electricity consumption.
3. Life cycle CO_2 ($LCCO_2$) is an indicator that converts GHG emissions into CO_2. The assessment takes into account not only the amount of CO_2 generated to produce the electricity needed for operation, but also the impact of refrigerant leakage and the CO_2 generated in the production of equipment and refrigerants.

REFERENCES

1. Hitachi Research Institute Product Passport. “1. What is Product Passport”, “Fig. 1 Overall picture of Product Passport,” 2023. https://www.hitachi-hri.com/research/researchreport/short/k130.html
2. Ministry of the Environment/Mitsubishi Research Institute, Inc. “Effective SEA and Case Analysis”, “1.3 Information on Effective SEA,” 2003, http://assess.env.go.jp/files/0_db/seika/0184_01/jp.pdf
3. Mitsui & Co, Ltd. “LCA Plus decarbonisation platform—Towards a clean society created through decarbonization,” “Product LCA—Emissions per unit of product and service,” 2023. https://lp.lcaplus-pf.com/
4. Daikin Industries, Ltd. “Daikin Group Sustainability Report 2023”, p.41, 2023, https://www.daikin.co.jp/csr/environment
5. Waseda University Open Innovation Strategy Research Institute. Next-generation heat pump technology strategy research consortium, “User technology analysis,” 2023, https://waseda-heatpump.com/ユーザー技術分析/
6. Japan Electro-Heat Center. Comprehensive site for waste heat and unused heat utilization Industrial heat pumps.com, “Simultaneous cooling and heating supply system,” 2023, https://sangyo-hp.jeh-center.org/heatpump_system2.html

7. Council on Competitiveness-Nippon (COCN). Final report of the project in 2022, "[Construction and social implementation of Ambient Energy Platform]—Towards early realisation of an integrated EMS including heat", p.15, Fig. 4, 2023, http://www.cocn.jp/report/bb8a485bdac95a274de373dfbf3c1876e38bc1d2.pdf
8. Nihon Business Publishing Co., Ltd. Environmental Glossary "Demand response," 2017, https://www.kankyo-business.jp/dictionary/002811.php
9. Oji Engineering Co., Ltd. "Energy saving measures for air conditioning equipment ", "Transition of air conditioner use and power consumption," 2023, https://homepage.ojieng.co.jp/solution/product/energy_saving_ measures/
10. Ministry of the Environment. "Global warming prevention and next generation refrigerant initiatives by Japan Refrigeration and Air Conditioning Industry Association," p. 20, 2011.

5 Heat Pump Technology for a Healthier Lifestyle and a Stable Food System

OVERVIEW

The areas in which the creation of new value by next-generation heat pump technology is expected to be significant include the living environment, food system, industry, and the electricity supply and demand system (see Figure 5.1).

In this chapter, we will give examples of how next-generation heat pump technology can contribute to the needs and issues of the living environment and our food system, which are indispensable for sustaining human life, and how a prosperous future society can be realized in these areas.

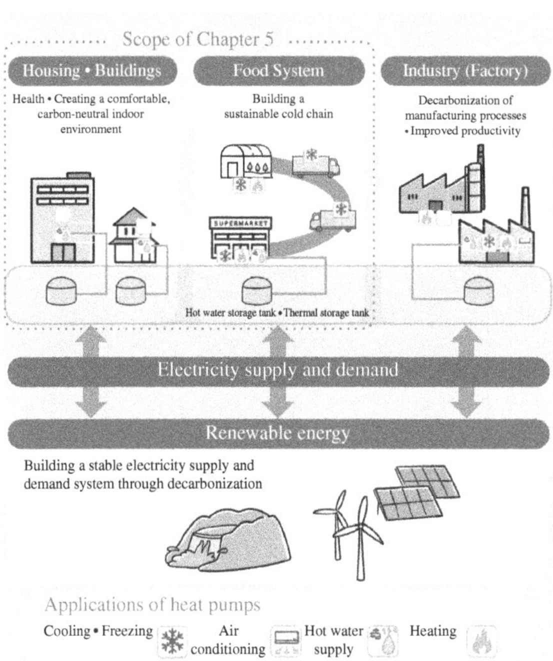

FIGURE 5.1 The role of next-generation heat pump technology to protect life.

The above are all visions of the future that need to be realized on a global scale, so it is important that they are applied and deployed not only domestically but also internationally.

5.1 CURRENT STATUS OF THE LIVING ENVIRONMENT: WELLNESS

5.1.1 Concept of Wellness and Expectations for Its Realization

Globally, the concept of "wellness"[1] [1], which takes the conventional concept of health[2] more broadly and aims for a more positive, spiritually and socially enriched, and energetic daily life is spreading. In Japan, people are increasingly interested in extending their healthy life expectancy and living a fulfilling life amid a declining and aging population. In addition, health management, which considers the support and promotion of employee health as a management issue, is becoming more widespread among companies.

The living environment, which is the foundation of people's lives, is one of the most important factors in improving wellness, and the indoor thermal environment has a particularly large impact on people's health and comfort. As shown in Figure 5.2, the thermal environment is determined by four environmental factors (temperature, humidity, airflow, and radiation) and two human factors (metabolic rate and clothing).

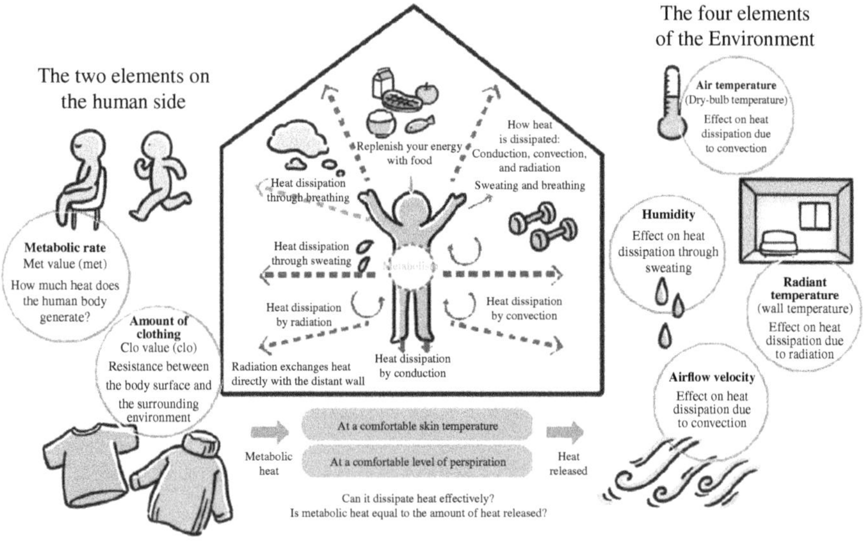

FIGURE 5.2 Six elements that make up the thermal environment. (From Masayuki Mae, "The science of comfort/Thinking about comfort in winter" [2].)

5.1.2 Factors Influencing Wellness

For a thermal environment to be comfortable, the amount of metabolic heat from the human body and the amount of heat dissipated from the skin surface must be in balance, as determined by the six aforementioned factors.

First, a body's metabolic rate determines the amount of metabolic heat that must be dissipated from the human body. In addition, the amount of clothing worn determines the surface temperature of the clothing, and the heat exchange (heat release) between this surface temperature and the surrounding environment determines the amount of heat released. The routes of heat dissipation to the surrounding environment include convection currents in the surrounding air, radiation to surrounding objects, and perspiration, and the four abovementioned environmental factors affect these. For example, if humidity is high, sweat does not evaporate and the amount of heat radiated by perspiration is small, making it easier to feel hot. If airflow is good, a lot of heat is radiated by convection, making it easier to feel cold.

Thus, to achieve a comfortable thermal environment, temperature, humidity, and airflow are important factors. Moreover, it is necessary to prevent heat from flowing in and out of the building by improving insulation performance, and to control these factors appropriately using air conditioning equipment and systems that utilize heat pump technology.

Ventilation has become even more important since the outbreak of the COVID-19 pandemic. Clusters of outbreaks have occurred, and it is well known that adequate ventilation has a significant control effect, as shown in Figure 5.3.

On the other hand, it is also true that energy losses due to ventilation are very significant. If normal ventilation and air conditioning continue even in situations where there are few occupants, this would be a significant step backwards from decarbonization.

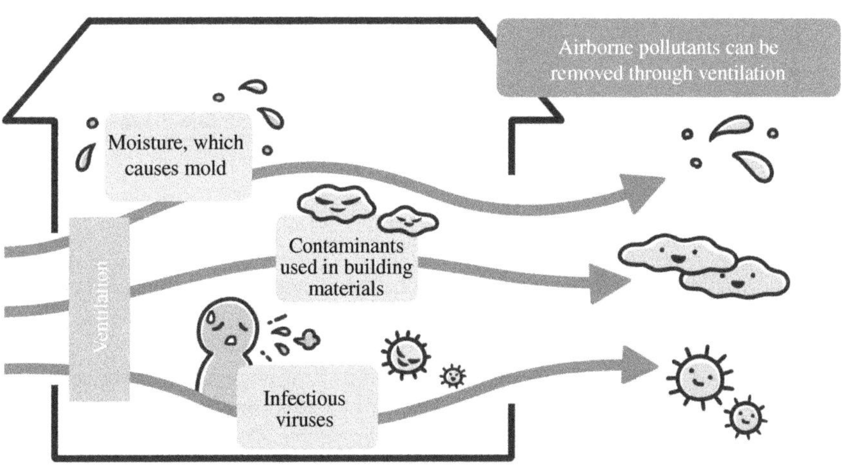

FIGURE 5.3 Importance of ventilation. (From Sekisui Industrial Corporation, "We teach you how to ventilate well in homes, offices and shops" [3].)

In addition, emission reductions in residential housing and commercial buildings, which emit approximately 30% of the total greenhouse gas (GHG) emissions in Japan as a whole, are essential to achieve carbon neutrality, as discussed in Chapter 2. Therefore, it is necessary to achieve a comfortable heat environment as described above, while also achieving energy saving and decarbonization.

5.2 CURRENT STATE OF THE LIVING ENVIRONMENT: DECARBONIZATION

5.2.1 Residential Energy Usage

5.2.1.1 Japanese Homes that Do Not Meet Insulation Performance Standards

Figure 5.4 shows the insulation performance standards for Japanese and international homes. The horizontal axis is the heating degree day[3], an indicator of low outdoor temperatures in winter, with larger numbers indicating colder temperatures and smaller numbers indicating warmer temperatures. The vertical axis, average heat transfer coefficient of outer skin of houses (average heat transfer coefficient of building envelope (UA value)), indicates the ease of heat escape, and the smaller the number,

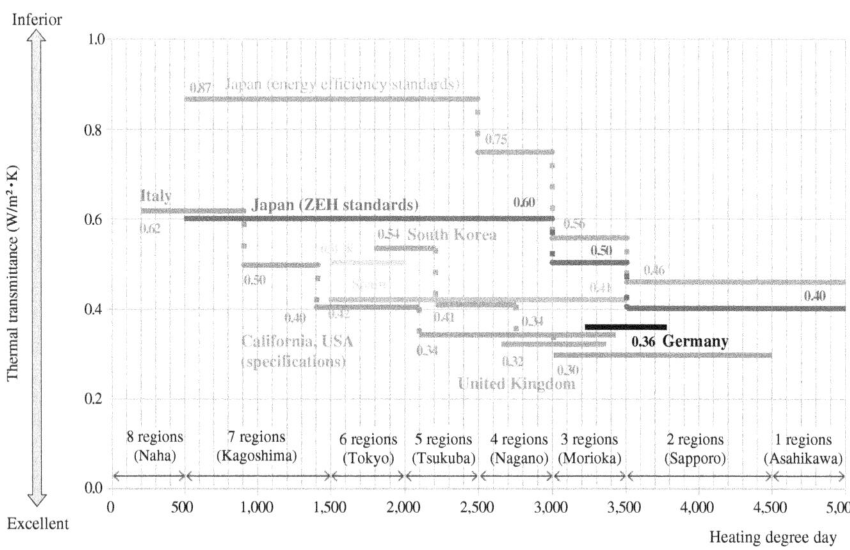

FIGURE 5.4 International comparison of standards for average heat transfer coefficient of building envelope (UA value) of houses. *Note:* ※In Spain, the energy efficiency standards are classified into five regional categories, but the figures above are only for the regional category to which Madrid belongs. (From the Ministry of Land, Infrastructure, Transport and Tourism's, Future Energy Saving Measures for Houses and Buildings (Third Report) and the Building Standards System (Fourth Report) [5]; Based on Nomura Research Institute, FY2021 Survey on Energy Saving Regulations and Standards for Houses and Buildings in Overseas Countries.)

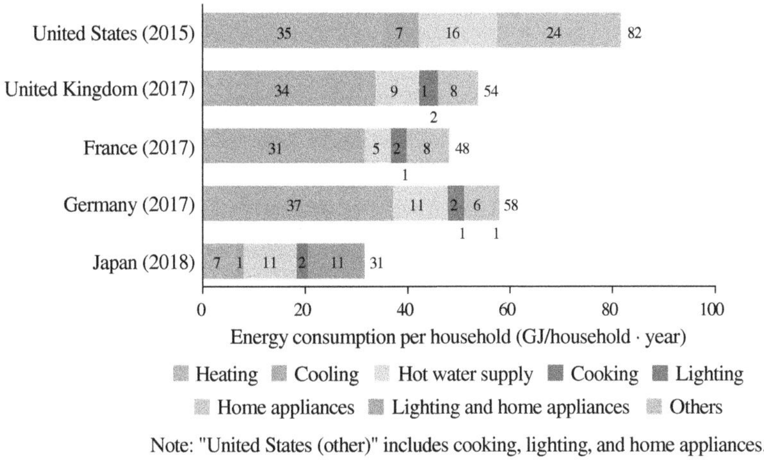

FIGURE 5.5 International comparison of energy consumption per household by use. (From Ministry of the Environment, White Paper on Environment, Sound Material-Cycle Society and Biodiversity 2020 [6].)

the higher the thermal insulation. It can be seen that Japan's current energy saving standards are lower than those overseas, especially in warmer regions. On the other hand, according to data from the Ministry of Land, Infrastructure, Transport and Tourism, around 90% of the total housing stock did not meet these current standards by 2020, of which around 30%, their insulation performance is not considered high. [4].

5.2.1.2 Low Insulation Performance and Heating Practices in Japan

Not only are Japanese houses poorly insulated, but their heating levels are also low. Figure 5.5 shows an international comparison of energy consumption per household by use, which shows that energy consumption for heating in Japan is very low compared to Europe and the USA.

Although it should be noted that the comparison is based on different conditions such as climate, house size, and culture, the main reason for this difference is that in other countries, the main method of operation is to heat the entire house or living room for a long period of time (continuous whole house operation or continuous living room operation). In Japan, on the other hand, it is the norm to heat a room intermittently, that is, only the living room or bedroom is heated during the necessary time period.

5.2.1.3 Regional Differences in Air Conditioning and Heat Pump Water Heater Usage

According to a statistical survey by the Ministry of the Environment, looking into equipment used for heating (Figure 5.6), air conditioning systems account for 30–40% of primary heating equipment in all regions except cold regions. In cold regions, especially in Hokkaido, the percentage of households using kerosene heaters is high and while air conditioner is used as the main heating equipment, the percentage of households using it is very low, at 1% in detached houses and 3% in

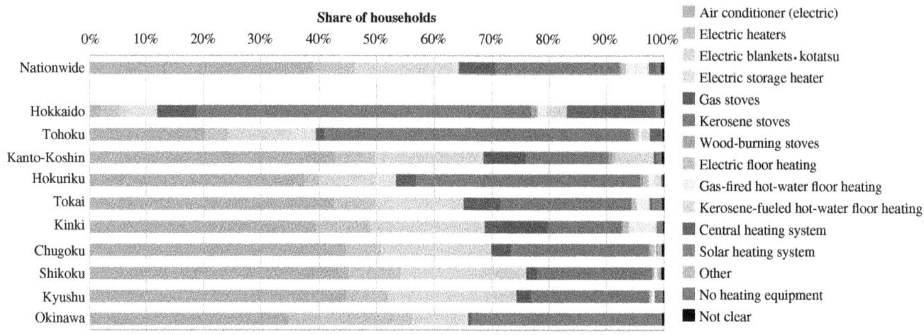

FIGURE 5.6 Most commonly used heating appliances in homes by region (FY2021). (From Ministry of the Environment, "Statistical Survey on CO_2 Emissions from the Household Sector in FY2021" [8].)

apartment buildings. According to the same survey, only around 30% of households in Hokkaido use air conditioner (air conditioner penetration rate), partly due to low cooling demand in summer, whereas more than 90% of households in the whole of Japan use air conditioner [7].

One of the reasons for the low uptake of air conditioner in cold regions is frost build-up on outdoor units. Air conditioning units remove heat from the outside air on the surface of the heat exchanger in the outdoor unit during heating operation, so the temperature inside the heat exchanger must be lower than the temperature of the outside air. Therefore, when the outside temperature is very low and the temperature inside the heat exchanger is below freezing, moisture in the air adheres to the surface of the heat exchanger as frost, inhibiting heat exchange with the outside air and reducing the heating capacity,

In addition, a separate energy input may be required to thaw the frost (in air conditioning for cold regions, performance at low temperatures has been considerably improved).

In addition, the comfort level of users differs between kerosene heaters, which transmit heat by radiation type[4] and convection type[5] air conditioning systems. It can be inferred that households accustomed to paraffin heaters may be resistant to switching to air conditioners.

The same applies to water heating: gas water heaters and heat pump water heaters have a high share in detached houses in warmer regions, while gas water heaters have a high share in colder regions, especially in Hokkaido, with heat pump water heaters accounting for less than 5% (see Figure 5.7).

Gas water heaters are the predominant type of water heater in housing complexes, regardless of region, and heat pump water heaters have a share of less than 5% in all regions, which is the norm for this type of water heater.

Compared with heat pump water heaters, gas water heaters are inexpensive, and their advantage lies in the fact that they are capable of instantaneous water heating and do not require the installation of a hot water storage tank, which means that installation space is less restricted. Heat pump water heaters are therefore subject to the same problems as air conditioners in cold regions in terms of frost build-up

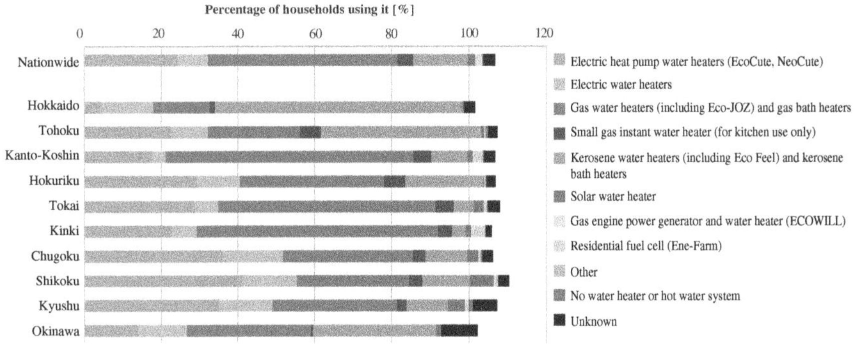

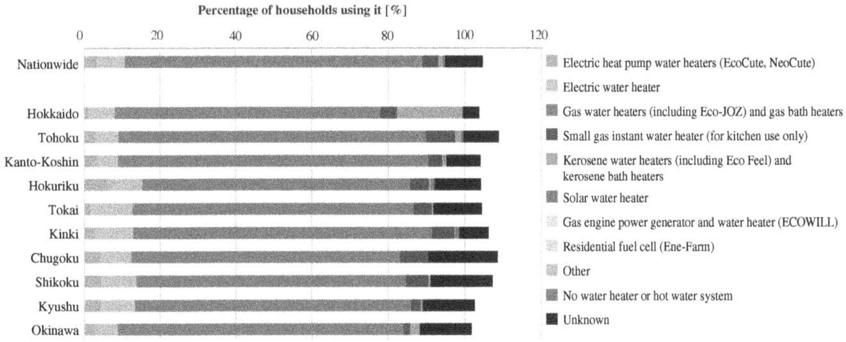

FIGURE 5.7 Water heaters and water heating systems used in houses by region and building type. (Top) Detached houses. (Bottom) Apartment buildings. *Note:* As there are households using more than one system, the total percentage of the number of households exceeds 100%. (From Ministry of the Environment, "Statistical Survey on CO_2 Emissions from the Household Sector in FY2021" [9].)

and thawing issues, as well as in terms of installation space restrictions in housing complexes, and this has been a disincentive to their widespread use.

5.2.2 Energy Usage in Buildings

For buildings other than houses, according to data from the Ministry of Land, Infrastructure, Transport and Tourism, as of FY2020, the equivalent of no thermal insulation accounted for about 30% of the total stock, and, like houses, their insulation performance is fairly low [7].

The breakdown of energy consumption in the building sector (commercial sector) by application (FY2020) is shown in Figure 5.8, with air conditioning (heating and cooling) accounting for approximately 27%, hot water supply 13%, kitchen 10%, and power/other 50%.

Of the above applications, the equipment used in air conditioning is mainly commercial heat pump equipment. On the other hand, equipment used for water heating includes heat pump water heaters as well as combustion water heaters such

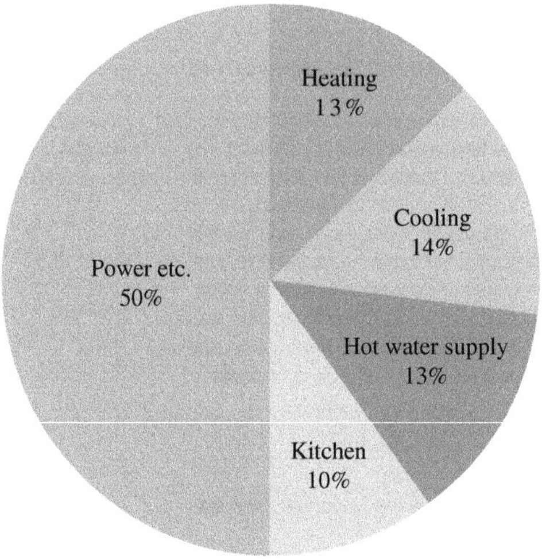

FIGURE 5.8 Breakdown of energy consumption by application in the commercial sector. (From the Institute of Energy Economics, Japan and Riko Tosho "EDMC Handbook of Energy & Economic Statistic" [10].)

as gas-fired hot water boilers and oil-fired hot water boilers, but at present combustion water heaters dominate the market share, with heat pump water heaters accounting for less than 1% [11].

5.3 CHALLENGES AND NEEDS OF THE LIVING ENVIRONMENT

5.3.1 Housing Performance in Japan

5.3.1.1 Thermal Efficiency of Housing

As mentioned in Section 5.2, Japanese housing struggles to achieve a comfortable heat environment due to its low envelope performance and a way of operating air conditioning (i.e. intermittent, for particular rooms, as opposed to constant, for the entire home) that differs from that in other countries. The World Health Organization (WHO) strongly recommends that the temperature in houses should be 18°C or higher in winter from a health perspective, but a survey of the actual room temperature in Japanese houses showed that the average temperature in living rooms was 16.7°C and in dressing rooms 12.8°C (Figure 5.9).

This inability to achieve a comfortable thermal environment has resulted in many indoor accidents caused by heatstroke in summer and temperature shock in winter (it is estimated that 28,000 people a year suffer from heatstroke indoors [13] and 19,000 people a year die suddenly while bathing indoor [14]). As the population ages further, it will become even more important to achieve a comfortable indoor environment to prevent these health problems.

As with housing, the first prerequisite for the needs of buildings other than houses is to achieve a comfortable indoor environment. Although there are few cases of heatstroke, temperature shock or other health hazards in buildings, the need for indoor

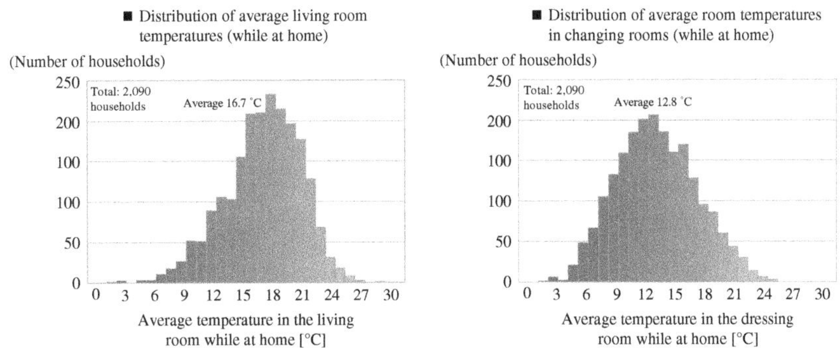

FIGURE 5.9 Distribution of average indoor temperature. (Left) Living room. (Right) Dressing room. (From the Better Living Foundation, "Warm living guide for healthy living" [12].)

environments that contribute to improving the health and intellectual productivity of employees is increasing due to the growing awareness of corporate health management and other factors.

As mentioned above, adequate measures need to be taken to prevent infectious diseases. However, of course, there must be no regression from the goal of decarbonization. For this reason, it will also be necessary to be able to adjust ventilation levels, for example, in the event of an infectious disease outbreak.

There is a need to transform the way equipment and systems that consumer energy are used in homes and other buildings, not only to support wellness, but also from the perspective of carbon neutrality.

5.3.1.2 The Need to Design Homes for Heat Pump Technology

As discussed in Section 5.2, combustion-based appliances using gas or kerosene are still widely used for residential heating and hot water applications in cold regions, while gas water heaters are the mainstay in apartment buildings, regardless of the region. As maximum electrification is required for decarbonization, it is important to overcome the challenges of current heat pump technology (frosting, installation space constraints, etc.) and move away from combustion equipment through technological development.

The same applies to buildings other than houses, where heat pump technology is already widely used for air conditioning, but combustion water heaters such as gas-fired hot water boilers and oil-fired hot water boilers currently have an overwhelming share of the market for hot water supply. In addition to the perspective of energy saving and moving toward net zero energy buildings (ZEBs) and net zero energy houses (ZEHs)[6] (Figure 5.10), the conversion from combustion water heaters to heat pump water heaters is desired from the perspective of promoting electrification, which is indispensable for decarbonization.

However, it is essential to implement measures for low global warming potential (GWP) refrigerants in conjunction with the conversion to heat pump appliances. For example, leakage measures could be taken on the assumption that flammable

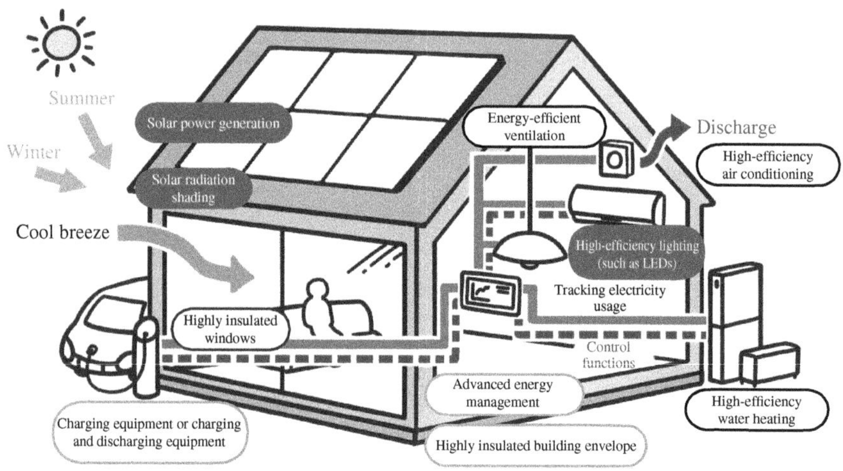

FIGURE 5.10 How a ZEH house works. (From the Sustainable Open Innovation Initiative, "ZEH subsidy by the Ministry of Economy, Trade and Industry and the Ministry of the Environment in 2023" [15].)

refrigerants such as hydrocarbons are used, or hot water heating systems, which are becoming increasingly popular in Europe, could be used.

Note that refrigerant measures are also required for existing heat pump equipment. Currently, refrigerant conversion to R32, which has a lower GWP, has just progressed, but further measures are inevitable. It will be necessary to either use flammable refrigerants such as hydrocarbons in direct expansion types with limited amounts of refrigerant, or to switch to inter-expansion types using chillers and other equipment (including the hot water heating systems mentioned above) that use cold and hot water instead of refrigerant for heat transfer. In either case, stricter management will be required, with the former requiring measures to ensure safety, such as refrigerant leakage prevention, and the latter requiring water leakage prevention. If chillers are to be used, the efficiency of the system will be reduced, but it may be improved by combining air conditioning and hot water supply in one system.

5.4 VISION OF THE FUTURE LIVING ENVIRONMENT CREATED BY HEAT PUMP TECHNOLOGY

5.4.1 Simultaneous Wellness and Decarbonization

As already mentioned, current housing and buildings mainly face the following challenges/needs

1. The low envelope performance of houses and air conditioning operation systems that differ from those in other countries do not allow for a comfortable heat environment

2. In buildings other than houses, awareness of the indoor environment is increasing due to the spread of health management, and there is a growing need for advanced control and ventilation, including not only temperature control but also humidity and air flow
3. In addition to the wellness perspective mentioned above, the carbon neutral perspective is also important

   Heat pump equipment is required to be widely used for heating and hot water supply in residential buildings in cold regions, and for hot water supply in housing complexes and buildings. However, it is essential that measures for refrigerants are implemented in conjunction. Ventilation has become very important, including for infection control, but unnecessary ventilation can also hinder decarbonization, as a lot of energy is lost.

Until now, energy saving has only been pursued in living environments, with little consideration given to comfort and health. In fact, comfort has been a major concern, as it inevitably requires a lot of energy to achieve.

In the future, there will be a need for living environments that support both wellness and decarbonization at the same time. In contrast, as discussed in Section 5.3, overall optimization is possible in conjunction and coordination with the performance of the house/building itself and related equipment and facilities, as the envelope performance is expected to improve in the future due to measures such as stricter energy efficiency regulations for houses and buildings.

In addition, by overcoming technical issues (use of flammable refrigerants, frosting, installation space restrictions, etc.), the introduction of next-generation heat pump technology that can be applied to cold regions, housing complexes, and commercial buildings can be promoted to meet the above challenges and needs.

A concrete vision of the future of housing and buildings is shown in Figure 5.11. Next-generation heat pump technology (e.g. whole-house air conditioning, desiccant air conditioning) is being used to achieve a higher level of wellness, not only in terms of temperature but also in terms of humidity and air flow, in line with the improved envelope performance of housings and buildings. From an energy perspective, overall system optimization is achieved through the selection and configuration of equipment, including air conditioning, hot water supply, and refrigeration and ventilation equipment.

5.4.2 Effective Use of Next-Generation Heat Pump Technology Using DX

In the operational phase, the use of Internet of Things (IoT) and digital transformation (DX), etc. enables heat to be visualized as data based on indoor temperature, humidity, and CO_2 concentration, as well as enabling optimum operation control based on electricity consumption data, thereby realizing both a comfortable indoor environment and saving energy. Control is not carried out for each individual piece of equipment, but multiple heat pump installations used for air conditioning, ventilation, hot water supply, and refrigeration and freezing equipment are controlled together by integrating the heat source, resulting in further energy savings. Furthermore, overall

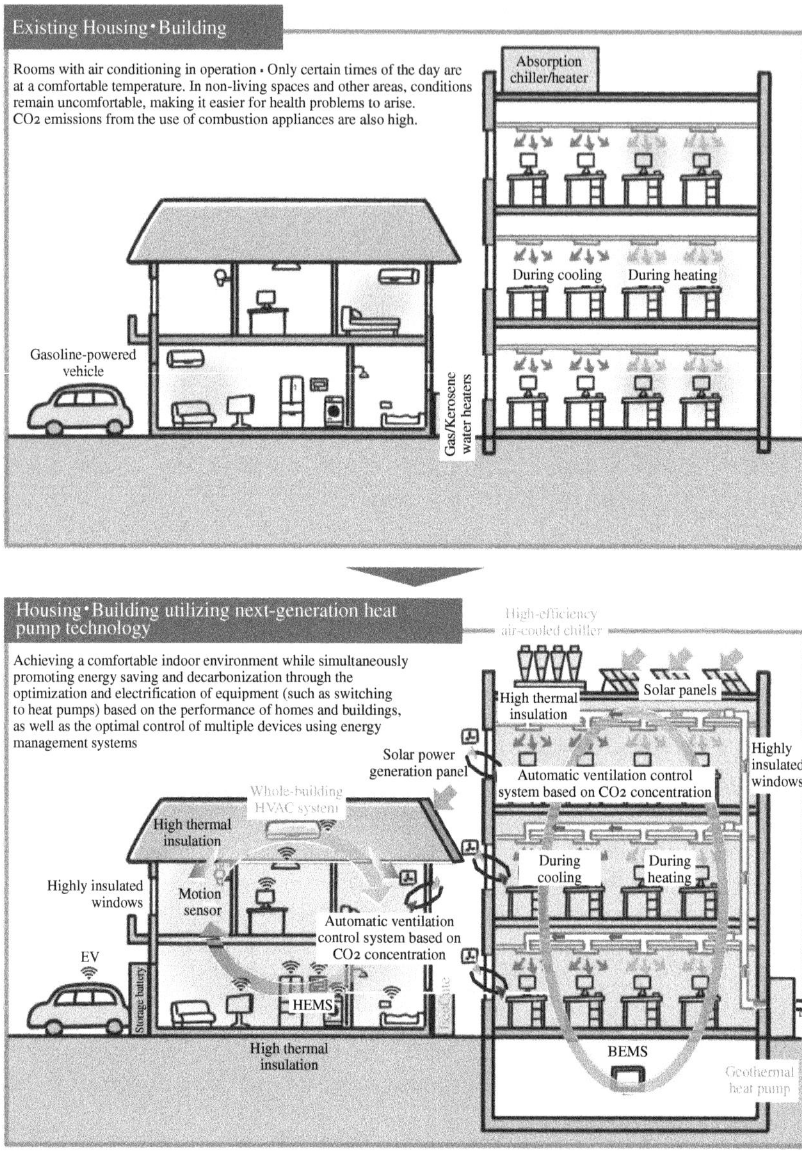

FIGURE 5.11 Housing/building with air conditioning versus those with next-generation heat pump technology.

optimization can be achieved by building an energy management system for the house or building as a whole, including on-site solar power, storage batteries, thermal storage, and electric vehicles (EVs), and by using heat pump technology as a means of improving the self-consumption rate of on-site solar power, etc. This maximizes the benefits for consumers.

With regard to refrigerants, leak detection systems for flammable refrigerants and integrated control of multiple equipments will be equipped using IoT and digital technology, to achieve both energy savings and refrigerant countermeasures. Depending on the situation, consideration may also need to be given to reviving water piping.

5.5 CURRENT STATUS OF FOOD SYSTEMS

5.5.1 CO_2 Emissions

The food system refers to the series of supply chains from the production of agricultural, forestry, and fishery products, through food processing, distribution, and sales until the food reaches the consumer and is finally consumed or disposed of. While the food system in Japan provides consumers with high-quality, high-value-added products, it faces sustainability challenges such as chronically low self-sufficiency, large-scale natural disasters, responses to climate change and the weakening of the production base due to a decrease in farmers.

Recent discussions have also pointed out that the food system has a significant impact on the balance between emissions and absorption of GHGs that cause climate change, due to methane contained in cattle burps, CO_2 emissions from the use of feed grain, land and water required for animal rearing, and methane generated during the breakdown of waste food in landfill, as a result of the expansion of animal production. The Food and Agriculture Organization of the United Nations (FAO) estimates that global food system-related GHG emissions are equivalent to one-third of total anthropogenic GHG emissions (see Figure 5.12).

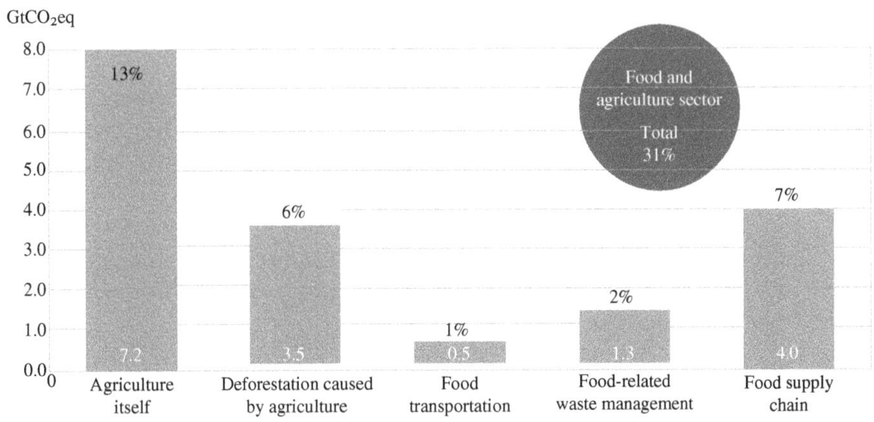

FIGURE 5.12 Global food system-related GHG emissions. *Note:* Percentages are in relation to GHGs of anthropogenic origin. Rounding of fractions has been carried out, which means that the total number and the breakdown total do not match. Data based on 2019 figures. (From Mitsubishi Research Institute Inc., "What is 'food tech' being promoted for? Expectations as a climate change mitigation measure" [16].)

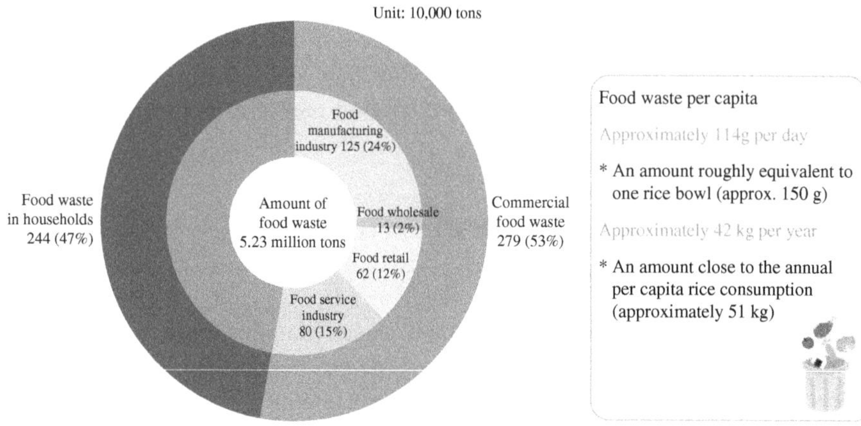

FIGURE 5.13 Amount and breakdown of food waste generated (estimated figures for FY2021). (From Ministry of Agriculture, Forestry and Fisheries, "What is food waste?" [18].)

5.5.2 Food Waste

While the challenges described in the previous section are significant, the amount of food that is wasted is significant: according to an FAO report, approximately 900 million tonnes of food, equivalent to about 20% of the world's food production, is wasted every year.

In Japan, as shown in Figure 5.13, 5.23 million tonnes of food is wasted, and the economic loss[7] due to such food waste is estimated to be around 4–6 trillion yen [17]. Although progress in food loss prevention has led to a significant reduction compared to the 9.8 million tonnes of food waste as of FY2000, further reductions are needed, despite there is a sense that the level has bottomed out in recent years. In addition, as the reduction of food waste is very important globally, not only from the perspective of reducing GHGs and other environmental impacts, but also from the perspective of responding to food crises caused by future population growth and avoiding economic losses, efforts should be made to contribute to the reduction of food waste not only in Japan, but also around the world.

Against this backdrop, there is a strong need to build a sustainable food system that achieves increased productivity and stable food supply, as well as decarbonization, by rationalizing and streamlining the entire supply chain and reducing food waste, not only in Japan but also in the wider world.

5.6 CURRENT STATUS OF COLD CHAINS

Cold chain refers to a system within the food chain in which food and other products are controlled and distributed at the appropriate temperature (low temperature, frozen, or refrigerated) and humidity from the point of production to the consumer, as shown in Figure 5.14. Cold chains play an important role in the food system and heat pump technology is used extensively. This section therefore focuses on the cold chain, explaining its current status, challenges, and needs, as well as the vision for the future.

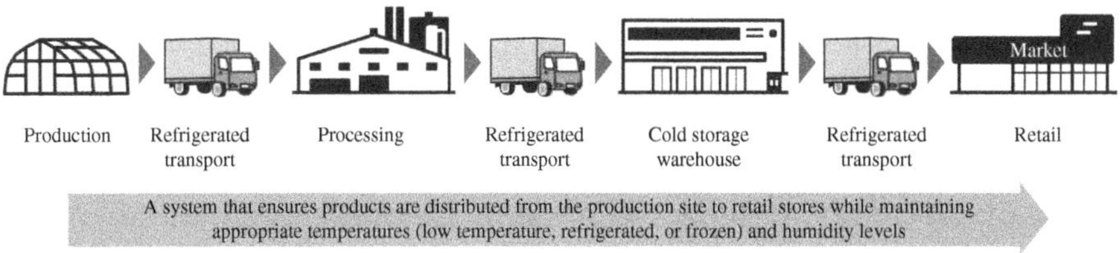

FIGURE 5.14 The cold chain. (Based on various sources.)

The cold chain is expected to contribute not only to the provision of safe and secure food, but also to the improvement of food self-sufficiency. In each of these processes (production, processing, transportation, storage, sales, etc.), freezing and refrigeration equipment (e.g. refrigeration unit) is used that incorporates heat pump technology. The impact of heat pump technology on the cold chain is therefore significant. In particular, it is also true that each process in the distribution chain requires refrigeration and freezing equipment to be operating constantly, which entails considerable energy consumption. This section describes the current status of equipment usage and energy consumption in the cold chain, focusing on logistics warehouse, transportation and distribution, and retail outlets, which are the processes that follow intermediate distribution, where the temperature and quality control of food products is particularly problematic.

5.6.1 Temperature Control in Logistics Warehouses (Refrigerated Warehouses)

A logistics warehouse is a warehouse that not only stores goods but is also equipped with service functions such as cargo handling, packaging, and distribution processing, as well as an information system to efficiently perform these services. In the logistics sector, the temperature at which goods are stored is often divided into frozen (–18°C or lower), refrigerated (–18–10°C), and room temperature (10–20°C). Logistics warehouses that store goods in frozen or refrigerated conditions are called refrigerated warehouses.

According to data from the Japan Refrigerated Warehouses Association, most of the energy usage in refrigerated warehouses is electricity required to power refrigeration and freezing equipment, with the actual electricity consumption in FY2021 being 1.93 billion kWh [19]. Recently, energy efficiency has improved as refrigerated warehouses are being rebuilt; old sites are being decommissioned; new, larger refrigerated warehouses are being built; and energy-saving refrigeration and freezing equipment is being adopted. In addition, an increasing number of refrigerated warehouses are now installing solar power on their roofs to meet the electricity demand. Roofs of refrigerated warehouses often have a large surface area, and there are examples of Megawatt-scale solar installations.

Hydrochlorofluorocarbons (HCFCs) (R22) used to be the main refrigerant used in refrigeration/freezing equipment, but as restrictions on fluorocarbons (FCs) refrigerants have strengthened, the Ministry of the Environment's subsidy program has provided a boost, and conversion to natural refrigerants has been progressing,

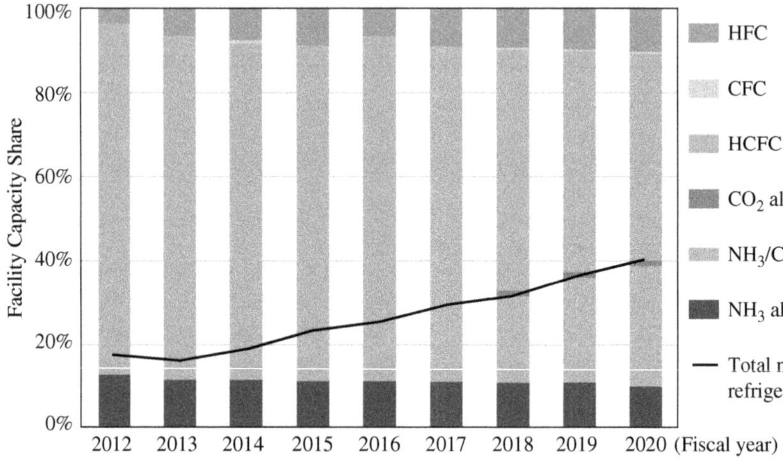

FIGURE 5.15 Refrigerants used in refrigerated warehouses of member companies of Japan Association of Refrigerated Warehouses. (From Shigekatsu Koganemaru (Japan Association of Refrigerated Warehouses), "Efforts to Address Global Environmental Issues by JARW" [20].)

particularly in recent years. Ammonia/CO_2 systems, in which CO_2 is cooled by an ammonia refrigerator and CO_2 brine is fed into the refrigerator, are on the increase (Figure 5.15).

An example of an ammonia/CO_2 refrigeration system is shown on the left in Figure 5.16. Although ammonia has high refrigeration cycle efficiency, it has safety issues (toxicity, odor, flammability, etc.), but by using the ammonia/CO_2 method, safety is dramatically improved while maintaining high efficiency, and it has been widely adopted. In recent years, CO_2 condensing units such as those shown on the right in Figure 5.16 have been developed and increasingly adopted by small and medium-sized business operators as replacements for their relatively small air-cooled

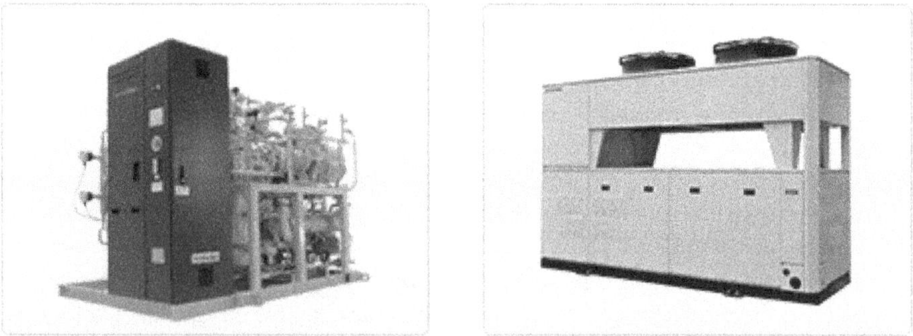

FIGURE 5.16 Ammonia/CO_2 method chiller and air-cooled CO_2 condensing unit. (From Mayekawa Manufacturing Corporation, (left) "High-efficiency natural refrigerant chiller NewTon" [21]; (right) "CO_2 condensing unit COPEL" [22].)

R22 refrigeration units, with the share rising rapidly from 1.3% in FY2018 to 4.2% in FY2020.

On the other hand, as can be seen from Figure 5.15, HCFC (R22) still account for around half of the overall market share, and there is a need for continued conversion to low GWP refrigerants, including natural refrigerants.

5.6.2 Temperature Control in Retail Outlets

In retail outlets such as supermarkets and convenience stores, condensing units and showcases are used to keep food and beverages at temperatures ranging from –25–8°C. Air conditioning equipment, condensing units, and showcases are estimated to account for about half of the total electricity consumption in retail outlets.[8] [23]. As with logistics warehouses, there has been an increase in the number of retail shops introducing roof-mounted solar panels to meet their electricity needs in recent years and, according to data from the Japan Franchise Association (JFA), of which convenience stores are members, 13,620 out of 56,448 member shops had installed solar power as of FY2021 [24].

The type of showcase is selected according to the storage temperature range of the food and the way it is displayed, with multi-stage open showcases often being selected for products in the refrigerated temperature range. Products in the frozen temperature range are displayed vertically and are used as reach-in types with doors and flat open showcases with openings facing upwards, although flat open showcases have been on the increase in recent years, and there are also dual types combining vertical and flat showcases. In addition, as shown in Figure 5.17, in recent years,

FIGURE 5.17 Example of a natural refrigerant showcase with doors. (From Lawson Inc., “Lawson’s first to completely ‘non-fluorocarbonize’ all refrigeration equipment used in its shops, aiming to prevent global warming and use highly efficient energy—also implementing power-saving control initiatives through IoT of equipment” [25].)

an increasing number of types have been fitted with doors in both the freezing and refrigeration temperature ranges from the perspective of saving energy, and doors that can be added later are also provided, but the market penetration rate is low.

Condensing units and showcases in retail outlets often have long installation pipes, and 16% of refrigerant is leaked annually while operating. The leakage impact is also high because many units still use R404A refrigerant, which still has a high GWP as a whole.

5.6.2.1 Temperature Control in Transportation and Delivery

In order to maintain low temperatures during the transportation and delivery phase, it is necessary to insulate the surrounding environment while in transit. Compared to warehouses and shops, the time required for cooling is shorter and the cooling area can be switched during transfers, etc. In the past, insulation or the addition of a cold storage agent was mainly used to prolong the cooling time. However, the increase in transportation distances, the need to uphold food safety measures, and the development of pharmaceuticals, the quality of which is assured on the basis of temperature control, have led to the demand for more rigorous temperature control. In response, temperature control using refrigeration equipment has become the mainstream method for road, rail, and ship transportation. However, even today, in light vehicles such as rear-driven cars and bicycles, where space and weight restrictions are severe, or in aviation, where heat dissipation is difficult because the vehicle is placed in an enclosed space, a system in which a cold storage agent is added to the heat insulation is often used.

In trucks, refrigeration units are often installed on a vehicle-by-vehicle basis and, as shown in Figure 5.18, there are three types: sub-engine type with a diesel engine, direct-coupled type using the engine power of the truck, and electric type that operates from the truck's power supply or from an on-board battery.

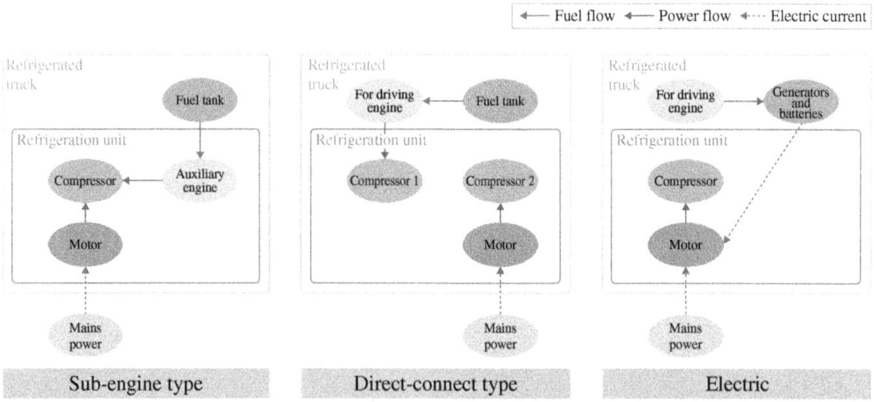

FIGURE 5.18 Energy flow of a refrigeration unit for transportation. (From Japanese Society of Refrigerating and Air Conditioning Engineers, "Refrigeration Cycle Control (Japanese Society of Refrigerating and Air Conditioning Engineers Technical Book Series)" [26].)

FIGURE 5.19 Cold storage box with compact refrigeration unit. (From Yamato Holdings Corporation, "Energy and Climate—Mitigating Climate Change" [27].)

In the electric type, as trucks have become more electrically powered in recent years, a system that supplies power to the refrigeration unit from the truck's running batteries has also been developed. The conditions is becoming more conducive to improving the total energy efficiency of the vehicle, including the use of renewable energy sources.

At the bottom end of the cold chain, known as the last mile, cold storage boxes with cold storage agent were often installed not only in vehicles with cold maintenance functions as described above, but also in regular small cars and light vehicles that do not have heat maintenance functions. In recent years, an increasing number of operations have replaced these with cold storage boxes equipped with ultra-compact refrigeration units (Figure 5.19), a trend that ensures reliable temperature control all the way to the end of the cold chain.

5.7 COLD CHAIN CHALLENGES AND REQUIREMENTS

5.7.1 The Increasingly Complex Cold Chain Distribution Channel

As discussed in Section 5.6, maintaining a cold chain requires the constant operation of freezing and refrigeration equipment, which consumes a considerable amount of energy, so in addition to promoting the introduction of existing energy-saving equipment (e.g. showcases with doors in retail shops), further improvements in equipment efficiency and energy saving are needed. From the perspective of decarbonization, in addition to energy saving, it is also necessary to promote the use of electricity from renewable energy and to address the issue of refrigerants, which are essential for freezing and refrigeration equipment. In recent years, progress has been made in the conversion to low GWP refrigerants, but when looking at the market stock as a whole, there is still a large amount of equipment using refrigerants with high GWP, and it is necessary to continue to promote the conversion to low GWP refrigerants, as well as measures to prevent leakage and recovery from existing equipment.

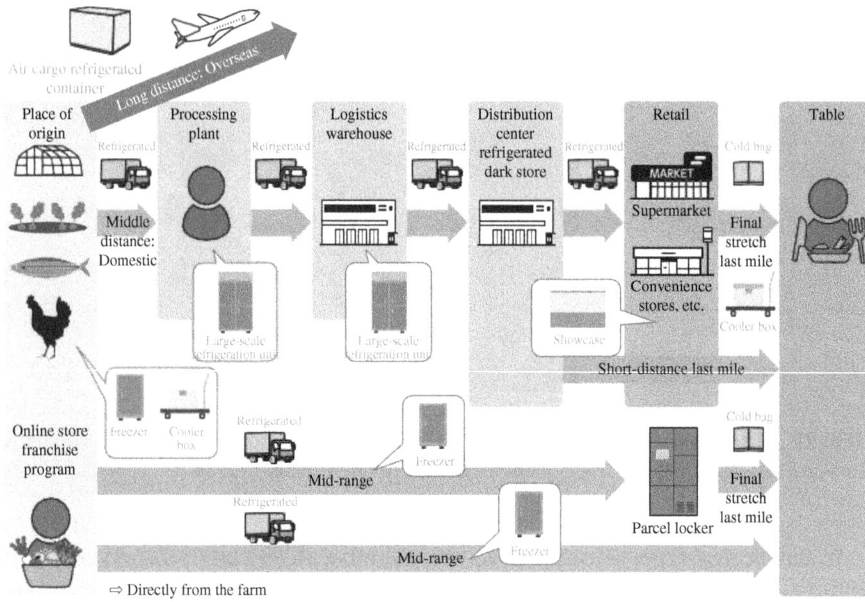

FIGURE 5.20 Increasing complexity of cold chain distribution processes and diversification of equipment used. (Based on multiple sources.)

In addition, as online shopping and e-commerce become more widespread, consumer needs are becoming more diverse and sophisticated, and the cold chain distribution channel is becoming more complex, with an increasing number of products being delivered directly from production sites and logistics warehouses to consumers, rather than through shops (Figure 5.20).

Refrigerated warehouses are not the traditional type that store products near seafood markets or food manufacturers, but are increasingly logistics warehouses that store, sort, and process a wide variety of products between production and consumption centers. There is also an increasing number of facilities known as dark stores, which act as delivery points between warehouses and shops, where pick-up operations are carried out. In light of this change, the refrigeration and freezing equipment used in the cold chain is also becoming more diverse, particularly for transportation.

5.7.2 Challenges Surrounding the Cold Chain

With the increasing complexity of the distribution channel, various players are involved in the process of delivering food products from the production area to the consumer, and it is becoming increasingly difficult to carry out temperature and humidity control throughout the entire distribution process as the equipment used becomes increasingly diversified. On the other hand, consumers' demand for freshness and quality of food is increasing, and the need for appropriate temperature

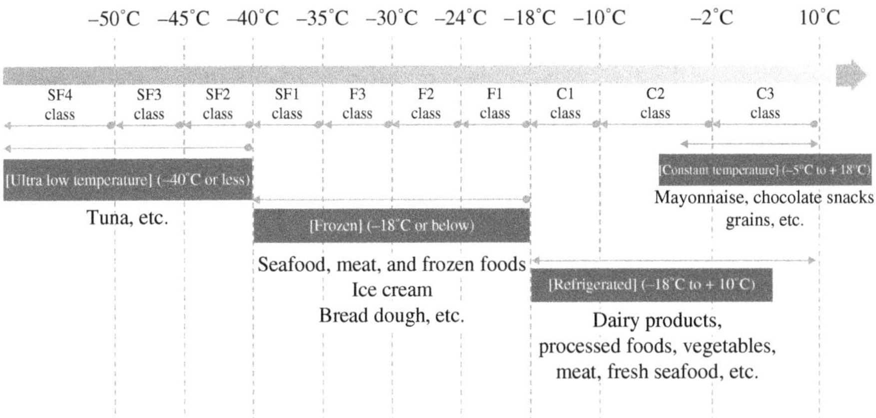

FIGURE 5.21 Food storage temperature range. (From Japan Association of Refrigerated Warehouses, "About storage temperature ranges" [28].)

control throughout the entire cold chain distribution process is expected to increase in the future, as is the need to reduce food waste (Figure 5.21).

However, the lower the temperature, the lower the energy efficiency of refrigeration and freezing equipment, which can lead to increased energy usage. Moreover, the refrigerants that can be used in low temperature areas have a higher GWP, making it more important to prevent refrigerant leakage and develop new low GWP refrigerants in addition to increasing efficiency.

In addition, labor shortages are a serious issue in the cold chain. The number of home deliveries of items including foodstuffs is increasing steadily year by year, and exceeded 4.9 billion units in FY2021, according to data published by the Ministry of Land, Infrastructure, Transport and Tourism [29]. The increase is due to the growing number of home deliveries, such as direct delivery of products to individual homes due to the proliferation of online shopping and e-commerce, and home delivery services by local supermarkets. This trend is expected to continue in the future.

5.7.3 The Need to Construct a Tailored Cold Chain

As the labor shortage is expected to become even more acute in the future, the need for more efficient transportation and delivery and easier facility management is expected to increase. For example, truck transportation will be adapted to improve transport efficiency and unmanned operation, and refrigeration equipment for transportation will be required to comply with these requirements. The shift from road to rail transportation and other means of transportation that require less labor, or a combination of the two, may increase the importance of container-type transport refrigeration units that are not tied to the means of transportation. In addition, from the perspective of curbing redistribution, which has become one of the challenges in the logistics sector, there is likely to be a growing need for the installation of receiving lockers with freezing and refrigeration functions on the consumer side.

In addition, users of equipment that uses fluorocarbon refrigerants are obligated, under the Fluorocarbon Emission Control Law, to conduct regular inspections and properly manage the equipment to prevent refrigerant leaks. There are, however, many sites where there are no specialist equipment operators for whom this responsibility is a major burden. Therefore, it is expected that the introduction of systems that automate leakage management and the conversion to non-fluorocarbon refrigerants will be highly sought after.

5.8 VISION OF THE FUTURE COLD CHAIN CREATED BY HEAT PUMP TECHNOLOGY

5.8.1 Establishing a Cold Chain Centered on Next-Generation Heat Pump Technologies

As described in Section 5.2.2, the current cold chain mainly faces the following challenges and needs

1. As refrigeration and freezing equipment must be kept in constant operation, it consumes a considerable amount of energy, and there is a need to promote the introduction of energy-saving equipment, further increase the efficiency of equipment and energy efficiency through energy management, and utilize renewable energy
2. Looking at the market stock as a whole, there is still a large amount of equipment using refrigerants with a high GWP, and there is a need for low GWP conversion and measures to prevent refrigerant leakage
3. The distribution channel is becoming more complex as consumer needs become more diverse and sophisticated, and the refrigeration and freezing equipment used is becoming more diversified, making it more difficult to control temperature, humidity, etc. throughout the entire process, which can lead to food waste
4. Growing labor shortages have increased the need for more efficient transportation and delivery, and easier management of equipment and refrigerant

In contrast, if next-generation heat pump technology is used, which enables overall optimization based on centralized management of the relevant data, the above challenges and needs can be met at the same time.

A concrete vision of the future of the cold chain is shown in Figure 5.22. By constructing a system that manages data for the entire cold chain, centered on refrigeration and freezing equipment using next-generation heat pump technology, information on the temperature and humidity of foodstuffs in the cold chain, information on refrigerant management for refrigeration and freezing equipment, and other related information (energy consumption, CO_2 emissions, order quantities, production quantities, etc.) will be stored, shared, and utilized as digital data.

This enables optimum temperature and humidity control throughout the entire processing and distribution stages, which helps to maintain the freshness and quality of food, leading to improved consumer satisfaction, as well as significantly extending the shelf life and consumption life of food, as well as reducing food waste.

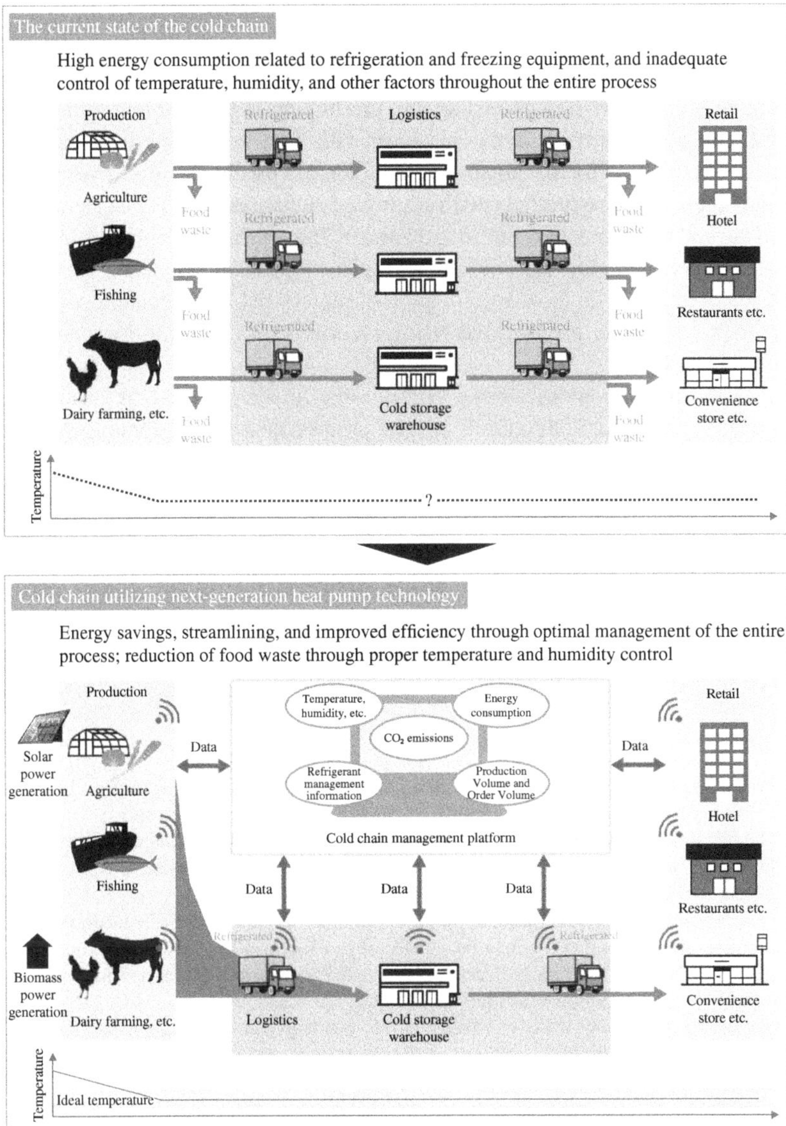

FIGURE 5.22 Current cold chain versus cold chain realized by next-generation heat pump technology.

5.8.2 Deployment of Next-Generation Heat Pump Technology in the Food System

As described above, next-generation heat pump technology enables a visualization of energy consumption, CO_2 emissions, and refrigerant leakage in the entire cold chain, while also enabling optimization of the selection and operation of refrigeration and freezing equipment on this basis. At the same time, equipment and refrigerant management can be automated, reducing the time and cost required for these tasks. In addition, the system can contribute to the rationalization and efficiency of the entire process by precisely matching order quantities with production quantities, thereby reducing waste in food manufacturing and production to the absolute minimum.

Thus, the establishment of a cold chain with next-generation heat pump technology at its core is one of Japan's models for a sustainable food system, which combines increased productivity with improved sustainability. The system can also contribute not only to Japan but also to the world by being deployed in ASEAN countries, where demand for cold chain logistics services has been increasing in recent years, in line with the diversification of dietary habits associated with population growth, economic growth, and rising income levels, and the expansion of the e-commerce market.

Although this section describes a vision of the future centered on the cold chain, as described in Chapter 1, Section 7, next-generation heat pump technology will also change the future of horticultural facilities at the production stage.

Decarbonization can be achieved by converting from the current mainstream fuel oil heaters to next-generation heat pump technology as the heat source for temperature control in horticultural facilities. In addition to this, it will enable advanced control of not only temperature, but also humidity, to control the growth and development of crops. This will lead to stabilized production and reduced food waste by adjusting the timing of shipments, thereby contributing to a stable supply of food.

NOTES

1. Wellness: Definitions of wellness include "the active pursuit of activities, choices and lifestyles that lead to overall health" by the Global Wellness Institute.
2. Health: The Preamble to the Charter of the World Health Organisation (1946) defines health as "a state of complete physical, mental and social well-being and not merely the absence of disease or infirmity."
3. Heating degree day: An indicator used to calculate the amount of heat required to heat a house or other structure for one winter. It is calculated by adding up the difference between the statistical daily average outdoor temperature and the heating temperature during the period when heating is required in the area.
4. Radiation type: Heat transfer caused by electromagnetic waves such as infrared radiation. No heat medium is required, and heat transfer occurs both in a vacuum and at a distance, so if there is a hot object, it feels warmer in that direction. Walls and floors can also be heated, but it takes time to heat the entire space.
5. Convection type: Heat transfer by fluid flow caused by temperature differences. The entire space can be warmed quickly by convection of warm air, but the walls and floors are not directly heated, so it takes time to warm them.
6. ZEB (zero energy building)/ZEH (zero energy house): Buildings and houses in which the same amount of energy is generated by renewable energy and other sources as is consumed, making energy consumption virtually zero.
7. Economic loss due to food waste: An example calculation by the Tokyo Institute of Technology and the National Institute for Environmental Studies estimates that the economic loss due to food waste in 2015 was 4,672.2 billion yen.
8. Electricity consumption of retail outlets: According to the breakdown of electricity consumption by industry sector shown in the Energy-Saving and Electricity Conservation Leaflet for businesses prepared by the Agency for Natural Resources and Energy, wholesale and retail outlets account for approximately 40–50% and food supermarkets

for approximately 70–75% in summer, depending on the area; in winter, wholesale and retail outlets account for approximately 20–40% and food supermarkets, depending on the area, wholesale/retail outlets account for about 20–40% and food supermarkets for about 60–65%.

REFERENCES

1. Global Wellness Institute, "What is Wellness?," 2023. https://globalwellnessinstitute.org/what-is-wellness/
2. Masayuki Mae/Sassokusha Inc., "Masayuki Mae's Science of 'Igokochi' (comfort)", Replan web magazine, vol. 008, 2018, https://www.replan.ne.jp/articles/6401/3/
3. Sekisui Industry Corporation, "We teach you how to ventilate well in houses, offices and shops," 2020, https://www.sekisui-ind.co.jp/column/other/post-2163
4. Ministry of Land, Infrastructure, Transport and Tourism, Expert Committee on Energy Efficiency Strategies for Housing and Buildings in Pursuit of a Decarbonized Society, "Estimation of the composition ratio of new construction and stock of housing and buildings by energy-saving performance," 2023, https://www.mlit.go.jp/jutakukentiku/house/content/001631903.pdf
5. Ministry of Land, Infrastructure, Transport and Tourism, Council for Social Infrastructure Development, Future Energy Saving Measures for Housing and Buildings (Third Report) and the Future Direction of the Building Standards System (Fourth Report), Annex 3, p. 35, 2022, https://www.mlit.go.jp/report/press/content/001487807.pdf
6. Ministry of the Environment, White Paper on the Environment, Sound Material-Cycle Society and Biodiversity 2020, Chapter 3, Section 1 Lifestyle Innovation for a Decarbonized and Sustainable Society Figure 3-1-7, 2020, https://www.env.go.jp/policy/hakusyo/r02/.%20html/hj20010301.html#n1_3_1
7. Ministry of the Environment, "Summary of the Survey Results on CO_2 Emissions in the Household Sector in FY2021 (Confirmed Figures)," Table 2-1 by Basic Items (Household, House, Equipment Usage, etc.)—Quantity of Equipment Used (TV, Refrigerator, Air Conditioner, Washing Machine, Clothes Dryer, Bathroom Dryer, Dishwasher, Dish Dryer, Microwave Oven, Gas Oven), (2023), https://www.env.go.jp/content/000122573.pdf
8. Ministry of the Environment, "Summary of the Survey Results on CO_2 Emissions in the Household Sector in FY2021 (Confirmed Figures)," Basic Items (Household, House, Equipment Usage, etc.)—Heating Usage (Temperature Setting Status of Heating Equipment Most Used, Heating Equipment Most Frequently Used, Weekday Use of Heating Equipment in the Cold Season, Presence or Absence of Heating for Pets, Number of Heating Rooms), 2023, https://www.env.go.jp/content/000122573.pdf
9. Ministry of the Environment, "FY2021 Statistical Survey on CO_2 Emissions in the Household Sector," Basic Items (Households, Houses, Equipment Usage, etc.)—Type of Equipment (Water Heaters, Hot Water Systems, Stoves), 2023, https://www.env.go.jp/content/000122573.pdf
10. Institute of Energy Economics, Japan and Riko Tosho "EDMC Handbook of Energy & Economic Statistic," 2022.
11. Heat Pump and Thermal Storage Technology Center of Japan, Japan Electro-Heat Centre, "Report on the Prospects for the Spread of Electrification in 2022," 2023.
12. Better Living Foundation, "Warm living guide for healthy living: Your health from the temperature of your home," p. 1, 2019, https://www.onnetsu-forum.jp/file/pamphlet.pdf
13. Ministry of Internal Affairs and Communications, Fire and Disaster Management Agency, "Emergency medical transportation status due to heatstroke in 2022 (May to September)," 2022, https://www.fdma.go.jp/disaster/heatstroke/items/r4/heatstroke_geppou_202205-09.pdf

14. Masaru Suzuki, et al., "Sudden death phenomenon while bathing in Japan," Circulation Journal (2017). https://pubmed.ncbi.nlm.nih.gov/28392545/
15. Sustainable Open Innovation Initiative (SII), "ZEH subsidies by the Ministry of Economy, Trade and Industry and the Ministry of the Environment in 2023," 2023, https://sii.or.jp/moe_zeh05/uploads/zeh05_pamphlet2.pdf
16. Mitsubishi Research Institute Inc, "What is 'Foodtech' being promoted for? Expectations as a climate change mitigation measure Part 1," (2022), https://www.mri.co.jp/knowledge/column/20221222.html
17. Yosuke Munai (Institute of Science Tokyo)/Toshihiko Masui and Yuko Kanamori (National Institute for Environmental Studies), "A study on assessment of environmental, economic and social impacts of food loss and loss reduction in Japan," 2020, https://www.erca.go.jp/suishinhi/kenkyuseika/pdf/symposium_r02_Munesue.pdf
18. Ministry of Agriculture, Forestry and Fisheries, "What is food loss?," 2023, https://www.maff.go.jp/j/shokusan/recycle/syoku_loss/161227_4.html
19. Japan Association of Refrigerated Warehouses, "Keidanren (Japan Business Federation) Carbon Neutral Action Plan FY2022 Follow-up Result: Individual Industry Version: Vision of the refrigerated warehouse industry towards carbon neutrality in 2050 (basic policy, etc.)," p. 4, 2022, https://www.keidanren.or.jp/policy/2022/095_kobetsu39.pdf
20. Shigekatsu Koganemaru (Japan Association of Refrigerated Warehouses, "The Efforts to Address Global Environmental Issues by JARW" (Atmosphere APAC Summit 2022 presentation material), 2022, https://drive.google.com/file/d/1uXdQLGmLLl9jIgZ0TlmiU7EfTvdvwdCG/view
21. Mayekawa Manufacturing Co. Ltd, "High-efficiency natural refrigerant chiller NewTon," (2023), https://www.mayekawa.co.jp/ja/products/detail/refrigeration/03/index.html
22. Mayekawa Manufacturing Co. Ltd, "CO_2 condensing unit COPEL," 2023, https://www.mayekawa.co.jp/ja/products/detail/refrigeration/06/index.html
23. Ministry of Economy, Trade and Industry, Agency for Natural Resources and Energy, Energy Conservation Portal Site, "Summer energy and power saving menu leaflet/Winter energy and power saving menu leaflet," 2023, https://www.enecho.meti.go.jp/category/saving_and_new/saving/media/index.html
24. Japan Franchise Association (JFA), "Measures to combat global warming in the convenience store industry—Carbon neutral action plan 2021 results report," p. 9, p.11, 2023, https://www.meti.go.jp/shingikai/sankoshin/sangyo_gijutsu/chikyu_kankyo/ryutsu_wg/pdf/2022_001_05_01.pdf
25. Lawson Inc., "Lawson's first completely 'fluorocarbon-free' refrigeration equipment used in stores. Aiming to prevent global warming and use highly efficient energy—Also implementing initiatives for power-saving control through the use of IoT for equipment," 2019, https://www.lawson.co.jp/company/news/detail/1387673_2504.html
26. Japanese Society of Refrigerating and Air Conditioning Engineers, Refrigeration Cycle Control (Technical Book Series of the Japanese Society of Refrigerating and Air Conditioning Engineers), p. 163, 2018.
27. Yamato Holdings Co. Ltd, "Energy and Climate—Mitigating Climate Change," 2018 https://www.yamato-hd.co.jp/csr/environment/climate_change.html
28. Japan Association of Refrigerated Warehouses, "Storage temperature ranges," 2018. https://www.jarw.or.jp/study/guardian/temperature
29. Ministry of Land, Infrastructure, Transport and Tourism, "FY2021 Parcel delivery service handling results," 2022, https://www.mlit.go.jp/report/press/jidosha04_hh_000255.html

6 How Heat Pump Technology Supports Industrial Development and Stable Energy Supply

OVERVIEW

Chapter 5 explains how next-generation heat pump technology can make a significant contribution to the living environment and food system as a technology that supports healthier environments and a more secure cold chain.

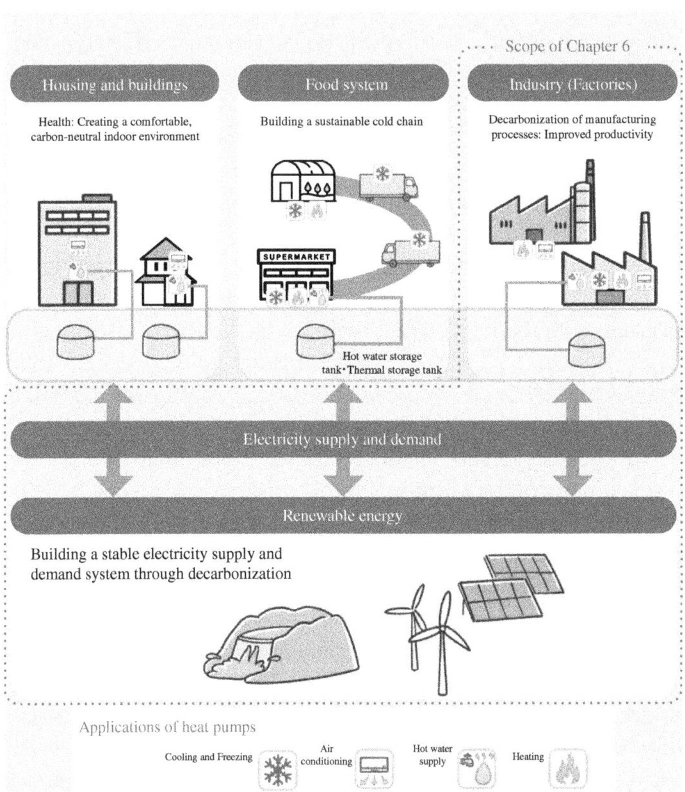

FIGURE 6.1 The role of next-generation heat pump technology in new value creation.

DOI: 10.1201/9781003666950-6

Here, as shown in Figure 6.1, the potential contribution to industry (manufacturing processes) and the stable supply of energy is considered as an area where new value is expected to be created by next-generation heat pump technology.

This chapter explains how, as a lot of heat is utilized in the manufacturing process, heat pump technology will play a major role in meeting the strong demand for decarbonization through electrification of the manufacturing industry. It is also explained that heat pump technology plays an important role in energy security with demand response[1] (DR) and virtual power plants (VPP) as one of the decentralized energy resources on the demand side.

6.1 THERMAL ENERGY TECHNOLOGY IN MANUFACTURING PROCESSES

6.1.1 Decarbonization in Industry

As mentioned in Chapter 2, Section 2.1, decarbonization efforts in the industrial sector [1], which emits about one-third of Japan's total greenhouse gas (GHG) emissions of 1.17 billion t-CO_2 (FY2021 value) (after electricity and heat allocation), are essential if Japan is to achieve the country's stated goal of carbon neutrality. There is also rapidly mounting pressure on businesses to decarbonize, not only from central government as a national target and policy, but also from surrounding stakeholders (Figure 6.2).

Companies with global operations, for example, are accelerating their moves toward decarbonization throughout their supply chains in compliance with reduction targets consistent with the 1.5°C target required by the Paris Agreement, such as the Science-Based Targets (SBT)[2]. The trend is becoming increasingly common, with large companies as well as small and medium-sized enterprises (SMEs) required to take action.

Delays in addressing decarbonization can lead to a range of risks for company operations, including contract termination from suppliers, reduced competitiveness of products, and difficulties in accessing finance. Conversely, if companies are able to move quickly toward decarbonization, e.g. by transforming their energy and manufacturing processes, they will be able to maintain and strengthen their competitiveness in the global marketplace.

In addition to decarbonization, another macro trend in the business environment is the development of digitalization, as discussed in Chapter 2, Section 2.6. It is crucial for companies to utilize digital technologies such as the Internet of Things (IoT), digital twin[3], and artificial intelligence (AI) for digital transformation (DX) in order to survive and grow in the future. This section therefore focuses in particular on the industrial sector.

This section describes the current status, challenges and needs, and future of heat utilization technology in manufacturing processes (especially in the low temperature

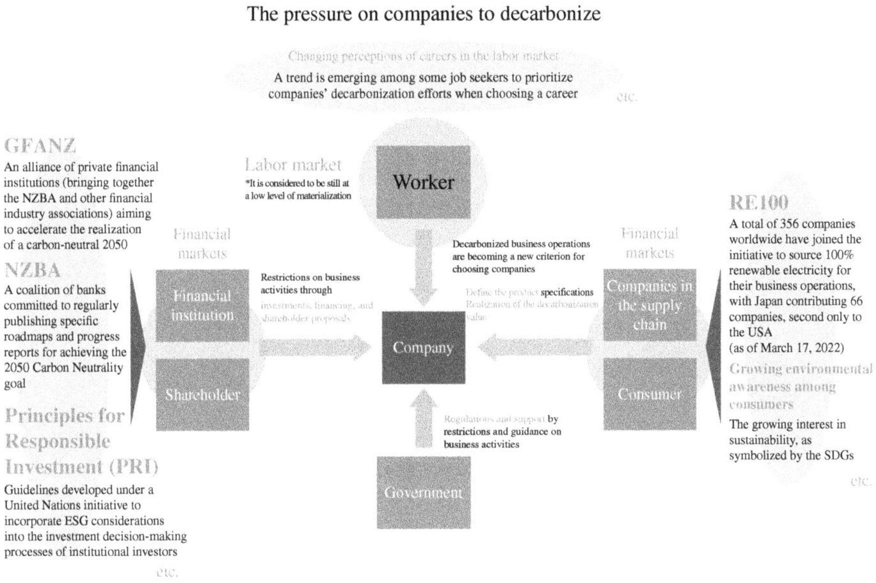

FIGURE 6.2 Decarbonization pressures on businesses. (From Industrial Science and Technology Policy and Environment Bureau, Agency for Natural Resources and Energy, and the Ministry of Economy, Trade and Industry, Interim Clean Energy Strategy [2]).

range below 200°C). This accounts for a particularly large share of emissions from the industrial sector and is where heat pump technology is expected to be utilized, mainly from the perspective of decarbonization and DX.

6.1.2 Current Status of Manufacturing Processes

Figure 6.3 shows an example of the estimated heat demand by temperature range in the industrial sector in a survey conducted by the Agency for Natural Resources and Energy. Although a wide range of temperatures exist, the main sources of heat demand are for steam and hot water.

The demand for lower temperatures is the largest, accounting for more than a quarter of the total demand in the low temperature zone of 20–200°C, which mainly corresponds to steam and hot water. It can be seen that the demand for cooler temperatures is mainly covered by electricity (centrifugal chiller, chilling units, and other heat pumps), whereas the steam and hot water demand in the volume zone is almost exclusively covered by fuel (steam and hot water boilers, and other fuel-fired equipment). The share of combustion heating by industrial furnaces is also overwhelmingly high for direct heating demand.

Therefore, electrification of steam and hot water systems and direct heating with decarbonized electricity and conversion to decarbonized fuels (hydrogen, ammonia, and synthetic fuels) is crucial for the decarbonization of the industrial sector. In particular, steam and hot water systems in the low temperature range is an area

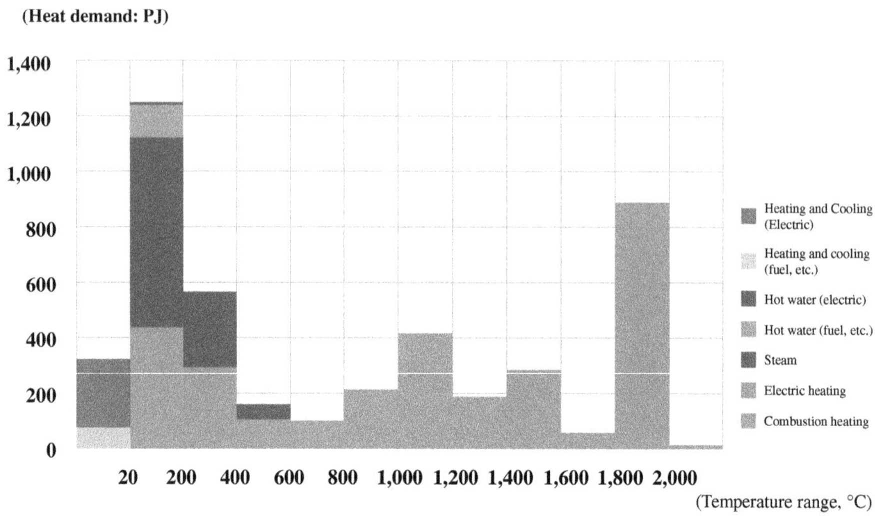

FIGURE 6.3 Heat demand by temperature range in the industrial sector. (From Mitsubishi Research Institute Inc., "Basic Survey on the Promotion of the Introduction of New Energy, etc. in FY2017 (Survey on the Supply and Demand of Heat and the Characteristics of Heat Supply Equipment)" [3]). *Note:* For steam reception, the breakdown between electricity and fuel etc. is not shown, but another chart in the same report shows that it is almost entirely covered by fuel etc. (i.e. boilers).

where electrification through conversion from boilers is expected, given the development of heat pumps for process temperature control in factories, etc. that are already in progress.

However, as the current shipment status of industrial boilers and industrial heat pumps shows (Figure 6.4), the proportion of industrial boilers is still overwhelmingly higher and the speed of adoption of indust rial heat pumps has not increased.

As mentioned in Chapter 3, there are issues behind this, such as the fact that, at present, the applicable range is limited to 120°C or lower (some systems can be applied up to 165°C).

6.1.3 DX in Manufacturing Processes

There are several survey examples regarding the status of DX-related initiatives in the manufacturing industry: in a survey commissioned by the Ministry of Internal Affairs and Communications in 2020, 22.8% of the manufacturing industry reported that they were "implementing" DX.

In a survey conducted by the Teikoku Databank Corporation in 2021, 14.7% of the manufacturing sector reported that they understood the meaning of the term DX and were working on it. Although there are variations, it is estimated that only 10–20% of companies are already working on DX, and it cannot be said that the initiatives have made sufficient progress.

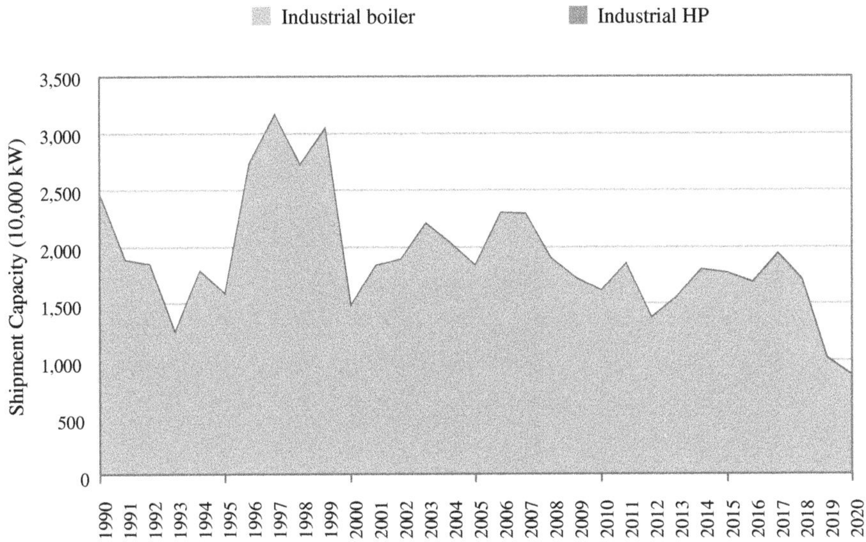

FIGURE 6.4 Transition of shipment capacity of industrial boilers and industrial heat pumps. (From Heat Pump & Thermal Storage Technology Center of Japan/Japan Electro-Heat Center, Survey Report on the Prospect of Electrification in FY2020 [4]).

The status of IoT and digital twin use in Figures 6.5 and 6.6 shows that Japan is not making as much progress as the USA, with around 20% of companies implementing IoT company-wide or in part, and around 20% using digital twin in situations such as manufacturing processes, factories, and supply chains. The proportion of companies using IoT in the whole or part of the company is around 20%, while the proportion of companies using digital twin in manufacturing processes, factories and supply chains is only around 5-10%.

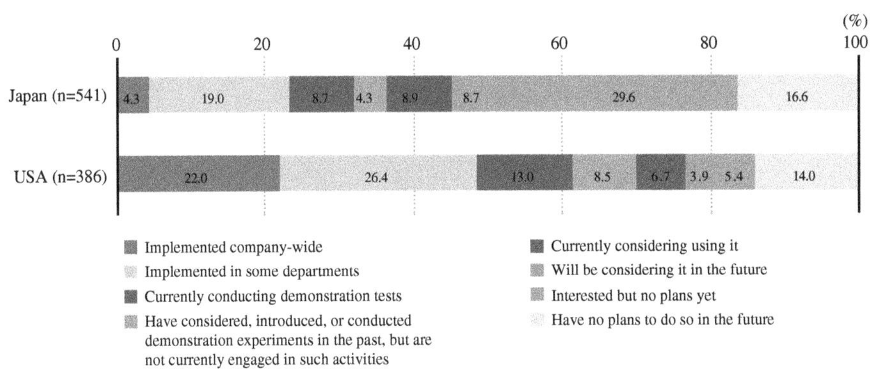

FIGURE 6.5 Status of IoT uptake. (From Information-technology Promotion Agency (IPA), DX White Paper 2023 [5]).

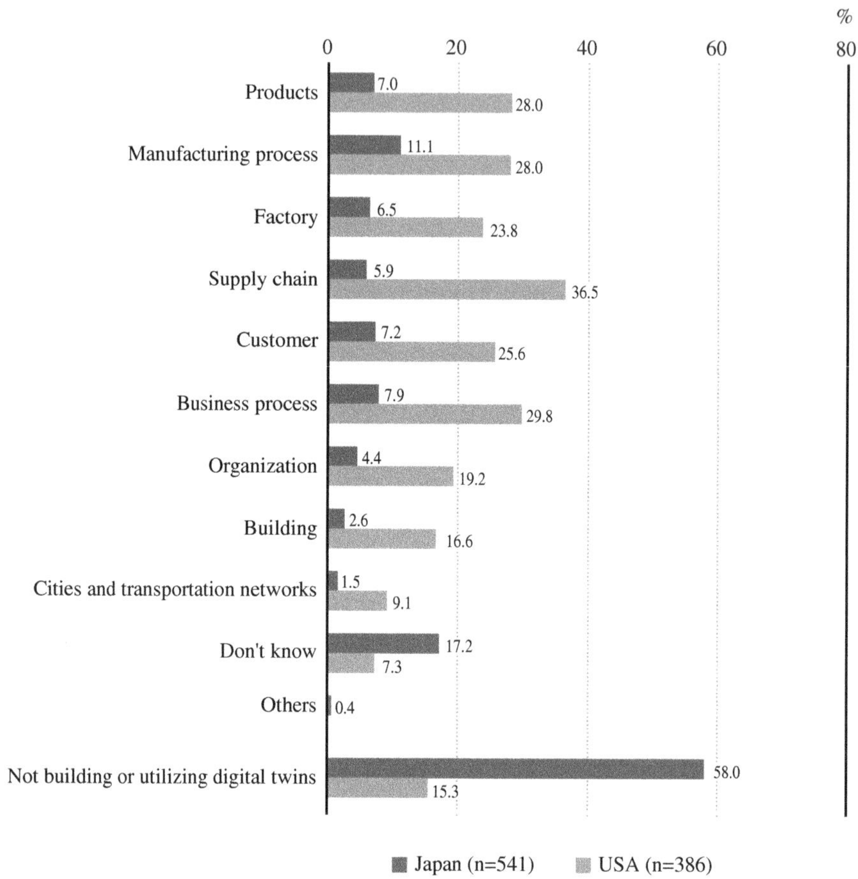

FIGURE 6.6 Status of digital twin construction and utilization. (From IPA, DX White Paper 2023 [5]).

6.2 CHALLENGES AND NEEDS FACING MANUFACTURING PROCESSES

6.2.1 Current Reliance on Fossil Fuels

As discussed in Chapter 2, Section 2.2, the current industrial sector in Japan is largely dependent on fossil-fuelled combustion boilers and industrial furnaces to meet the majority of its heat demand.

In addition to electrification, the use of decarbonized fuels (hydrogen, ammonia, synthetic methane, etc.) is being explored as a means of decarbonizing heat systems in the industrial sector. While it takes a certain period of time to establish such supply chains of decarbonized fuels, continued use of fossil fuels will, at least in the meantime, result in carbon lock-in[4].

Energy conservation and fuel switching

Future measures based on heat demand by temperature range in the industrial sector

- In promoting decarbonization, the effective use of heat—including improving thermal efficiency and utilizing unused heat—remains crucial.
- For low-temperature steam demand, heat pumps are a promising option, but their limited temperature range, lack of widespread deployment, and high costs remain challenges.
  Therefore, these issues will be addressed through the development and demonstration of high-efficiency heat pumps (expansion of temperature range) and through capital investment support to promote wider adoption.
- On the other hand, for the medium- and high-temperature ranges where the introduction of heat pumps is difficult, necessary support measures will be provided for appropriate options—such as electrification or the shift to non-fossil fuels—after taking into account medium- to long-term power supply structures and costs.

Temperature range		Possible Approaches to Decarbonizing Heat Sources	Time
Overall		• Improving heat utilization efficiency and utilizing unused heat, promoting collaborative energy conservation, and strengthening support for small and medium-sized enterprises	From the near term
	200°C or less (steam)	• Support for heat pump adoption and technology development	From the near term
	Around 100–1,500°C	• Decarbonization of the power supply + electrification	From 2030 onward
		• Utilization of non-fossil fuels such as hydrogen, ammonia, synthetic methane, and biomass fuels/Support based on price differentials compared to existing fuels	From the near term—2030 and beyond
		• Shift to natural gas → Transition to non-fossil fuels/Support based on price difference with existing fuels	From the near term—2030 and beyond
	Around 2000°C	• Technological innovations (such as hydrogen-reduced ironmaking)/support based on the price differential with existing fuels	From 2030 onward

FIGURE 6.7 Direction for decarbonization of heat systems in the industrial sector. (From Industrial Science and Technology Policy and Environment Bureau, Agency for Natural Resources and Energy, and Ministry of Economy, Trade and Industry, Interim Clean Energy Strategy [6]).

In addition, it is difficult to produce such fuels on a large scale domestically, from economic and other perspectives, which forces a certain degree of dependence on imports from abroad. From the perspective of energy security, for steam and hot water demand in low-temperature ranges, electrification technology, i.e., industrial heat pumps, is already established. It is necessary to transition to industrial heat pumps as soon as possible.

The Interim Clean Energy Strategy (published May 2022) states that, for low temperature heat demand in the industrial sector, conversion to heat pumps is a promising option, and that support should be provided for the development and demonstration of technology to overcome the current issues (e.g. limited applicability) and for capital investment to expand the introduction of such systems (Figure 6.7).

6.2.2 Importance of Heat Pump Technology in De-Fossilization

The use of boilers as a heat source for steam and hot water poses a number of challenges, in addition to GHGs and other emissions associated with the combustion of fossil fuels. For example, in factories, high-temperature, high-pressure steam is usually generated intensively in one boiler room and supplied to the production

TABLE 6.1
Heat Loss from Boiler to Production Equipment in Factories

Location	Heat loss (loss rate to heat input)	Main sources of loss
(1) Boiler	Several % ~ 20%	Exhaust gas, heat release, blow, excess air ratio
(2) Steam piping	Several dozen % ~ 50%	Heat dissipation, steam leakage, excessive design
(3) Production equipment	Several % ~ 20%	Condensate, heat radiation, steam leakage
(1) + (2) + (3)	30 ~ 70%	

Source: Based on New Energy and Industrial Technology Development Organization (NEDO), "Energy saving can be done! Industrial Heat Pump Exposition" [7].

facilities through steam pipes extending several hundred meters, but a lot of heat is lost in the process. Although there are differences depending on the type of industry, type of production equipment, and aging of equipment, 30–70% of heat is lost in relation to the fuel input (see Table 6.1), which means that fuel input and associated costs are wasted.

In addition, when steam pipes are lengthy, it takes several minutes before steam reaches the production facilities from the boiler, making it difficult to supply the right amount of steam at the right time according to the conditions of the production facilities. This is also a problem in terms of controllability. In contrast, heat pumps are characterized by their high controllability, and by making use of this feature, it is expected that the controllability of the entire production process can be improved and stabilized.

6.2.3 Use of Digital Technology

In addition, DX and digitalization-related initiatives in the manufacturing sector have not progressed sufficiently compared to those in the USA and other countries, as discussed in Chapter 6, Section 6.1.1. On the other hand, the source of value creation in business is already shifting to the use of data and digital technologies.

In order for businesses to survive the future competitive marketplace, various types of data generated by business activities should be collected and accumulated, and design development, production management, and other processes should be carried out efficiently using digital technologies such as IoT, digital twin, and AI. Alongside this, it is necessary to create higher added value and increase competitiveness.

As the decarbonization described above needs to be addressed across the entire manufacturing process and supply chain, the use of digital technologies is becoming essential to visualize efforts and share this information among stakeholders beyond organizational and corporate boundaries.

6.3 VISION OF FUTURE MANUFACTURING PROCESSES CREATED BY HEAT PUMP TECHNOLOGY

6.3.1 Decarbonization of Manufacturing Processes with Heat Pump Technology

As discussed in Chapter 6, Section 6.1, current manufacturing processes in the industrial sector, especially those that use steam and hot water, face the following main challenges

1. While decarbonization is becoming an essential issue for business management, the majority of heat demand in current manufacturing processes is still met by combustion boilers using fossil fuels
2. The method of supplying steam produced by boilers to production facilities not only causes significant heat loss in the process but also poses controllability issues
3. Insufficient progress has been made in digitalization and DX, which are essential for the future growth and competitiveness of enterprises, and the use of digital technologies such as IoT and digital twin has not progressed

The above issues can, however, be resolved by promoting the conversion to next-generation industrial heat pumps that can expand the scope of application (e.g. higher temperatures) and enable centralized management of information and overall optimization on this basis throughout the manufacturing process.

A concrete vision of the future of the manufacturing process is shown in Figure 6.8. By switching from a centralized heat source system, where heat is supplied from a conventional boiler to the production facilities, to a decentralized installation of next-generation heat pump technology, the heat losses caused by piping etc. can be significantly reduced. In addition to the effective utilization of air heat and waste heat from factories etc., which have not been utilized up to now, the integrated, simultaneous use of hot and cold thermal energy (when there is also demand for cold) can be achieved. Moreover, the excellent features of next-generation heat pump technology also bring about significant energy savings.

In addition, as heat pumps do not involve the combustion of fossil fuels and can instead use electricity from renewable energy sources, for example, they can reduce the associated GHG emissions to virtually zero, making a significant contribution to decarbonization.

6.3.2 Promoting the Use of Heat Pumps Using Digital Technology

The importance of DR on the demand side is increasing with the aim of further expanding the use of renewable energy, and there are moves to actively evaluate and give preferential treatment to businesses that respond to DR under the Energy Conservation Law and other systems (see Section 6.4 below).

Electricity companies are also increasingly offering contract forms and electricity rate options for consumers who support DR, as shown in Figure 6.9.

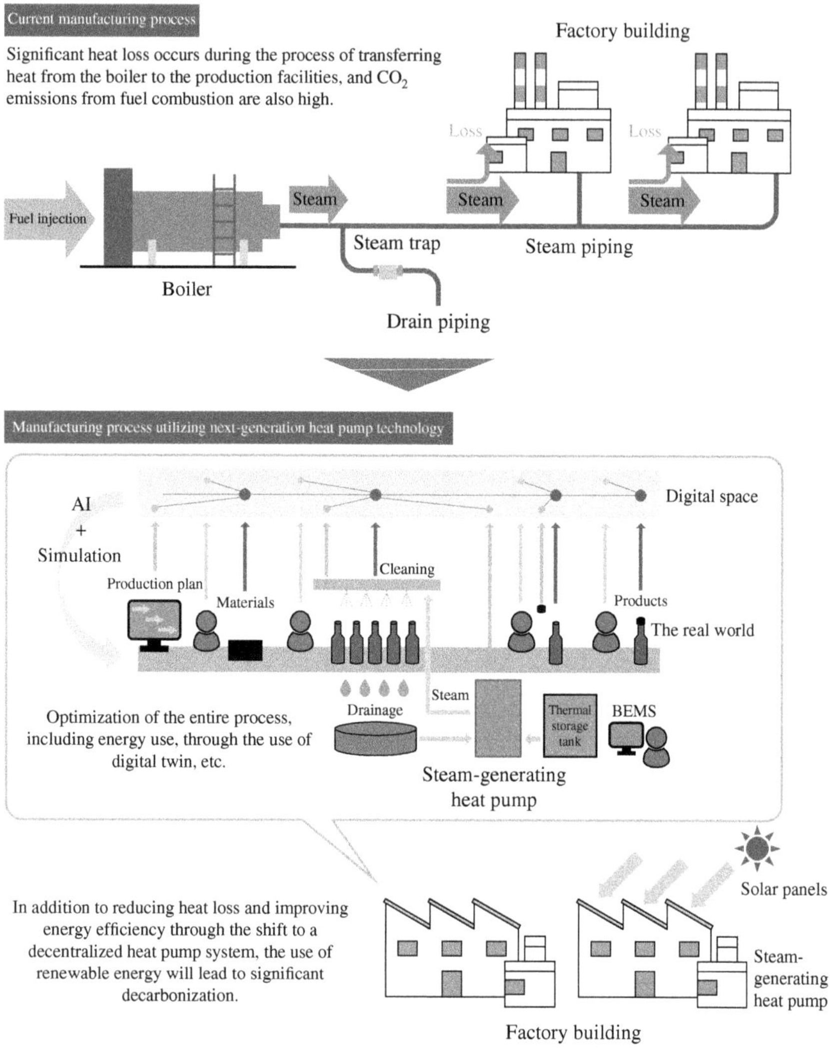

FIGURE 6.8 The current manufacturing process versus one realized by next-generation heat pump technology.

By converting from boilers to industrial heat pumps and setting up a heat supply system that is compatible with DR, including a heat storage system, it is possible to enjoy financial benefits such as reduced energy costs and rewards for DR performance. If the boiler-based heat supply system is to continue, it will be necessary to switch to decarbonized fuels, but, as mentioned above, this will require reliance on imports from abroad, which will lead to significant price fluctuation uncertainty. In contrast, electrifying the heat supply system with industrial heat pumps allows the use of electricity, which is expected to become cheaper as renewable energy becomes more widely used, resulting in less price fluctuation than fuel.

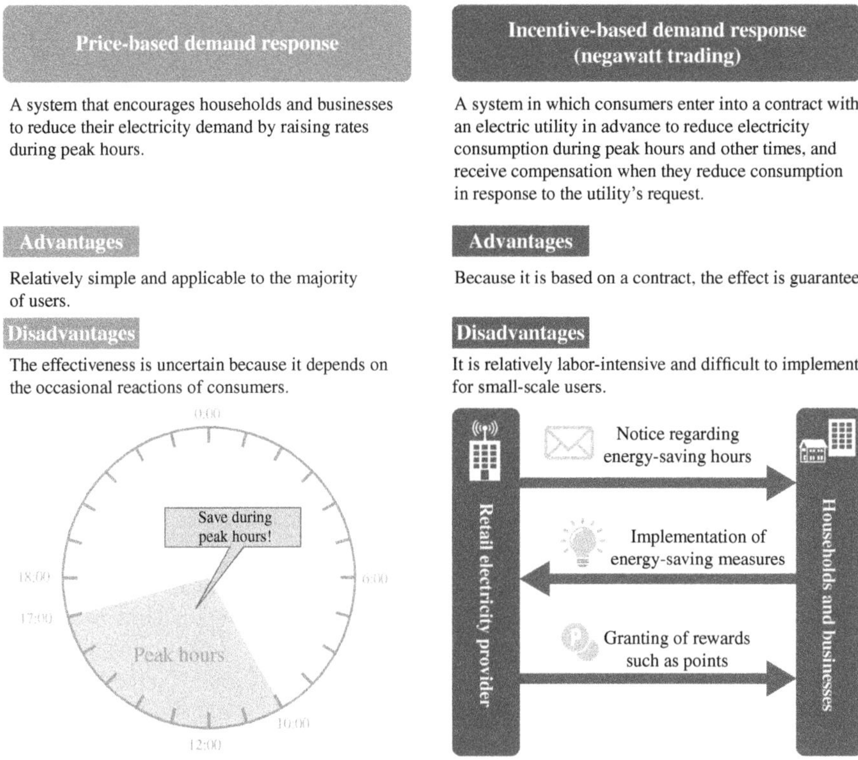

FIGURE 6.9 The DR mechanism. (From Ministry of Economy, Trade and Industry and Agency for Natural Resources and Energy, "What is demand response? Types of demand response" [8]).

As described above, the conversion from boilers to industrial heat pumps will lead to a reduction in energy costs and mitigation of fluctuations, which greatly affects competitiveness, and will lead to improved competitiveness.

Another advantage of next-generation industrial heat pumps is their controllability. As the cold and hot water withdrawal capacity, temperature, etc. can be accurately controlled using inverter control technology, which is one of Japan's strengths, it is possible to optimize operation to meet the heat demand of diverse needs, leading to improved controllability and stability of the entire production process.

In addition, the system is highly compatible with the digitalization and IoT of manufacturing processes, such as in factories. Therefore, the use of digital technology enables the construction of control and management systems that are also linked to other facilities and equipment in the factory, thereby realizing the optimum configuration and operation of the entire manufacturing process. This will lead to improved operational efficiency and productivity, and the production of high-quality products, and will contribute to improving the competitiveness of the Japanese manufacturing industry.

6.4 CURRENT STATUS OF THE POWER SUPPLY AND DEMAND SYSTEM

6.4.1 Building a Stable Power System with a View to Decarbonization

6.4.1.1 What Is a Power System?

A power system is a system comprising various elements related to the supply of electricity, including generation facilities such as hydro, thermal, solar, wind, and nuclear power; distribution facilities such as transmission lines, substations, and distribution lines; as well as operation and control systems that protect and control the supply of electricity and the occurrence of faults. In Japan, the ten electric power companies (former general electricity utilities) were responsible for the integrated management and operation of all aspects of electricity supply, from generation, transmission, and distribution to retail sales, but under the electricity system reform they were separated into power generation, general transmission and distribution, and retail electricity utilities respectively. Of these, the general transmission and distribution companies are responsible for the operation and control of the entire power system to ensure a stable supply of electricity.

For a stable supply of electricity, as shown in Figure 6.10, it is necessary to constantly balance electricity supply and demand. If this supply and demand balance is disrupted, the electricity supply will no longer be stable, resulting in the triggering of safety devices, which can cause power stations to shut down and in some cases lead to large-scale power outages.

For this reason, general transmission and distribution companies, which are obliged to maintain system frequency under the Electricity Business Act, maintain the supply–demand balance by, among other things, running power sources and other sources traded on the market in line with demand and adjusting the power sources as

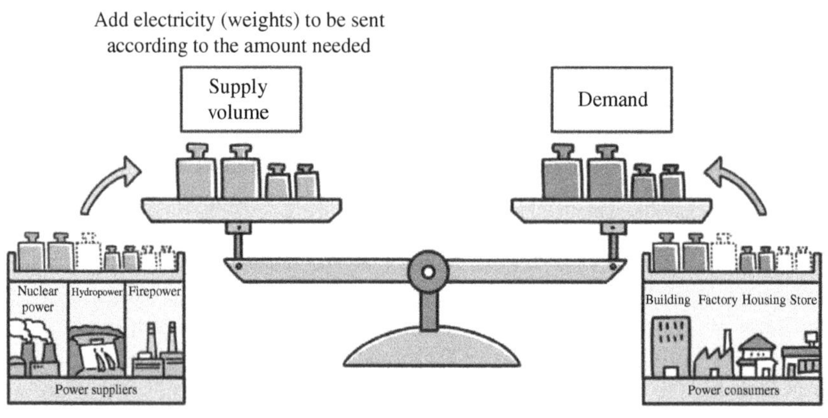

FIGURE 6.10 The balance of electricity supply and demand. (From Ministry of Economy, Trade and Industry and the Agency for Natural Resources and Energy, "What is the keyword 'balance of electricity supply and demand' for stable supply of electricity? Let's experience it with a game" [9]).

necessary for a stable supply of electricity to accommodate the constantly changing demand. The capacity of the systems and mechanisms that control generation facilities, demand facilities, etc. and their equivalents is referred to as regulating power.

6.4.1.2 Building a Power System for the Mass Introduction of Renewable Energy

The mass introduction of renewable energies is essential for the realization of carbon neutrality, but this will lead to the decentralization of power sources such as small-scale solar power, as well as to greater fluctuations than at present in power supply and demand due to variable renewable energy sources such as solar power and wind power, which will hinder the stable supply of electricity. There are also concerns that this will increase integration costs (e.g. grid stabilization costs). Therefore, there is a need to build a power system that achieves stable supply while ensuring decarbonization by securing the necessary regulating power to increase the flexibility of the power system.

Until now, large thermal power plants have played the role of regulating power, but in the future, as regulating power is also required to be decarbonized, initiatives to secure new regulating power, such as DR and VPP (where decentralized energy resources on the demand side are highly controlled in line with the supply side) are becoming increasingly widespread. Heat pump technology and thermal storage systems used in combination with heat pump technology are typical demand-side energy resources, and their use in DR and VPP can contribute to the stabilization of the power system, thereby contributing to the further spread and expansion of renewable energy.

This section describes the current state of the power system and the challenges and needs for the further mass introduction of renewable energy in more detail, as well as the future that is expected to be realized through the use of heat pump technology.

6.4.2 Current State of the Japanese Power System

Figure 6.11 shows the current status of electricity generation by power source in Japan, where renewable energy has been expanding since the start of the feed-in tariff scheme in FY2012, and its share in the power supply mix increased to 20% by FY2020.

With this expansion of renewable energy, grid constraints are becoming increasingly apparent. As shown in Figure 6.12, there are three types of grid constraints: constraints on the overall supply–demand balance within an area, constraints related to transmission capacity, and grid constraints associated with fluctuations.

6.4.2.1 Constraints on the Overall Supply–Demand Balance Within an Area

Constraints on the overall supply–demand balance within an area refers to the generation of electricity exceeding demand, resulting in a surplus. As mentioned above, the amount of electricity generated and the amount of electricity demanded must always be balanced to ensure a stable supply of electricity. Therefore, when a surplus occurs, the supply–demand balance is adjusted by curtailing the output of thermal power generation and utilizing pumped storage and inter-regional interconnection lines, etc. If there is still a surplus, the output of renewable energy is controlled.

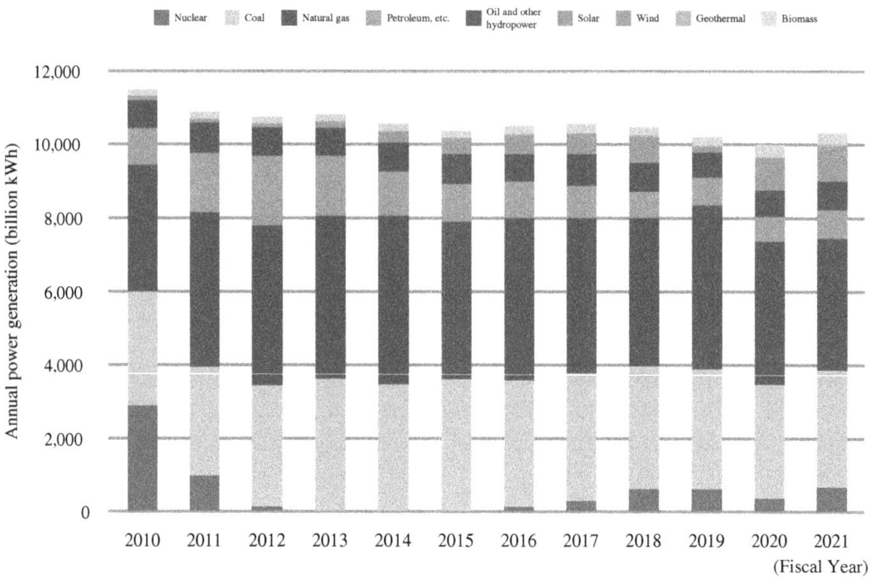

FIGURE 6.11 Changes in electricity generated by power source. (From Ministry of Economy, Trade and Industry, and the Agency for Natural Resources and Energy, Comprehensive Energy Statistics—Time Series (Reference Table) [10]).

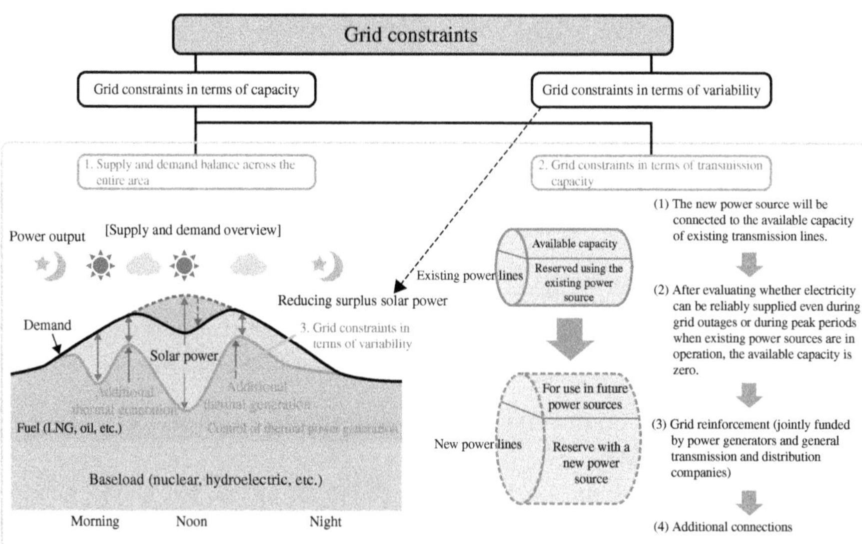

FIGURE 6.12 Types of grid constraints associated with the expansion of renewable energy deployment. (From Ministry of Economy, Trade and Industry and the Agency for Natural Resources and Energy, "Policy Issues in the Era of Large-Scale Introduction of Renewable Energy and the State of Next-Generation Power Networks" [11]).

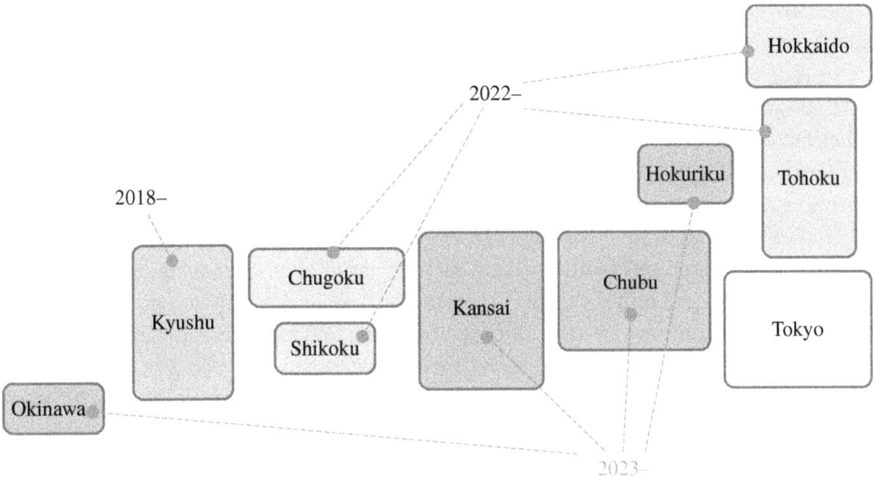

FIGURE 6.13 Status of implementation of renewable energy output control. (From Ministry of Economy, Trade and Industry and the Agency for Natural Resources and Energy, "Toward the Control of Renewable Energy Output Control" [12]).

Renewable energy output control was first implemented in the Kyushu area in 2018, followed by the Shikoku, Tohoku, Chugoku, Hokkaido, Okinawa, Chubu, Hokuriku, and Kansai areas, resulting in the Tokyo area being the only one where it has not been implemented as of July 2023 (Figure 6.13).

Along with the expansion of these implementation zones, the total amount of renewable energy output control is also on the rise, increasing from approximately 100 million kWh in FY2018 to approximately 600 million kWh nationwide in FY2022.

6.4.2.2 Constraints Related to Transmission Capacity

Constraints related to transmission capacity refers to the lack of transmission capacity for renewable energy generators and others to grid interconnection[5]. Normally, if there is an application from a power producer for an agreement to connect to the grid, capacity is reserved for transmission within the transmission capacity in the order in which the applications are received, but if there is no more space available, new transmission facilities are built according to the wishes of the applicant, and connections are made within the increased capacity.

However, as it costs a lot of money and takes a lot of time to increase transmission capacity, efforts are being made to make maximum use of the existing grid in order to connect at low cost and in a short time.

6.4.2.3 Grid Constraints Associated with Fluctuations

Grid constraints associated with fluctuations refers to the lack of regulating power to control the electricity supply and demand balance in order to ensure a stable supply of electricity. As mentioned above, general transmission and distribution companies, which are responsible for the stable supply of electricity, manage day-to-day supply and demand with the regulating power they have secured in advance.

In recent years, as the spread of naturally variable renewable energies such as solar power and wind power has increased, output fluctuations due to weather and other factors have grown, and it has become increasingly important to ensure that the regulating power to follow output fluctuations is available, but currently the majority of this is dependent on thermal power generation and other sources.

In parallel with these developments, decentralized energy resources (e.g. energy-creating equipment and facilities such as roof-mounted solar panels and fuel cells, and energy storage equipment and facilities such as heat pump equipment, thermal storage air conditioning systems, storage batteries and electric vehicles) introduced on the demand side are also becoming more widespread, and this has been accompanied by a growing trend. In addition to the conventional flow of electricity from the higher-level grid to the lower-level grid, many flows of electricity from the lower system to the upper system are also becoming apparent.

6.5 CHALLENGES AND NEEDS FACING POWER SYSTEMS

6.5.1 Challenges Facing the Power System

As mentioned in Section 6.2.1, the challenges facing Japan's power system include the increasing integration costs (e.g. grid stabilization costs) to accept renewable energy into the power system, as grid constraints have become more apparent with the increasing penetration of renewable energy.

In order to achieve carbon neutrality in the future, it will be necessary to increase the flexibility of the electricity supply and demand system by, for example, efficiently and effectively securing adjustment power in response to fluctuations and strengthening measures to maintain the power supply and demand balance, in order to ensure a stable supply and to further expand the use of renewable energy.

In this context, there is a growing need to utilize decentralized energy resources on the demand side, which have become increasingly popular in recent years, to improve the flexibility of the power system. Although each of the decentralized energy resources on the demand side are small in scale, they can be used as a regulating force for electricity supply and demand balance by aggregators bundling them together through advanced energy management technology utilizing the IoT, and controlling them remotely and in an integrated manner (Figure 6.14). This mechanism is called a VPP, as it allows multiple decentralized energy resources to function as if they were a single power plant. As a means of realizing VPP, controlling decentralized energy resources on the customer side and changing the pattern of electricity demand is called DR.

6.5.2 Distributed Energy Services Using VPP and DR

Heat pump water heaters with hot water storage tanks and thermal storage air conditioning systems that combine heat pumps and thermal storage systems are among the decentralized energy resources that are expected to be utilized for DR and VPP as energy storage equipment and facilities that can flexibly shift operating hours. Some related initiatives are already underway.

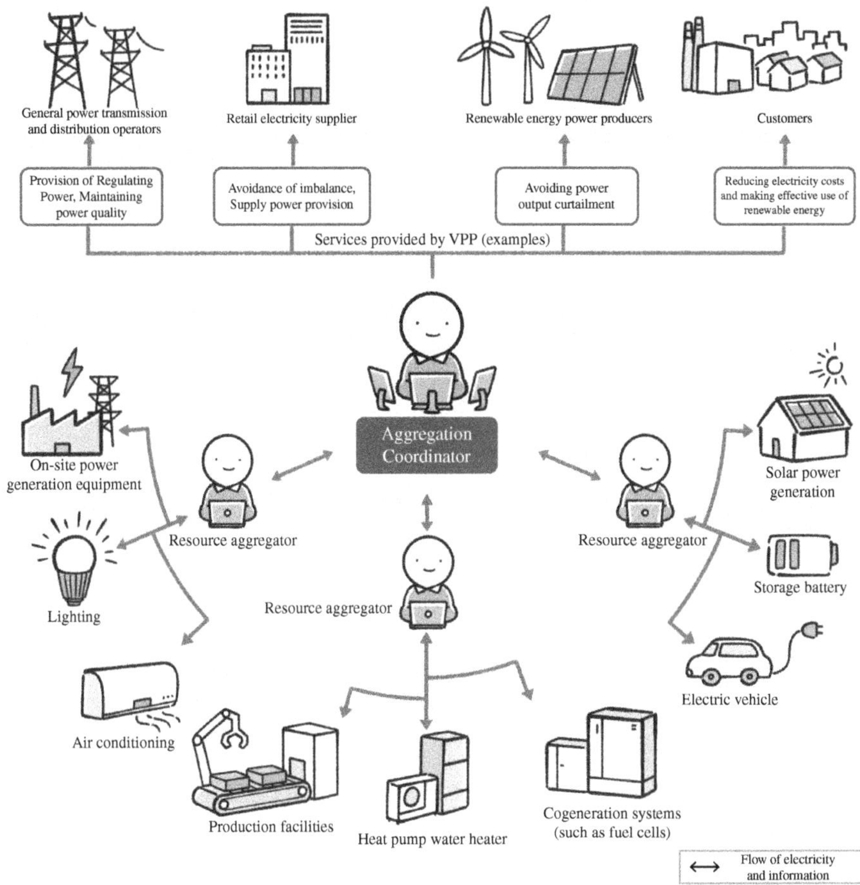

FIGURE 6.14 How VPP work. (From Ministry of Economy, Trade and Industry and the Agency for Natural Resources and Energy, "What are VPP and DR?" [13]).

For example, with regard to heat pump water heaters, the Ohisama Eco-Cute, which uses surplus electricity from solar power generation to boil water during the daytime, has seen some uptake in homes with residential solar power.

More recently, efforts have also begun to utilize DR to increase the flexibility of the power system: from December 2022, Hokuriku Electric Power has begun to use the heat pump water heaters of users of its leasing service by remote control, in return for using them as a means of adjusting electricity demand (Figure 6.15). The company has started a DR service for which it pays compensation.

In addition, for air conditioning, a thermal storage air conditioning system combined with a thermal storage system (water storage tank, ice storage tank, system to store heat in the building thermal mass), etc., makes it possible to separate air conditioning use time from heat pump operation time, thus ensuring the comfort of the occupants and others, while also responding to supply and demand adjustments (Figure 6.16). The usefulness of this system has already been confirmed in demonstration projects.

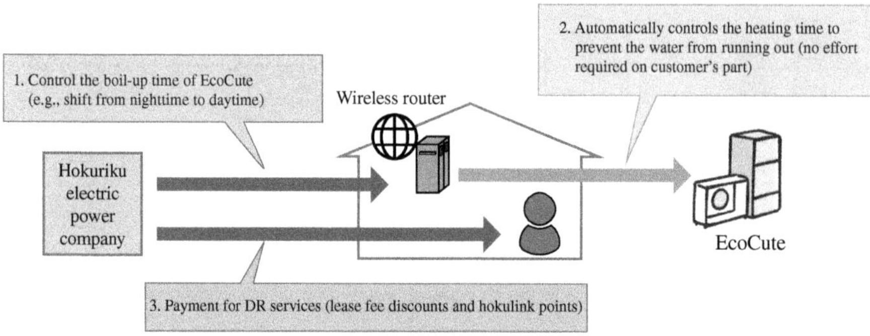

FIGURE 6.15 Hokuriku Electric Power Company's DR service using EcoCute. (From Hokuriku Electric Power Company and Hokuriku Electric Power Living Service Co. Ltd, "Launch of 'Easy Cute' leasing service for Eco-Cute" [14]).

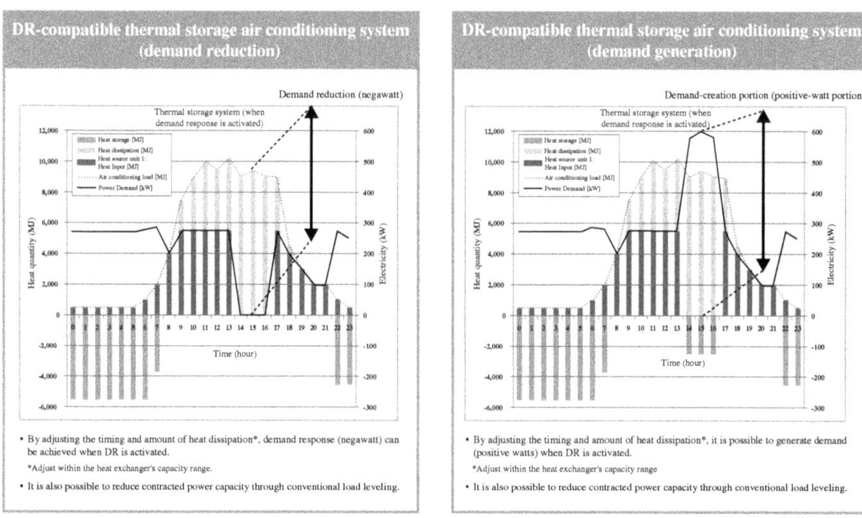

FIGURE 6.16 DR using a thermal storage air conditioning system. (From Heat Pump & Thermal Storage Center, "'Heat Pump & Thermal Storage Technology Center of Japan" [15]).

Other examples of DR-compatible refrigerated warehouses have also emerged in the cold chain, such as the case shown in Figure 6.17: A logistics warehouse in Fukuoka City that is one of the largest refrigerated warehouses in Japan, where a fully automatic control system for the refrigeration units has been installed, enabling the warehouse to respond to DR without the need for human intervention.

6.5.3 Constructing Digital Systems that Meet Needs

While DR initiatives are progressing in various sectors in this way, discussions are underway in the country on the creation of subsidies to support the introduction of DR-compatible and other measures, and on the expansion of DR-compatible resources, using the framework of the Energy Conservation Law among others.

FIGURE 6.17 Example of a refrigerated warehouse with fully automatically controlled DR implemented. (From Mayekawa Manufacturing Corporation, "Implementation of Demand Response with fully automated control in one of the largest refrigerated warehouses in Japan—Enabling participation in the public offering of coordination power through unmanned operation" [16]).

Specifically, in the household sector, it is being discussed that air conditioners and heat pump water heaters, which account for a large proportion of household electricity consumption, should be equipped with remote control functions (DR-compatible equipment), using the framework of the Top Runner Program.

In addition, the revised Energy Conservation Law, which came into force in April 2023, made it mandatory for large-scale consumers to submit periodic reports on their efforts to optimize electricity demand. Furthermore, an additional framework for evaluating DR performance is being considered for the content of these periodic reports, and the provision of incentives such as public announcements. Preferential treatment in the form of related subsidies for operators with excellent DR performance is also being considered. In this way, there is a growing demand for DR not only in the consumer sector, but also in the industrial sector.

In addition, the expansion of DR and VPP initiatives that utilize decentralized resources, for which needs have been increasing in recent years, is expected to contribute to the stabilization of power supply and demand, and the expansion of the use of renewable energy.

6.6 VISION OF THE FUTURE ELECTRICITY SUPPLY AND DEMAND SYSTEM CREATED BY HEAT PUMP TECHNOLOGY

6.6.1 Power Systems Created by Heat Pump Technology

As described in Section 6.2.2, the current electricity supply and demand system mainly faces the following challenges

1. Grid constraints are becoming more apparent with the expansion of renewable energy, and the integration costs of accepting renewable energy into the electricity system are increasing
2. The need to maintain a stable supply of electricity and to further massively introduce renewable energies, for which it is necessary to secure the ability to adjust and increase the flexibility of the power system, and there is a growing need for DR and VPP as one of these initiatives

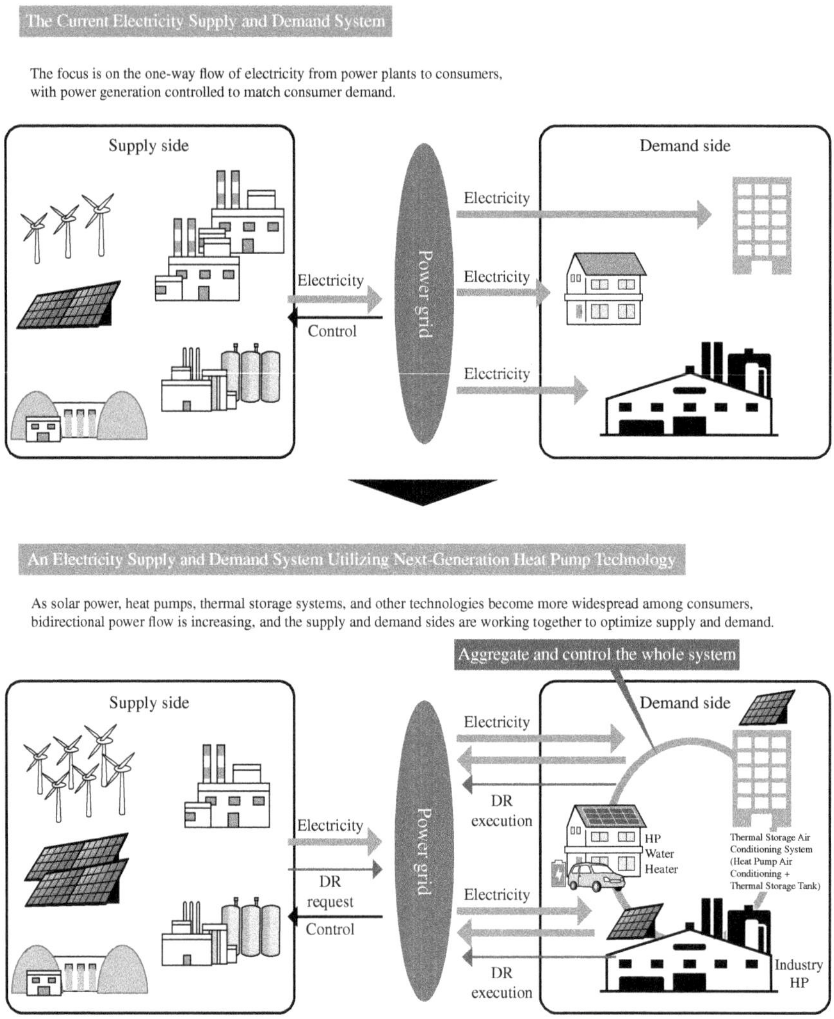

FIGURE 6.18 The current electricity supply and demand system versus a power system realized by next-generation heat pump technology.

In response, the above issues and needs can be addressed by utilizing next-generation heat pump technology, which enables flexible control according to the level of electricity supply.

A concrete vision of a future electricity supply and demand system is shown in Figure 6.18. First, by utilizing next-generation heat pump technology and thermal storage systems used in combination with DR and VPP, surplus power can be absorbed and used as a regulating force when dealing with surplus power or a lack of regulating power as a result of the increased diffusion of renewable energy. This contributes to the effective use of renewable energy and the further expansion of renewable energy in the future.

6.6.2 Reducing Power Generation Costs Using Heat Pump Technology

DR and VPP can also lower demand during peak hours or shift it to other times of the day, thereby reducing the number of power generation facilities that operate infrequently to meet peak demand, which occurs only for a short period of time. In many cases, the required demand is met during peak hours by firing up power sources with high fuel costs, so by reducing peak demand, it is also possible to reduce the use of these additional sources.

Such measures can reduce power generation costs, leading to a more economical power system.

In addition, while peak power sources such as thermal power and pumped storage power have been used as regulating power to ensure a stable supply of electricity, DR/VPP use heat pumps, etc. as regulating power, thereby reducing capital costs and stabilizing the grid at a lower cost. This is expected to reduce capital costs and stabilize the grid at low cost.

There are also benefits for consumers as they can receive compensation accordingly for participating in VPP and responding to DR.

NOTES

1. Demand response (DR): A method whereby the demand side of energy wisely changes its consumption patterns in response to power adjustment orders from the supply side.
2. Science-Based Targets (SBT) are greenhouse gas emission reduction targets set by companies, consistent with the levels required by the Paris Agreement, with a target reduction of 4.2%/year or more and a target year of five to 10 years ahead of the base year.
3. Digital twin: A concept proposed by Michael Greaves of the University of Michigan, USA, in 2002. A technology that collects information from the real world through IoT and other means and reproduces the real world in a virtual space based on the transmitted data. It is called a digital twin in the sense that it constructs a counter (twin) environment of the real world in a virtual space.
4. Carbon lock-in: Equipment with high emissions is maintained for its useful life and continues to emit emissions.
5. Grid interconnection: The connection of electricity generated to the electricity grid for flow to the transmission and distribution lines of a general transmission and distribution utility or distribution company.

REFERENCES

1. Ministry of the Environment, "Greenhouse gases emissions and absorptions in FY2021 (FY2021) (confirmed values)," p. 1, p. 7, 2023, https://www.env.go.jp/content/000128750.pdf
2. Ministry of Economy, Trade and Industry, Industrial Science and Technology Policy and Environment Bureau/Agency for Natural Resources and Energy, "Interim Clean Energy Strategy Arrangement," p. 48, 2022, https://www.enecho.meti.go.jp/committee/council/basic_policy_subcommittee/carbon_neutral/report_20220519_01.pdf
3. Mitsubishi Research Institute, Inc., "FY2017, Basic Survey on the Promotion of the Introduction of New Energy, etc. (Survey on Heat Supply and Demand and Characteristics of Heat Supply Equipment)," (Research report commissioned by Agency for Natural Resources and Energy), p. 88, 2018, https://dl.ndl.go.jp/view/prepareDownload?itemId=info%3Andljp%2Fpid%2F11222688&contentNo=1

4. Heat Pump and Heat Storage Technology Center of Japan/Japan Electro Heat Center, "Survey Report on the Prospect of Electrification in FY2020," p. 115, (2022), https://www.hptcj.or.jp/Portals/0/data0/press_topics/令和4年年度/R4Houkokusyo.pdf
5. Information-technology Promotion Agency, Japan (IPA), DX White Paper 2023, p. 33/35, (February 2023), https://www.ipa.go.jp/publish/wp-dx/gmcbt8000000botk-att/000108041.pdf
6. Ministry of Economy, Trade and Industry, Industrial Science and Technology Policy and Environment Bureau/Agency for Natural Resources and Energy, "Interim Clean Energy Strategy Arrangement," p.27, (2022), https://www.enecho.meti.go.jp/committee/council/basic_policy_subcommittee/carbon_neutral/report_20220519_01.pdf
7. New Energy and Industrial Technology Development Organization (NEDO), "Energy saving can be done! Industrial Heat Pump Exposition," p. 2, (2020), https://www.nedo.go.jp/content/100925495.pdf
8. Ministry of Economy, Trade and Industry, Agency for Natural Resources and Energy, "What is Demand Response?: Types of Demand Response," (2023), https://www.enecho.meti.go.jp/category/electricity_and_gas/electricity_measures/dr/dr.html
9. Ministry of Economy, Trade and Industry, Agency for Natural Resources and Energy, "What is the keyword 'balance of electricity supply and demand' for stable supply of electricity? Let's experience it with a game," 2023, https://www.enecho.meti.go.jp/about/special/johoteikyo/balance_game.html
10. Ministry of Economy, Trade and Industry, Agency for Natural Resources and Energy, "Comprehensive Energy Statistics—Time Series (Reference Table)," (April 21, Reiwa 5), (2023), https://www.enecho.meti.go.jp/statistics/total_energy/
11. Ministry of Economy, Trade and Industry, Agency for Natural Resources and Energy, Subcommittee on Large-Scale Introduction of Renewable Energy and Next-Generation Power Networks (1st meeting), Document 3 "Policy Issues in the Era of Large-Scale Introduction of Renewable Energy and the State of Next-Generation Power Networks," p. 39, (2017), https://www.meti.go.jp/shingikai/enecho/denryoku_gas/saisei_kano/pdf/001_03_00.pdf
12. Ministry of Economy, Trade and Industry, Agency for Natural Resources and Energy, Subcommittee on Large-Scale Introduction of Renewable Energy and Next-Generation Power Networks (52nd meeting), Document 3 "Toward Controlling the Output of Renewable Energy," p. 4, (2023), https://www.meti.go.jp/shingikai/enecho/denryoku_gas/saisei_kano/pdf/052_03_00.pdf
13. Ministry of Economy, Trade and Industry, Agency for Natural Resources and Energy, "What are VPP and DR?," (2023). https://www.enecho.meti.go.jp/category/saving_and_new/advanced_systems/vpp_dr/about.html
14. Hokuriku Electric Power Company and Hokuriku Electric Power Living Service Co., Ltd., News Release, "Launch of Eco-Cute's Lease Service 'Easy Cute'," p. 1, (2022), https://www.rikuden.co.jp/press/attach/22121402.pdf
15. Heat Pump & Thermal Storage Technology Center of Japan, news release, "'Heat pump and thermal storage system' the trump card to achieve carbon neutrality—'Heat Pump and Thermal Storage Month' has started," p. 11, (2023), https://www.hptcj.or.jp/Portals/0/data0/event/symposium/2023/News%20Release_all.pdf
16. TEPCO Energy Partner, Inc., News Release, "Implementation of Demand Response by Fully Automatic Control at One of Japan's Largest Refrigerated Warehouses—Unmanned Adjustment Ability Participation Becomes Possible," (2021), 2023, https://www.tepcoventures.co.jp/news/news-470/

7 Policies and Roadmaps for the Realization of Next-Generation Heat Pump Technology

OVERVIEW

This chapter presents a roadmap and policy recommendations for the realization of next-generation heat pump technology and how the future society should be ushered in.

FIGURE 7.1 Policies and roadmaps necessary for to realizing next-generation heat pump technology.

DOI: 10.1201/9781003666950-7

In realizing next-generation heat pump technology, it is necessary to develop it not only in Japan but also in the rest of the world, as a way of contributing to the resolution of social problems in the world as a whole and to develop it as a Japanese industry. From this perspective, a roadmap for international expansion will also be examined (Figure 7.1).

7.1 POLICY AND ROADMAP FOR NEXT-GENERATION HEAT PUMP TECHNOLOGY

Centralization of information and promotion of a comprehensive evaluation methodology are essential for the realization of next-generation heat pump technology. However, if the parties concerned only work on these issues in isolation, there is a risk that partially optimized models may compete with each other and not be achieved smoothly. In addition, in order to actually realize next-generation heat pump technology, it is necessary to gain acceptance and uptake in society.

7.1.1 Roadmap of the Social Situation Related to Next-Generation Heat Pump Technology

In order to realize next-generation heat pump technology, the relevant social conditions explained in Chapter 2 need to be fully considered. In particular, carbon neutrality, refrigerant conversion, circular economy, and digital transformation (DX) will have a significant impact. An important point is that these factors are closely interrelated.

A roadmap of the relevant social conditions is therefore shown in Figure 7.2 (for more detail, see Chapter 2).

7.1.2 Technology Roadmap for Next-Generation Heat Pump Technology

To solve individual issues, heat pump technology itself needs to be improved and upgraded.

Products with high comfort and energy-saving features will naturally be accepted by the market, but products that meet social demands, such as being environmentally friendly, will not be properly communicated to users and will be difficult to spread.

In order to address these non-technical issues, a centralized management platform for all information related to heat pump technology and a comprehensive evaluation methodology for the overall environmental impact and benefits (e.g. comfort) brought about by heat pump equipment throughout its life cycle need to be devised, and the optimization of the entire value chain needs to be pursued.

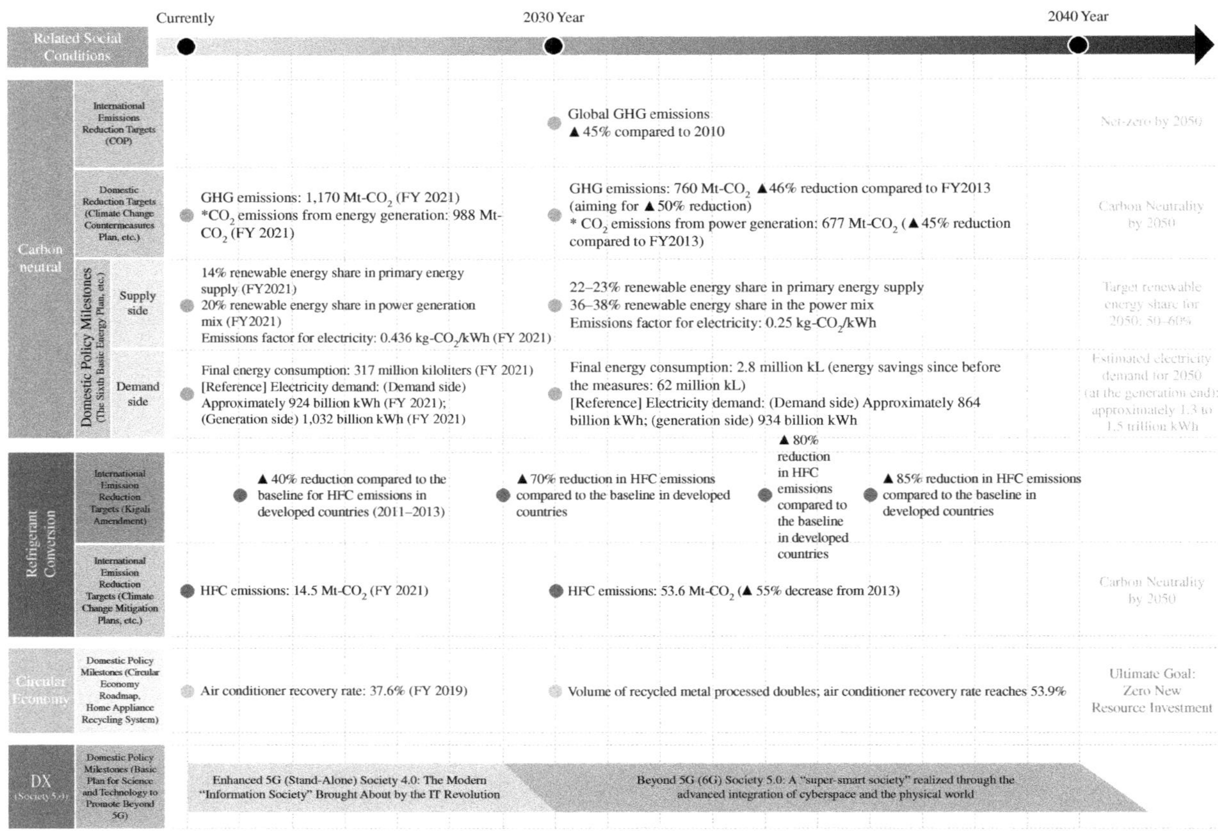

FIGURE 7.2 Roadmap of relevant social situations surrounding next-generation heat pump technology.

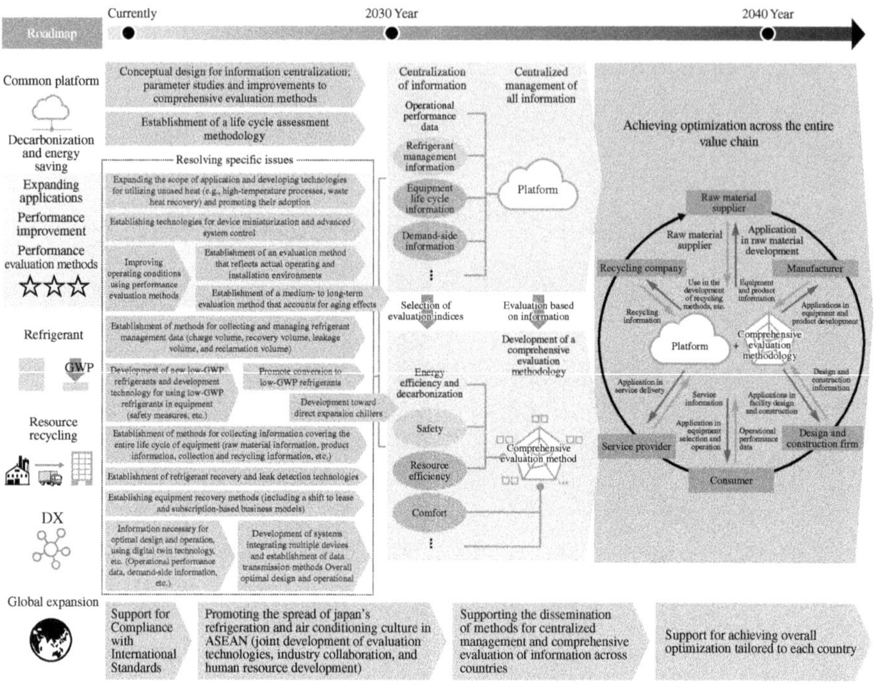

FIGURE 7.3 Technology roadmap for the realization of next-generation heat pump technology.

Figure 7.3 shows a technology roadmap of the necessary technological developments and other initiatives as a path from the resolution of individual issues to the realization of overall optimization.

In order to popularize heat pump technology in households and industry, it is necessary to aim for overall optimization of the value chain. To follow is a proposed timeline of steps to make this happen.

- 2020s
  Development and digitalization of elemental technologies and next-generation heat pump technology to solve individual issues (refrigerant conversion, decarbonization and energy saving, resource circulation, DX)
  Grand design for centralization of information management systems (platforms and comprehensive evaluation methodologies), model trials, etc.
- Late 2020s to early 2030s
  Development of equipment and equipments integrating elemental technologies
  Implementation of centralized information management platforms and comprehensive evaluation methodologies
- Mid-2030s onwards
  Realization of overall optimization of the value chain based on centralized information management platforms and comprehensive evaluation methodologies

It is proposed that in the 2020s, digitalization and the establishment of an information management system for each issue area should be carried out on an individual basis, and that preparations should be made for the establishment of a centralized management platform and comprehensive evaluation methodology that integrates the individual digitalized areas from the late 2020s to the early 2030s.

Specifically, in order to give shape to the final form of information centralization as early as possible, it is assumed that studies and discussions, such as conceptual design and parameter case studies of comprehensive evaluation methodologies, will be carried out from the 2020s.

In addition, in order to make the method actually usable, it is necessary to improve it through a number of practices. For this reason, a trial will be carried out based on various proposals throughout the 2020s to decide which method is most suitable. In order to establish a comprehensive evaluation method for a wider range of aspects, it is also important to establish a life cycle assessment methodology. It is therefore desirable to develop specific methods applicable to each type of equipment and to establish a method by around 2030, before the comprehensive evaluation method is fully deployed. On the platform of centralized management of information, it is necessary to adopt and implement evaluation methodologies for each application and aim to realize overall optimization using comprehensive evaluation methodologies throughout the value chain by the late 2030s.

Furthermore, while the above has been organized with initiatives in Japan in mind, the expansion of these initiatives to countries and regions outside Japan will contribute to solving various social issues, including decarbonization, across the world.

7.1.3 Roadmap for Future Society for Next-Generation Heat Pump Technology

The technology roadmap for a future society to be realized by next-generation heat pump technology is shown in Figure 7.4 to illustrate the changes from the current situation.

Next-generation heat pump technology, which has evolved in accordance with the technology roadmap in response to social demands such as carbon neutrality, DX, and circular economy, will be used in various locations and applications, causing a transformation of that social system in each place of use.

As explained in Chapters 5 and 6 and shown in Figure 7.4, the four areas (listed below) are where new value is expected to be created by next-generation heat pump technology in particular and are represented as examples of roadmaps for a future society. Figure 7.4 also shows the related trends (e.g. targets and prospects in related policies) that form the background against which the realization of these future society visions will be sought.

1. Housing and buildings: Creating healthy, comfortable, and decarbonized living environments
2. Food: Constructing sustainable food systems
3. Industry: Decarbonization and productivity improvement of manufacturing processes
4. Energy: Achieving a stable power supply and demand system

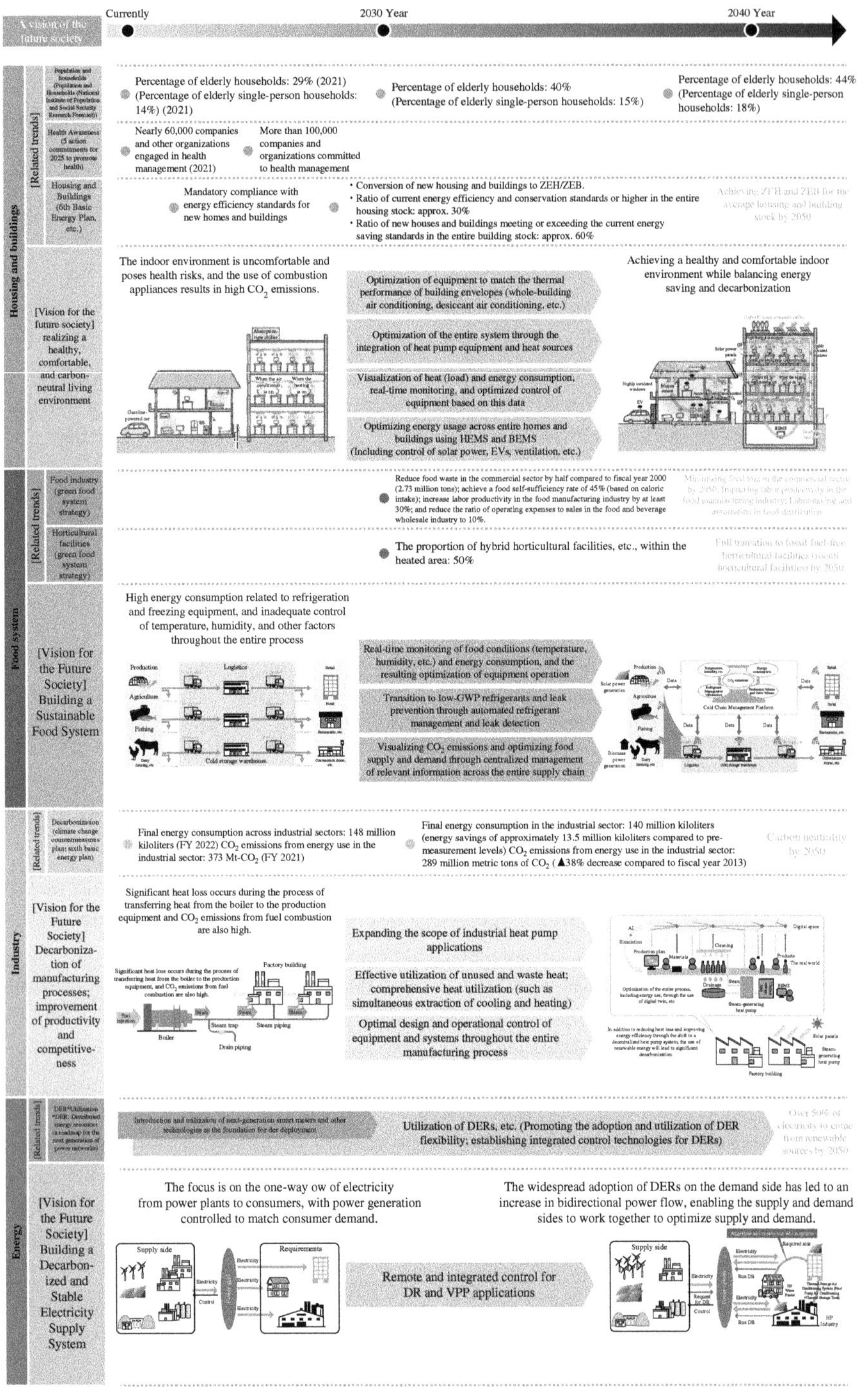

FIGURE 7.4 Roadmap for future society to be realized by next-generation heat pump technology.

7.1.4 Policy Recommendations for Next-Generation Heat Pump Technology Policy

Table 7.1 outlines the three policy guidelines (as shown in the Perspective column) which are deemed necessary for the realization of next-generation heat pump technology. Details are explained in the next section.

TABLE 7.1
List of Next-Generation Heat Pump Technology Policy Recommendations

Perspective	Recommendation	Details	Corresponding section
Policy direction for the realization of next-generation heat pump technology	1. Development of a common platform for realizing overall optimization	Presentation of a grand design for information centralization	Section 7.2
		Construction of an information platform	
		Construction of a comprehensive evaluation methodology	
	2. Improving decarbonization: Upgrading performance standards	Creation of standards for measuring actual operational performance	Section 7.3
		Creation of performance evaluations that take into account the load caused by the actual installation environment	
		Creation of medium- and long-term performance evaluation methodologies	
	3. Refrigerant measures	Development of training and qualification systems for technicians handling refrigerants	Section 7.4
		Development of safety standards	
		Development of waste disposal systems	
		Dissemination of safe disposal methods	
		Refrigerant management IT system to monitor refrigerant management status and ensure traceability Development of management IT systems	
	4. Resource circulation	Development a system for individual-level management of products and materials	Section 7.5
		Rule-making for leasing and subscription services	
	5. DX	Development and sharing of technological elements required for digitalization of heat as energy source	Section 7.6
		Rule-making and platform building for management of heat as energy source	
		Building digitalization indicators	

(Continued)

TABLE 7.1 *(Continued)*
List of Next-Generation Heat Pump Technology Policy Recommendations

Perspective	Recommendation	Details	Corresponding section
Direction of measures to promote next-generation heat pump technology	1. Promote the introduction of next-generation heat pump technology	Support for initial introduction	Section 7.7
		Support for use	
	2. Establishment of heat supply plans	Clarification of the treatment of thermal energy by heat pump technology as renewable energy	Section 7.8
		Clarification of the position of next-generation heat pump technology in heat supply plans	
	3. Promoting international expansion	Advancement of energy efficiency regulations	Section 7.9
		Upgrading of evaluation equipment	
		Advancement of equipment and refrigerant management	
		Human resource development exchange	
Sectoral policy guidelines	1. Housing and buildings	Development of performance evaluation methodology based on the installation environment	Section 7.10
		Development of a comprehensive evaluation methodology for heat pump technology, including from the viewpoint of wellness	
		Rule-making for homes compatible with next-generation refrigerants	
		More active support for the introduction of heat pump water heaters that contribute to the mass introduction of renewable energies	
	2. Food systems	Establishment and linkage of cold chain data management and heat pump equipment data management	Section 7.11
		Establishment and promotion of a methodology for CO_2 visualization that makes the contribution of the cold chain visible Subsidies for the introduction of heat pump equipment, e.g. for institutional horticulture, and creating public awareness	
	3. Industry	Subsidies for the installation of high temperature heat pump equipment	Section 7.12
		Obligation and incentivization of demand response	
	4. Power supply and demand	Demand response rule-making	Section 7.13
		Establishment of optimal demand response tariffs	
		Subsidies for the introduction of heat storage tanks	

7.2 DEVELOPMENT OF A COMMON PLATFORM TO REALIZE OVERALL OPTIMIZATION

In order to realize overall optimization of the value chain of heat pump technology, it is necessary to promote digitalization and develop a common platform, including a centralized information management platform to integrate digitalized information and a comprehensive evaluation methodology based on this platform. Specifically, the following initiatives should be undertaken

1. Realization of a grand design for information centralization
2. Establishment of an information platform
3. Development of a comprehensive evaluation methodology

7.2.1 Realization of a Grand Design for Centralization of Information

Various attempts at digitalization have been proposed in isolated initiatives, with remote monitoring systems and refrigerant management systems for equipment inspection and refrigerant process management already in operation.

However, these are not yet fully understood as several systems coexist and only some of the equipment and refrigerants are registered (Figure 7.5).

In centralizing information for overall optimization, it is important to consolidate the following items, which are expected to be developed in the future, as a whole.

1. Individual systems that manage the supply chain of equipment from manufacture to disposal

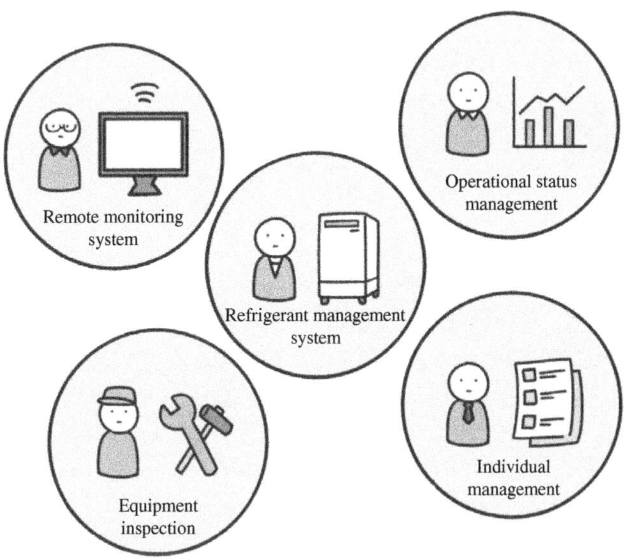

FIGURE 7.5 Data fragmentation due to individual construction.

2. An integrated operation system that manages the operating status, optimizing the operation of equipment in use to manage both the supply and demand of energy, while based on (1).
3. Refrigerant management IT systems to monitor the status of refrigerant and ensure traceability

If these are built in isolation, centralization of information will be difficult. The government needs to present a grand design at an early stage that will guide the development of systems accordingly (Figure 7.6).

In order to promote the use of next-generation heat pumps, it is necessary to show the specific direction of information centralization at an early stage and to proceed with the conceptual design.

In addition, centralization of information could be extended to include operating status management and refrigerant management, which have already begun in part, but most of the current equipment in the market is not covered. There is little incentive to use a new platform when it is already performing certain functions, and it may be more difficult to migrate from one platform to another. For this reason, it is desirable to design centralized management to encompass the whole system from the outset and to develop it accordingly. However, if it is assumed that the equipment in the market is also fully covered, the transition costs may be very high. Considering the possibility of gradual expansion, such as at the time of purchase or maintenance, the coverage will be gradually expanded after 2030.

Note that centralization of information does not necessarily mean centralized data management. It may be interlinked as decentralized data management[1] that can easily reflect the intentions of many actors and respond flexibly to institutional and market changes [1].

It is desirable to find appropriate ways of doing this in future developments.

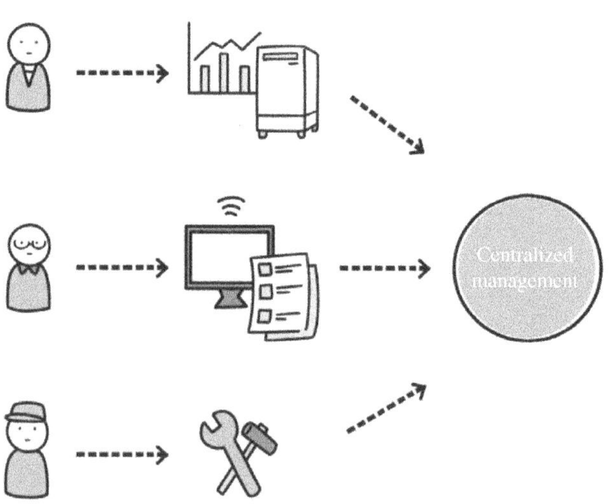

FIGURE 7.6 Information centralization.

7.2.2 Building Information Platforms

For information centralization, a common system could be created, but realistically, a platform would need to be established to link separately created systems so that they can operate with each other.

On such a platform, a heat pump passport framework should be created that can handle equipment data (e.g. operational status) in addition to information about the equipment (e.g. attributes), as handled in the digital product passport. In this framework, data items, formats, acquisition methods and accumulation methods will be standardized (Figure 7.7).

In launching this framework, the following items should be required by law and regulation

1. Registration of equipment and maintenance status
2. Registration of refrigerant management status
3. Recording of recycling fee payments, etc.

However, it would be good for the industry to actively propose systems for such records and platforms, as they go beyond the scope required by law and could be developed into a variety of services. The introduction of lease and subscription services would be effective in conjunction with a change in business structure, as it would be easier to manage under such a system.

7.2.3 The Need for a Comprehensive Evaluation Methodology

7.2.3.1 Improving Equipment Performance

As improvements in equipment performance themselves have been implemented in the past, the scope for future improvements is shrinking. Therefore, rather than improving the performance of individual equipment by enhancing performance of their internal components, the direction will be to improve the performance of the

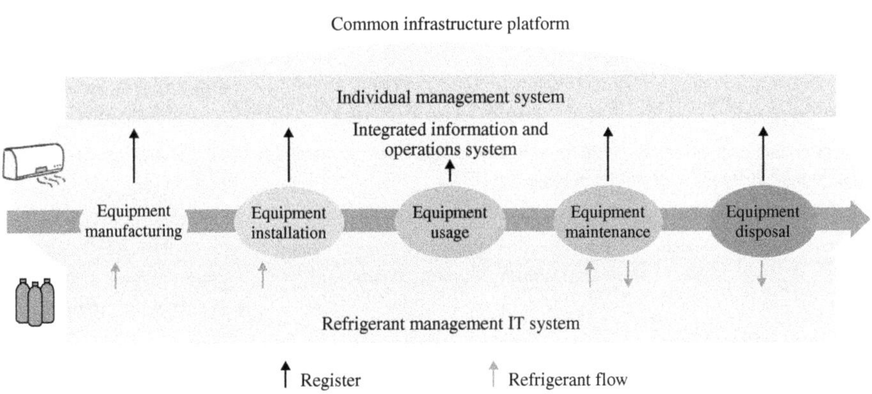

FIGURE 7.7 Concept of common infrastructure platform for equipment management and appropriate use.

system as a whole by improving overall system control. The technologies required for this are the downsizing of equipments and the establishment of advanced system control technologies by 2030.

7.2.3.2 Establishment of a Comprehensive Evaluation Methodology for Each Application

The comprehensive evaluation methodology for centralized management information can be considered as a comprehensive evaluation of both the heat pump technology as a system and the relevant social systems, which cannot necessarily be integrated into one.

As different situations and applications require different evaluation methodologies, it is necessary to develop a comprehensive evaluation methodology for each application (Table 7.2).

In assessing the environmental performance evaluation of heat pumps, it is necessary to assess not only the performance in terms of energy savings, but also in terms of life cycle assessment, taking into account the impact of refrigerants. So far, the Total Equivalent Warming Impact (TEWI)[2] and the Life Cycle Climate Performance (LCCP)[3] have been used for analysis. In the future, it will be necessary for the industry to establish and operate a life cycle assessment (LCA) method, including for the equipment itself.

In addition, energy efficiency and refrigerant conversion have so far been managed as separate policies and have not been handled in an integrated manner. In the future, it will be necessary to link the regulations of the Energy Conservation Law with the global warming potential (GWP) regulations for designated products under the Fluorocarbon Emissions Control Law (e.g. setting energy conservation standards in consideration of LCCP), taking LCA into account, for overall optimization (Figure 7.8).

TABLE 7.2
Example Uses Envisaged in a Comprehensive Evaluation plus Evaluation Methods

Assumed use(s)	Targets to be evaluated	Operation of evaluation indicators
Indication and evaluation of equipment attributes and performance	Energy efficiency, safety, environmental compatibility	Fixed at the time of product sale
Product selection by users	Energy efficiency, safety, environmental compatibility, comfort, health, etc.	Adjusted by the user based on information at the point of sale of the product
Equipment system design	Energy efficiency, safety, environmental compatibility, comfort, health, etc.	Set as design information for the installation site
Determination of operation and control methods	Energy efficiency, comfort, health	Real-time evaluation and dynamic changes

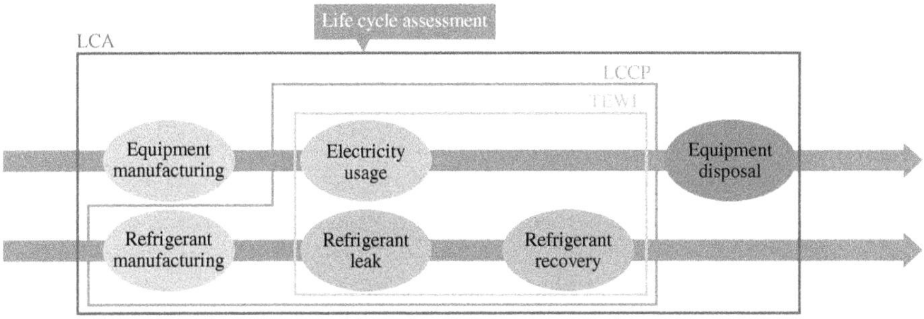

FIGURE 7.8 Extension to life cycle assessment.

Furthermore, there are other attributes of heat pumps besides performance, such as comfort and convenience. In order to consider the various aspects of heat pumps and select the most appropriate attributes, it is important to introduce a comprehensive evaluation methodology that includes these as well. This should be considered by the industry and widespread use also encouraged, for example by displaying this information together with the performance so that it is taken onboard more widely.

7.3 DECARBONIZATION AND ENERGY SAVING

7.3.1 Expanding the Use of Unused Thermal Energy with Heat Pump Technology

In order to decarbonization and save energy, heat pumps should be actively introduced to make more effective use of natural and untapped heat energy, such as solar heat (air heat) and factory exhaust heat.

It is important to expand the scope of areas where heat pumps are used as a substitute for combustion-based heat sources. The proliferation of high-temperature heat pumps and exhaust heat recovery is key, particularly for air conditioners and heat pump water heaters for cold climates and in the industrial sector. Specifically, two stages are required: the development of elemental technologies such as heat exchangers and compressors for commercialization, and product development, including the expansion of the product line for wider uptake. Realizing this by around 2030, together with unused heat utilization, will also contribute to the greenhouse gas (GHG) reduction targets for the 2030s.

7.3.2 Creation of Performance Evaluation Methodology under Actual Environmental Conditions

Currently, performance evaluations are carried out under certain environmental conditions that differ slightly from the actual operating environment. If equipment with high performance is used based on this performance evaluation criterion, energy savings may not necessarily be achieved under actual operating conditions.

For this reason, an international standard on next-generation performance is being discussed to improve the method of evaluating equipment performance under low load conditions; it is expected to be passed in 2026, but it is still only an improvement of the current evaluation method and has not yet led to a system that takes the actual operating environment into account.

In order to promote energy savings toward carbon neutrality, product selection based on more appropriate performance evaluation is a key factor. Discussions on the revision of international standards are expected even after their establishment, and it is hoped that Japan will propose the formulation of appropriate performance standards in the early 2030s to appropriately reflect technologies that contribute to energy savings in the actual operating environment. In addition to standardized equipment performance evaluation indicators being evaluated in tests, it will also be necessary to realize the following technologies by applying new digital solutions.

- Digital twin: Performance evaluation taking into account the load of the actual installation environment, e.g. simulation technology
- Internet of Things (IoT) data: Realization of medium- and long-term performance evaluations that trace changes in actual performance over time Medium- to long-term performance evaluation: Likely to take time to establish, as market data needs to be collected to evaluate changes over time, but other evaluation methodologies should be in place by around 2030

7.3.3 Creation of Medium- and Long-Term Evaluation Methodology

There need to be incentives for all individual events to move toward overall optimization of next-generation heat pumps at various levels of hierarchy.

To this end, it is necessary to establish an evaluation methodology that will guide the way toward the overall optimum.

As it is difficult to realize many elements at the same time, the industry needs to work toward the gradual extension of the following evaluation methodology, keeping the final form in mind (Figure 7.9).

1. Preparation of standards for measuring actual operational performance
2. Preparation of performance evaluations taking into account the load caused by the actual installation environment
3. Preparation of medium- and long-term performance evaluation methodologies

Even if an evaluation methodology is developed, it will not function if it is not adopted by the user. In order to promote the use of evaluation methodology, it is desirable to link it to the mechanism for using it. Therefore, it is necessary to promote the adoption of new energy saving performance evaluation methodologies, such as the one based on low load performance and automatic operation mode control in the Top Runner Standard of the Energy Conservation Law.

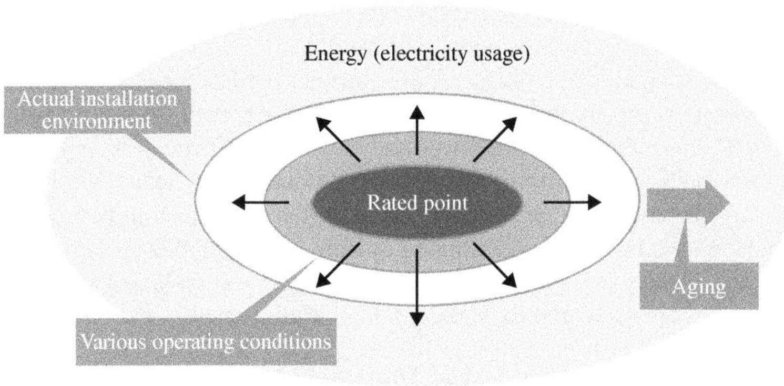

FIGURE 7.9 Expanding the scope of evaluation based on overall optimization.

7.4 INTERNATIONAL TRENDS AND POLICIES ON REFRIGERANT CONVERSION

7.4.1 Refrigerant Conversion Based on International Treaties

7.4.1.1 Kigali Amendment Regulating Alternative Fluorocarbons

The Kigali Amendment, which adds the reduction of production and consumption of alternative fluorocarbons (FCs) (such as hydrofluorocarbons (HFCs)) to the regulation of chlorofluorocarbons (CFCs) under the Montreal Protocol, was adopted at the 28th Meeting of the Parties (MOP28) to the Montreal Protocol held in Kigali, Rwanda, in October 2016.

In order to achieve carbon neutrality by 2050, it is essential to convert the refrigerants used in heat pumps from HFCs to lower GWP refrigerants. The ideal refrigerant would be one that, in addition to having a low GWP, would not significantly impair the energy-saving performance of heat pump technology and, moreover, would be safe to use with minimal risk during use, but there are currently many challenges. On the other hand, the Kigali Amendment calls for a 70% reduction in production and consumption compared to the baseline in 2029. Therefore, technological development needs to be pursued in the following timescale

- Early 2020s
  Develop technology for new low-GWP refrigerants
  Establish technologies for the use of equipment including safety measures for low GWP refrigerants, including natural refrigerants
- Late 2020s
  Disseminate low GWP refrigerant equipment.

Note that it is easy to adopt natural refrigerants (e.g. propane) or hydrofluoroolefin (HFO) for low GWP refrigerant conversion, for example, when the amount of refrigerant is small and the places where refrigerant circulates are limited, but if it is difficult, conventional direct expansion type equipment may be converted to inter-expansion type such as chillers to meet the need.

The Kigali Amendment calls for an even stricter reduction by 2036 than in 2029 of 85%. In order to ensure that the reduction target is achieved solely through the use of low GWP refrigerants, it will be necessary to start full-scale development toward the conversion to chillers from around 2030. As energy efficiency will deteriorate if direct expansion equipment is converted to chillers, it is also important to select refrigerants on the basis of developing equipment that combines multiple functions, such as water heating and air conditioning, rather than just replacing single purpose equipment. It is also essential to work with the building sector.

7.4.1.2 HFC Recovery from In-Use Equipment to Achieve Reduction Targets

For equipment using HFCs, of which there are already many in the city, GHG emissions can be reduced from heat pumps by taking measures to prevent leakage while operating and by ensuring recovery during servicing and disposal.

The aforementioned conversion to low GWP refrigerants will require a certain amount of time to replace equipment operating in the market and cannot necessarily be achieved in the same way for all models, considering optimization, including energy savings and safety. Eventually, it is likely that some HFC-based equipment will remain. However, it is important to promote the use of low GWP refrigerant equipment in order to minimize leakage during future use and reduce the uncollected amount during servicing and disposal.

7.4.1.3 The Need for Refrigerant Management in IT

It is also important to establish a proper refrigerant recycling market due to the increasing scarcity of refrigerants.

Therefore, in order to strongly promote measures against leakage, thorough recovery, and increased recycling, it is necessary to develop a refrigerant management IT system that can manage equipment and refrigerants throughout their life cycle (sales, use, maintenance, disposal, etc.) and to manage refrigerants throughout Japan. In doing so, this system should also be used to handle refrigerants that require careful management, such as natural refrigerants with flammability, which are required to comply with the aforementioned Kigali Amendment.

In order to realize a refrigerant management IT system, it is necessary to establish a method for collecting refrigerant management information (amount filled, amount recovered, amount leaked, and amount recycled). This will eventually be subject to information centralization and will be in place by around 2030, helping to minimize the amount of refrigerant used and emitted in all aspects of refrigerant conversion, emission control, and recovery/recycling, through monitoring of equipment use and other factors, and encouragement from system managers to equipment users.

7.4.2 Policy Direction through Refrigerant Conversion

Refrigerant conversion will progress in the future, as the Kigali Amendment has stipulated that the production and consumption of FCs should be reduced. Many of the refrigerants expected to be used as a result of the conversion are flammable,

Perspective	CFC refrigerant	Low-GWP refrigerants (such as natural refrigerants)	
Properties of refrigerants	Non-flammable (some materials are mildly flammable)	Many of them have properties such as flammability, toxicity, and high pressure	
Handling technology	Highly safe, easy to install, and can be handled by many facility operators	Installation, repair, and disposal require advanced technical expertise, particularly with regard to safety considerations. Additionally, when using a chiller to circulate water, improvements in workability are necessary.	Training of technicians establishment of a safe operating system
Legal regulations	Simplified compliance under the High-Pressure Gas Safety Law (e.g., For inert HFCs, a notification is required for refrigeration capacities of 20 tons or more)	Strict compliance requirements under the High-Pressure Gas Safety Law (e.g., For flammable gases (A3 and A2 refrigerants), a notification is required for refrigeration capacities of 3 tons or more)	Relaxation of restrictions following a thorough assessment of safety

FIGURE 7.10 Challenges and solutions for refrigerant conversion arising from next-generation heat pump technology.

toxic, etc., so it will be necessary to develop an environment in which they can be used safely.

In addition, if FCs are to be continuously used while being regenerated, thorough recovery and recycling of refrigerants will be necessary. To this end, industry and academic societies should undertake to (Figure 7.10):

1. Develop an education and qualification system for refrigerant handling technicians
2. Develop safety standards
3. Establish a waste disposal system
4. Disseminate safe disposal methods
5. Develop a refrigerant management IT system to monitor the status of refrigerant management and ensure traceability

In developing these and other measures, support from government is also essential. In addition, more robust legal measures for the recovery and recycling of refrigerants are also needed, as recovery is still obligatory at present, but many refrigerants are not yet recovered.

7.5 RESOURCE CIRCULATION FOR A CIRCULAR ECONOMY

7.5.1 Individual Management Systems to Track Information after Use

7.5.1.1 Necessity of Individual Management Systems

The current supply chain does not comprehensively track the use and disposal status of sold equipment due to lack of continuity in tracing goods from the point of sale to when the equipment has been used to the end of its lifespan. To realize a circular economy, it is important to ensure traceability of equipment and to know where equipment is in use and no longer in use.

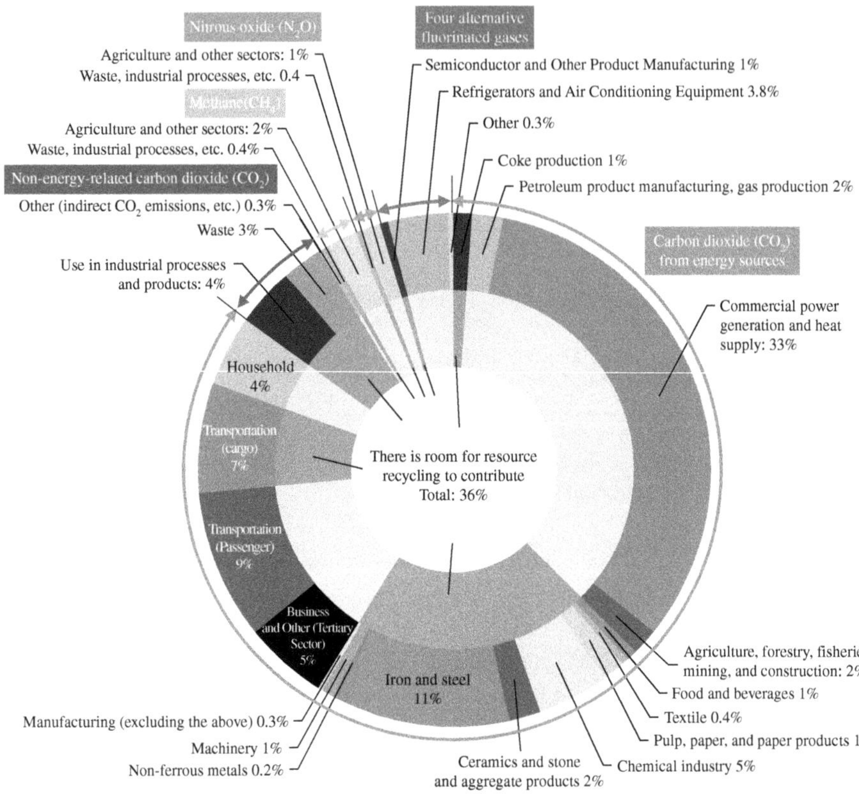

FIGURE 7.11 Proportion of sectors where there is scope for resource circulation to make a contribution. (From Ministry of the Environment, "The Fourth Basic Plan for the Promotion of a Recycling-Oriented Society and Outline of the Circular Economy Roadmap" [2]).

In order to realize a circular economy, it is necessary to develop an individual management system that can trace the entire life cycle of equipment (production, sale, use, disposal, etc.) by 2030, in conjunction with the measures outlined in the Ministry of the Environment's "Circular Economy Roadmap" [2] (Figure 7.11).

7.5.1.2 Points to Note on Resource Circulation of Refrigerants

In resource circulation of refrigerants, it is also important to ensure leakage prevention during use and recovery during servicing and disposal in order to increase the amount of recoverable refrigerant. It is desirable to establish refrigerant recovery and leak detection technologies by around 2030.

7.5.1.3 Information Centralization in the Circular Economy

The circular economy is not limited to the location of equipment through individual management systems. By promoting the centralization of information, such as the integrated understanding of equipment operating information, it is possible to aim for a supply chain that reduces the overall environmental impact, by measures such as increasing resource circulation and reducing GHG emissions, over the entire LCA.

In addition, in order to make these mechanisms function effectively, studies will also be conducted on methods that facilitate dismantling and re-use on the equipment side.

7.5.2 Measures to Promote Resource Circulation

In order to ensure that resource circulation is promoted, it is necessary to make visible the status of equipment and refrigerant use, and the path after use. The following guidelines can be considered as measures to achieve this

1. Developing a mechanism for individual management
2. Rule-making for leasing and subscription services

Any one of these measures should be introduced by the industry, to develop specific methods, and the government, to clarify the legal position of their introduction. At the same time, the public administration is expected to actively support technologies that facilitate the recovery of equipment and refrigerants.

7.5.2.1 Development of a System for Individual Management throughout the Life Cycle

Individual management of equipment throughout its life cycle, including its location, performance, material type, and amount used, will make the flow of resources visible and ensure more reliable resource circulation.

As such individual management requires information to be shared between a large number of entities, it is necessary to develop rules on how to convert data and the digital product passport required for heat pumps.

In addition, when individual management is carried out and equipment is connected, the use of information may be restricted in order to protect personal and confidential information such as administrator/user information. Technological solutions such as blockchain technology and quantum cryptography, which are expected to be utilized in the future, can be envisaged, but the government should also consider creating an institutional framework that balances the utilization and protection of such information so that it can be used effectively in society at large.

7.5.2.2 Rules for Leasing and Subscription Services

In order to enable the reliable recovery of products, a shift to businesses based on the sale of services, such as subscription services, is also a very effective means of ensuring the reliable recovery of products.

In order to promote such services, it is necessary to develop standard terms and conditions in the industry to ensure reliability of service content, as well as to study issues and solutions.

Note that subscription services are beginning to be deployed as equipment use services [3]. In addition to such services, it is also possible to view the service as heat utilization rather than equipment use, and to charge according to the amount of heat used.

7.6 DX REQUIRED FOR THE DEPLOYMENT OF HEAT PUMP TECHNOLOGY

7.6.1 DX Required for Digitalization of Heat

The digitalization of heat handled by heat pumps is not yet sufficiently advanced, which is a major barrier to the introduction of DX. However, the promotion of DX is essential for the realization of overall optimization.

For this reason, the necessary technological infrastructure must be developed as an industry and the status of efforts to encourage digitalization efforts should be visible, as described below.

Furthermore, it is necessary to clearly define what is feasible as an indicator of digitalization, that is, to categorize and show the difficulty level, thereby promoting a common understanding. Figure 7.12 is an example showing the classification of levels for automated driving in automobiles.

7.6.2 Development of Related Fields for the Introduction of DX

7.6.2.1 Carbon Neutrality

Electrification of combustion-based heat sources and optimization of energy supply are necessary to achieve carbon neutrality by 2050.

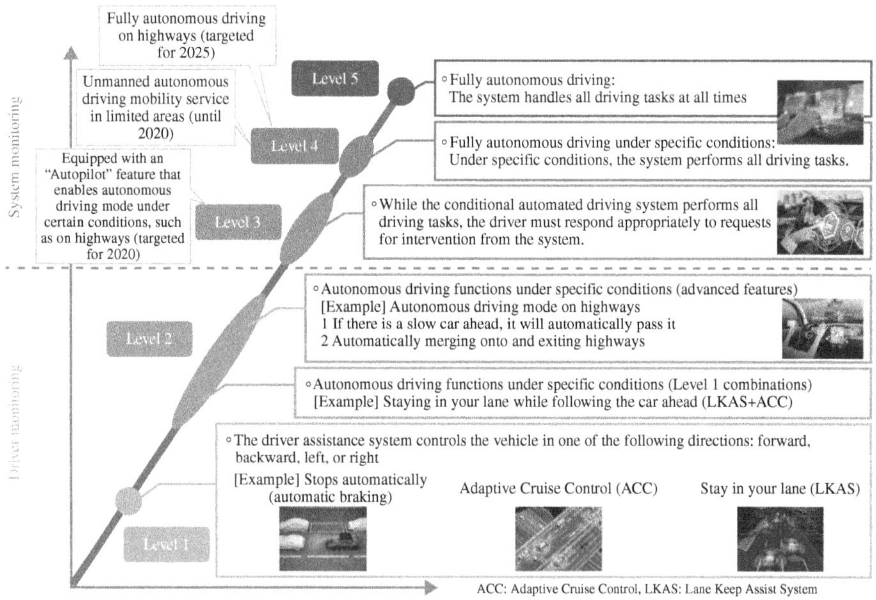

FIGURE 7.12 Level classification of automated vehicle operation. (From Ministry of Land, Infrastructure, Transport and Tourism, "Level classification of automated drive " [4]).

7.6.2.2 Communications Infrastructure

In order to establish Beyond 5G with the aim of achieving Society 5.0 in the 2030s, the Ministry of Internal Affairs and Communications has a policy to enhance 5G functions and create a Beyond 5G-ready environment during the 2020s.

Therefore, by about 2030, when we will enter a super-smart era with the realization of Society 5.0, heat pumps will also need to be optimally designed and operated based on actual operational performance evaluation and remote monitoring and control using digital twin technologies such as IoT and AI. To achieve this, the necessary information, along with methods for sending and receiving it, must be established in the early 2020s and implemented from the late 2020s.

7.6.2.3 Achieving Optimal Design and Operation of Heat Pumps

In order to promote next-generation heat pump technology in households and industry, it is expected that next-generation technologies will be combined from all perspectives, air conditioning and water heating will be optimized individually, systems will be integrated and system fusion will be achieved.

7.7 SUPPORT FOR THE INITIAL INTRODUCTION AND USE OF NEXT-GENERATION HEAT PUMP TECHNOLOGY

The introduction of next-generation heat pump technology requires environmental considerations such as refrigerant conversion and resource circulation, which is expected to increase the price of the equipment from the current level. Therefore, various measures to promote the introduction of new equipment will be necessary (Figure 7.13).

In order to promote the introduction of equipment, the following measures are recommended as necessary

1. Support for initial installation
2. Support for use
3. Clarification of the treatment of heat utilization by heat pump technology as renewable energy

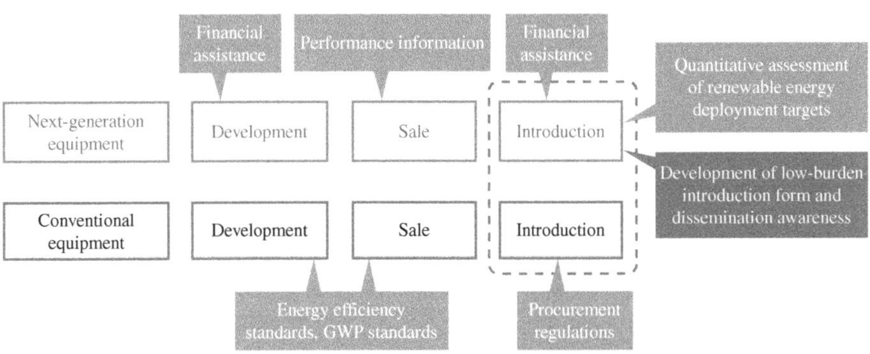

FIGURE 7.13 Approach to promoting the introduction of next-generation heat pump technology.

7.7.1 Support for Initial Installation

Support for initial installation is necessary to achieve society's overall goal of becoming carbon neutral and supporting a circular economy, and the following measures should be implemented by the public administration

1. Support for the provision of new performance information (e.g. legalization, financial support)
2. Financial support for the development of equipment for next-generation heat pump technology
3. Financial support for the introduction of next-generation heat pump technology

As for financial support for the introduction of next-generation equipment using next-generation heat pump technology, natural refrigerant equipment has been the target of support[4] with limited use in the past [5]. It would be desirable to broadly target the introduction of low GWP equipment for other applications, as well as the introduction of heat pumps combined with combustion systems and the conversion to net zero energy building (ZEB), which is a prerequisite for their introduction, so as to encourage their introduction, especially in the early stages of their uptake.

On the other hand, such financial support will increase uptake by compensating for price differences, and although price differences will narrow with them, some equipment will remain that will not ultimately eliminate the differences.

For this reason, support with financial costs, such as subsidies in the initial stage, cannot be implemented on a permanent basis. At a certain stage of market diffusion, consideration should be given to ensuring the competitiveness of next-generation equipment by strengthening regulations on the sale of and imposing a levy on conventional equipment[5] (Figure 7.14).

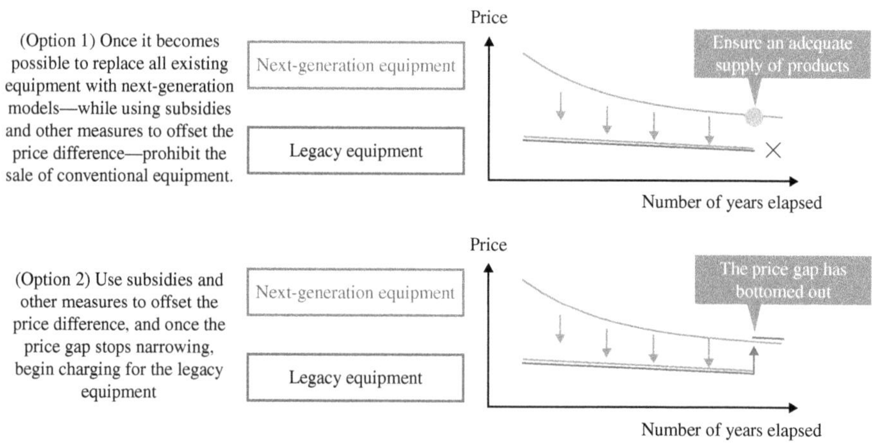

FIGURE 7.14 Two concepts of financial support for next-generation heat pump technology. *Note:* Example of a case where price differentials have to remain in the end.

7.7.2 Support for Use

As part of efforts to promote refrigerant conversion for low GWP and energy savings to reduce energy consumption, one type of financial support for next-generation heat pump technology is not only policies to encourage the installation of the heat pump equipment itself, but also to influence the energy costs during operation.

As part of the government's green transformation (GX)[6] policy, as well as the introduction of carbon pricing[7], the economic valorization of such a clear decarbonization will also be a major boost to the adoption of next-generation heat pump technology. A system should be established by the government to ensure that carbon pricing is also taken into account, and fairly compared and valued in the market.

7.8 TREATMENT OF HEAT PUMP HEAT AS RENEWABLE ENERGY

7.8.1 Incorporating Heat Pump Usage into Quantitative Assessment

Heat pump equipment contributes to renewable energy utilization and energy savings but is positioned in current policy mainly as energy-saving equipment and not explicitly as heat supply. For this reason, heat pump technology is treated from a regulatory perspective, such as with energy efficiency standards and refrigerant GWP standards, and it has not been shown what role it should play in the overall heat supply.

Currently, heat pumps are not quantitatively evaluated as a renewable energy source under the Energy Conservation Law and therefore do not contribute to the achievement of the renewable energy targets of the respective operators.

If the amount of heat pumps used as renewable energy could be included in the quantitative assessment in the future, it would promote the active introduction of heat pumps by operators. If the use of high temperature areas where heat pumps are not yet widely used can be actively, quantitatively assessed, this could trigger a shift away from combustion-based equipment (Figure 7.15).

7.8.2 Heat Supply Planning

In order to identify how the heat supply should be directed toward carbon neutrality, a heat supply plan has been proposed. Such a plan would clarify the significance of next-generation heat pump technology and allow for the formulation of how and in which specific areas it should be promoted. It would also require a system to promote the use of the technology, rather than simply using the existing system to regulate it (Table 7.3).

7.9 INTERNATIONAL EXPANSION OF HEAT PUMP TECHNOLOGY

7.9.1 Promoting International Expansion

Heat pump equipment, which is integral type[8] of equipment, is an area in which the Japanese manufacturing industry excels, and international expansion of next-generation heat pump technology should be actively promoted and deployed by

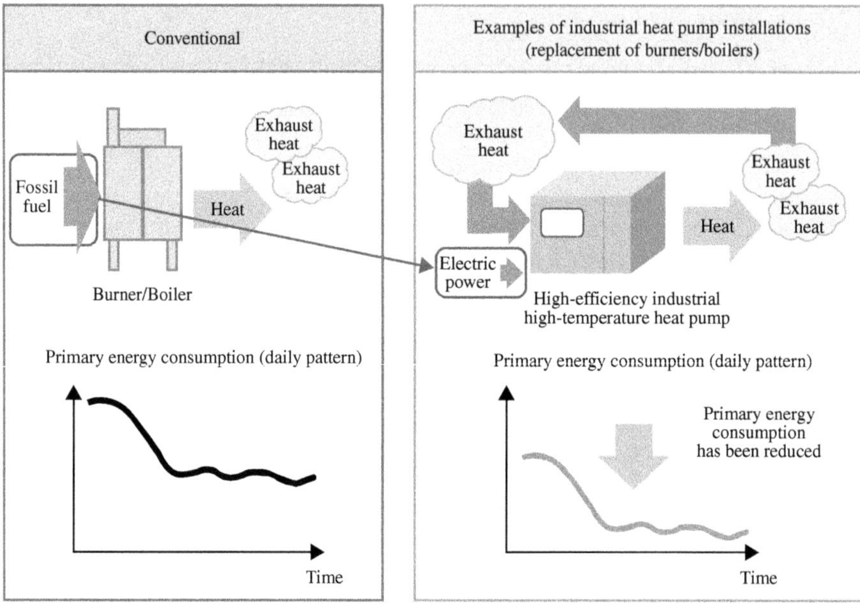

FIGURE 7.15 Illustration of quantitative evaluation incorporating heat pumps. (From New Energy and Industrial Technology Development Organization (NEDO), "Development of an industrial heat pump simulator that can quantitatively evaluate the effects of introducing heat pumps" [6]).

TABLE 7.3
Heat Supply Plan

Temperature range	Heat demand	Type of supply	
		Sector	**Percentage**
~20°C	●PJ	Cold heat (electricity)	95%
		Cold heat (fuel)	5%
20–200°C	◆PJ*	Combustion heating	10%
		Electric heating	10%
		Steam (fuel)	40%
		Steam (electric)	25%
		Hot water (fuel)	5%
		Hot water (electricity)	10%

Note: Heat pumps are used for cold heat (electricity), steam (electricity), and hot water (electricity). *PJ (petajoule) is a unit of energy equal to 10^{15} joules. It is commonly used to describe large-scale energy consumption or production

Japan through international exchange. Specific guidelines for this could include the following

1. More sophisticated energy efficiency regulations
2. Advancement of evaluation equipment
3. Advancement of equipment and refrigerant management
4. Human resource development and exchange

7.9.1.1 Advancement of Energy Efficiency Regulations

First, in order to promote next-generation heat pump technology, it is necessary to introduce high-performance heat pump equipment, and it is important to spread its use through; for example, its adoption in energy efficiency regulations established in various countries. In particular, as energy efficiency regulations are being introduced in the Association of Southeast Asian Nations (ASEAN) and are very similar to Japanese air conditioning regulations, intensive efforts should be made to standardize and promote the use of these regulations.

7.9.1.2 Upgrading of Evaluation Equipment

In addition, advanced evaluation equipment needs to be installed on site to enable performance evaluations of next-generation heat pump technology to be carried out.

7.9.1.3 Upgrading of Equipment and Refrigerant Management

In addition, systems for upgrading equipment and refrigerant management could be advanced only in developing countries where no such systems exist. While supporting them as leading examples, it is also proposed to build advanced models and to import them back to Japan.

7.9.1.4 Human Resource Development and Exchange

The foundation of all these initiatives is human resource development. It is necessary for the government and industry to work closely together to strategically develop exchanges with other countries by way of strengthening the competitiveness of Japan's manufacturing industry in the future. It would also contribute to improving the living environment and realizing a carbon neutral and circular economy throughout the world.

7.9.2 Different Applications and Expectations for Diffusion in Different Countries and Regions

The rate of uptake and application of heat pumps around the world varies widely, ranging from regions where heat pump technology is widely used to regions where they are more scarce, either due to the natural environment and cultural context of each region or policies and roadmaps for the realization of next-generation heat pump technology (see Section 7.3). Social conditions such as decarbonization with the aim of becoming carbon neutral need to be addressed globally, and next-generation heat pump technology has the potential to contribute globally to solving various social challenges.

Countries and regions where heat pump technology is expected to make a particular contribution include the cooling needs in the tropics and temperate areas, which are expected to expand in the future. The target countries are those where there is little awareness of ways to conserve energy, such as setting appropriate temperatures, and where air conditioning culture is not well entrenched. In addition, uptake in cold regions, where heating applications are not currently widespread, are also expected to expand, with Europe and the USA being targeted.

In the case of heating in the industrial sector, high-temperature heat pumps are expected to be adopted in countries that consume large amounts of fossil fuels, while refrigeration and freezing are expected to be deployed in countries in the ASEAN region, where cold chains will become more widespread in the future (Figure 7.16).

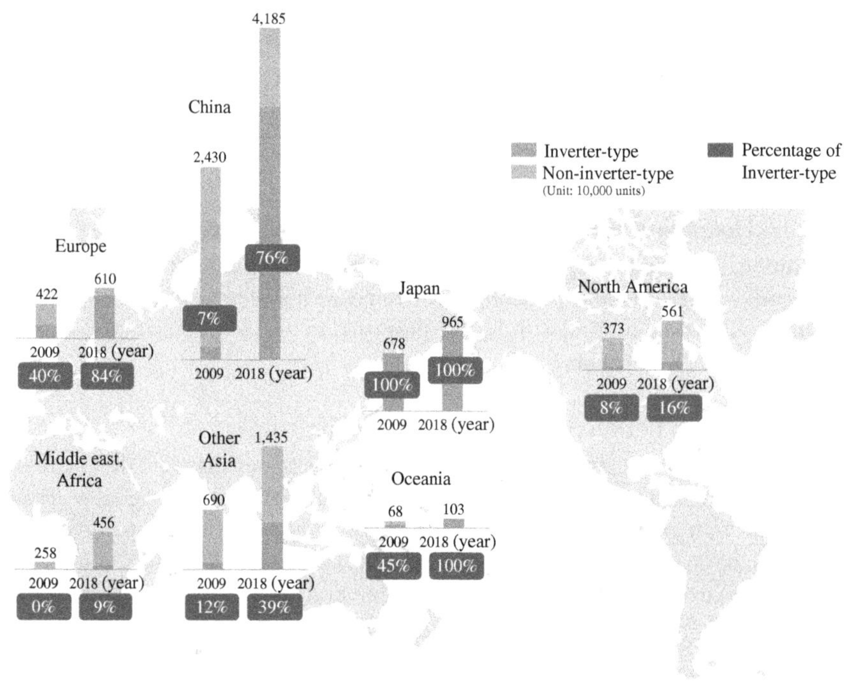

FIGURE 7.16 Residential air conditioning market demand in units and ratio of inverter air conditioner (2018). (From Daikin Industries, Ltd/Toyo Keizai Inc, "Air conditioning 'to increase by about three times by 2050' problem severity: essential in extreme heat, increasing demand worldwide and environmental concerns" [7]). *Note:* Prepared by Daikin Industries Ltd with reference to data from the Japan Refrigeration and Air Conditioning Industry Association. Residential air conditioning: Residential ductless air conditioners excluding wind and portable types. Residential ducted air conditioning is included only in North America.

7.9.3 Contribution of Heat Pump Technology in Energy Saving

With regard to the accurate evaluation of energy savings, to which heat pump technology is particularly expected to contribute, an international standard on next-generation performance is currently being discussed (see Section 3.7) with the aim of improving the evaluation method to take low load conditions into account. As the standard is expected to be passed around 2026, it is necessary to develop new international expansion standards. On this basis, efforts must be made to develop products suitable for each country and to establish a centralized information management platform and comprehensive evaluation methodology, taking into account differences in background and culture between countries.

In addition, by providing support, including for their operation, we can contribute to solving social issues in each country.

If the centralized information management platform and comprehensive evaluation methodology can be promptly expanded internationally in line with the dissemination of the centralized information management platform and comprehensive evaluation methodology aimed for domestic deployment in the 2030s, it will contribute to solving social issues worldwide. In advance, it will be important to promote and disseminate the Japanese refrigeration and air conditioning culture[9], which has advanced energy-saving and refrigerant countermeasures, as a foundation prior to overall optimization.

7.9.4 Compliance with New International Standards

Discussions are currently underway to revise the international standards for heat pump technology for air conditioning applications in order to correct the discrepancy between the actual operating environment and the test environment. The revision is expected to be reflected in the performance evaluation indicators of Japan and other countries, so it will be necessary to develop heat pump products that comply with the new international standards.

7.9.5 Development of Locally Fit Products

Each country has a different air conditioning culture and required product specifications. The development of locally fitted products is required, taking into account the background of each country. In developing locally fitted products, the development of equipment for the domestic market can be used as a reference.

Specifically, international expansion is expected in the development of air conditioner for cold regions, food supply systems with cold chains and high-temperature heat pumps for the industrial sector.

In addition, as the operation of equipment, such as appropriate temperature settings, has a significant impact on energy consumption, it will also be important to promote and develop the use of the equipment.

7.10 NEXT-GENERATION HEAT PUMP TECHNOLOGY POLICY FOR HOUSING AND BUILDINGS

7.10.1 Achieving a Healthy, Comfortable, and Decarbonized Living Environment

As discussed in Section 5.1, there is a growing need for healthier and more comfortable indoor environments in housing and buildings, which are the foundation of society and life, partly due to an aging population and an increasing number of companies addressing health issues. Decarbonization of housing and buildings in the private sector is also essential as a further step toward carbon neutrality. It is expected that regulatory measures and other measures will improve the energy efficiency of housing and buildings in the future (Table 7.4).

TABLE 7.4
Trends, Indicators, and Targets for Housing/Buildings and Wellness

Related trends	Examples of indicators	Current situation and targets	
Progress of aging population [8]	Ratio of elderly households	Current situation (2021)	30%
		2030	40%
		2040	44%
Growing health awareness [9]	Companies engaged in health management, etc.	Current situation (2021)	Fewer than 60,000 companies
		2025	More than 100,000 companies
Improved performance of housing and buildings [10, 11]	Policy targets for energy efficiency of housing and buildings	Current status (2023)	Compliance already obligatory for new buildings of medium size and above
		2025	Conformity obligatory for new housing and small-scale new buildings
		2030	Net zero energy housing (ZEH)/ZEB for new housing and buildings
		2050	Realization of ZEH/ZEB in the average housing and building stock

Source: Based on National Institute of Population and Social Security Research, "Future Estimates of the Number of Households in Japan (National Estimates) (2018)" [8]/Japan Health Council, "Five Declarations of Action for Health Promotion 2025" [9]/Ministry of Economy, Trade and Industry and Agency for Natural Resources and Energy, "The 6th Basic Energy Plan" [10]/National Land and Infrastructure Based on the Ministry of Land, Infrastructure, Transport and Tourism's "Outline of Energy-saving Measures in Houses and Buildings Towards a Decarbonized Society" [11].

In the future, the aim is to achieve a comfortable temperature in living and working environments and to optimize energy use by improving equipment and system configuration to reduce heat load and by establishing more advanced operation and control technology through improved performance of the houses and buildings themselves.

Next-generation heat pump technology advances according to the technology roadmap can fulfil the following functions to meet the above-mentioned needs

(1) Optimization of equipment to match the performance of the house/building (e.g. whole house air conditioning, desiccant air conditioning)
(2) Overall system optimization through integration of equipment and heat sources
(3) Visualization of heat load and optimized operation control of equipment based on this, taking comfort and other factors into account
(4) Overall optimization of energy use in housing and buildings through a home energy management system (HEMS) and a building energy management system (BEMS), etc.

As set out in the technology roadmap (evaluation methodology section), it is necessary to establish an evaluation methodology that reflects the actual operating and installation environment. If evaluations are carried out taking into account the heat load (temperature, humidity, etc.) in actual houses and buildings, it will be easier to distribute equipment (whole-house air conditioning, desiccant air conditioning, etc.) that provides a more comfortable indoor environment in houses and buildings whose performance is expected to improve in the future (i.e. point (1) above).

In addition, the use of DX, as mentioned in the roadmap's product technology section, will enable design optimization not only as isolated equipment for each application, but also as a system integrating multiple units and heat sources (i.e. (2)).

In addition, digital twins, for example, can be used to improve operational control at the actual operational stage. As mentioned above, once an evaluation methodology that takes into account the actual heat load (temperature, humidity, etc.) has been established, operation control based on this will also be possible (i.e. (3)). In addition, as HEMS and BEMS can also perform more advanced operation control, it will be possible to further the overall optimization of energy use in homes and buildings (i.e. (4)).

As described above, next-generation heat pump technology will enable the realization of the above functions, which will lead to a healthy, comfortable, and decarbonized living environment, as shown in Figure 7.17 (Figure 5.11 repeated).

7.10.2 Policy Recommendations for Housing and Buildings

In housing and buildings, heat pump technology are primarily used for air conditioning and hot water supply. With this in mind, the following policy guidelines for housing and buildings are recommended

1. Development of a performance evaluation methodology based on the installation environment
2. Development of a comprehensive evaluation methodology for heat pump technology from a wellness perspective

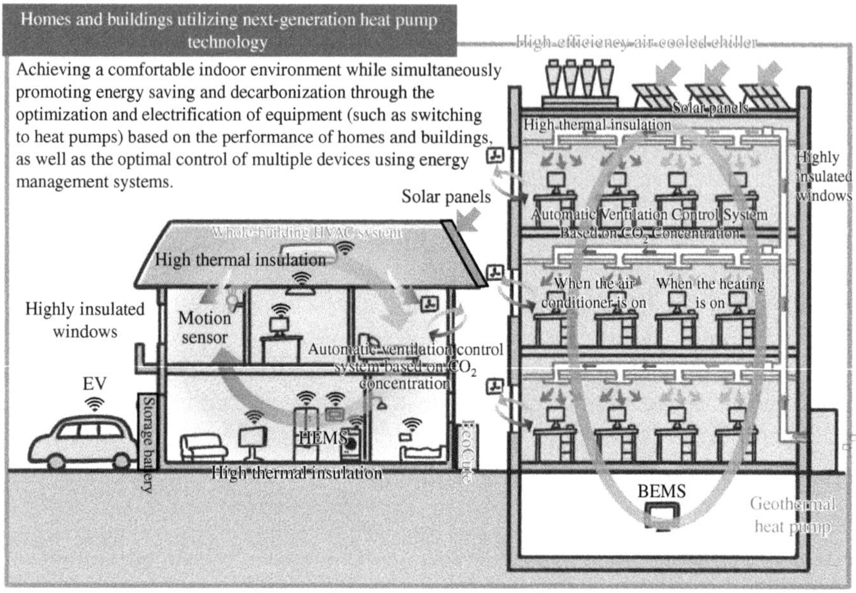

FIGURE 7.17 A living environment supported by next-generation heat pump technology.

3. Establishment of rules for homes compatible with next-generation refrigerants
4. Proactive support for heat pump water heaters to cope with the massive introduction of renewable energy

7.10.2.1 Development of Performance Evaluation Methodology Based on the Installation Environment

In the case of air conditioning, a performance impact assessment based on the installation environment needs to be established in cooperation with the building industry, as it is affected by the insulation and thermal storage properties of the building. The method is needed to work with existing systems such as the Building Energy Efficiency Act[10] and should be promoted so that better equipment and design can be selected.

7.10.2.2 Development of a Comprehensive Evaluation Methodology for Heat Pump Technology from a Wellness Perspective

In the case of air conditioning, in addition to energy-saving performance, comfort and wellness are also important. There is already a system for evaluating building spaces from a wellness perspective, such as the WELL certification[11], but this is an evaluation of the building, not of the heat pump technology. The industry needs to develop a comprehensive evaluation methodology for heat pump technology, including wellness considerations, so that the right heat pump technology can be selected, taking comfort and other factors into account.

7.10.2.3 Making Rules for Homes Compatible with Next-Generation Refrigerants

In the case of air conditioning, it is difficult to select a refrigerant with a GWP below 10, such as natural refrigerants, due to performance and safety considerations. What's more, it's necessary to consider measures against the use of refrigerants. In the case of chillers, appropriate rules need to be set by the industry, as the housing side has to be addressed, for example by installing water piping.

7.10.2.4 Active Support for Heat Pump Water Heaters that Contribute to the Mass Introduction of Renewable Energy

In the case of hot water supply, it is easy to adjust the operating hours of heat pump technology because the hot water can be kept warm in a storage tank and used without concern of the time of day. Taking advantage of this characteristic, equipment that heats water according to the time of day when solar power is being generated is also being introduced. Subsidies have been set up for heat pump water heaters that contribute to the mass introduction of such renewable energy, but they are not particularly favored in terms of subsidy amounts or rates [12]. As they also contribute to the regulation of power supply and demand, it is incumbent on the government to provide more active support.

7.11 NEXT-GENERATION HEAT PUMP TECHNOLOGY POLICY FOR THE FOOD SYSTEM

7.11.1 Constructing a Sustainable Cold Chain

Globally, the entire food supply chain is being streamlined and made more efficient, and food waste is being reduced (see Section 5.2), but there is still a strong need to build a food system that is both more productive and more sustainable. Japan, like many countries, has developed a strategy for constructing such a system, and specific targets and other measures have been set (Table 7.5).

Heat pump technology in food systems is mainly used in refrigeration and freezing installations. The cold chain, in which food and other products are managed and distributed at low temperatures from the production site to the consumer, is one of the core processing and distribution processes in the food system, as shown in Figure 7.18.

In order to realize the targets set out in Table 7.5, further reduction of food waste and more efficient and streamlined processes as a whole are required in the future.

Where next-generation heat pump technology advances in accordance with the technology roadmap, it can fulfil the following functions to meet the needs indicated:

(1) Provide data on temperature, humidity, etc. around food conditions, energy consumption, etc., and optimize operation control of equipment based on this data
(2) Help to prevent leakage through low GWP refrigerant conversion, automation of refrigerant management, and leak detection policies and roadmaps

TABLE 7.5
Trends, Indicators, and Targets for the Food System

Related trends	Indicator example	Current status and targets	
Rationalization and efficiency of food processing and distribution processes	Business food waste	Base year (FY2000)	5.47 million tonnes
		2030	2.73 million tonnes: halve compared to base year (FY2000)
		2050	Minimization of business food waste
	Food self-sufficiency ratio	Base year (FY2018)	37%
		2030	45% (based on heat supplied)
	Labor productivity in the food manufacturing industry	Base year (2018)	5149 thousand yen/person
		2030	6694 thousand yen/person: 30% improvement on base year (2018)
		2050	Further increase in labor productivity through the use of AI, etc.
	Ratio of expenses in the food and beverage wholesale industry	Base year (2016)	11.6%
		2030	10%
		2050	Further reduction through manpower saving and automation of food distribution
Zero emissions in agriculture, forestry, and fisheries	[Horticultural facilities] Ratio of hybrid horticultural facilities in heated area	Current status (2018)	10%
		2030	50%
		2050	Complete transition to fossil fuel-free facilities

Note: Hybrid horticultural facilities refer to horticultural facilities that do not depend solely on fossil fuels by utilizing technologies such as hybrid operation of heat pumps and fuel oil heaters.

Source: Based on the Ministry of Agriculture, Forestry and Fisheries, "Green Food System Strategy" [13], "'Green Food System Strategy' KPI 2030 Target Setting" [14] and "Food, Agriculture and Rural Areas Basic Plan" [15].

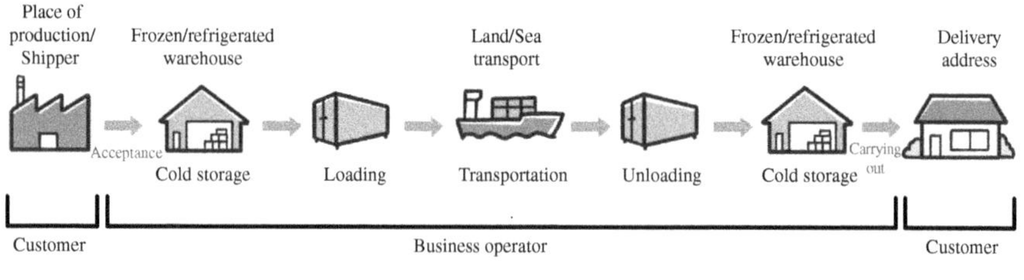

FIGURE 7.18 How the cold chain works. (Based on various sources.)

(3) Help to optimize foodstuff supply and demand through centralized management including other relevant information

Establish optimal operational control, etc., by utilizing evaluation methodologies that reflect the actual operating and installation environment as set out in the evaluation methodology section of the technology roadmap, and digital twin technology as set out in the 'product technology' section. By doing so, optimum operation control that takes into account the conditions of foodstuffs (temperature, humidity, etc.) at each site in the actual cold chain will be possible, which will maintain the freshness and quality of foodstuffs and lead to the extension of foodstuff expiry dates (1).

In addition, as set out in refrigerant conversion section of the roadmap, the conversion to low GWP refrigerants will also be promoted in cold chains where there remains a lot of equipment using refrigerants with high GWP. At the same time, the establishment of a method for collecting and managing refrigerant management information will automate refrigerant management and leak detection, enabling leakage countermeasures to be taken without time and cost (2).

Moreover, as set out under 'common platform' in the roadmap, a centralized management platform will be established that includes the above-mentioned temperature, humidity, energy consumption, refrigerant management information, as well as CO_2 emissions, production, and order volume transaction information etc. In this way, the CO_2 emissions of the entire cold chain can be examined and the supply and demand of foodstuffs can be optimized. Furthermore, it will also help to promote decarbonization initiatives through collaboration among stakeholders and reduce waste in food production and food production (3).

As described above, the aforementioned functions of next-generation heat pump technology will enable the realization of a sustainable food system with reduced food waste and increased efficiency and rationalization of the entire process, as shown in Figure 7.19 (Figure 5.22 reproduced).

7.11.2 Policy Recommendations for the Food System

In the food system, heat pumps are mainly used to develop the cold chain.

For this reason, the following policy directions are recommended in the food system

1. Establishment of data management and linkages between cold chain and heat pump equipment
2. Establishment and promotion of a system to capture data on CO_2 emissions in the cold chain
3. Subsidize and publicize the introduction of heat pumps in horticulture and other areas

7.11.2.1 Creating and Linking Data Management of the Cold Chain with Heat Pump Equipment

In next-generation heat pump technology, it is necessary to establish a data management system for the entire cold chain in order to manage the cold chain more meticulously.

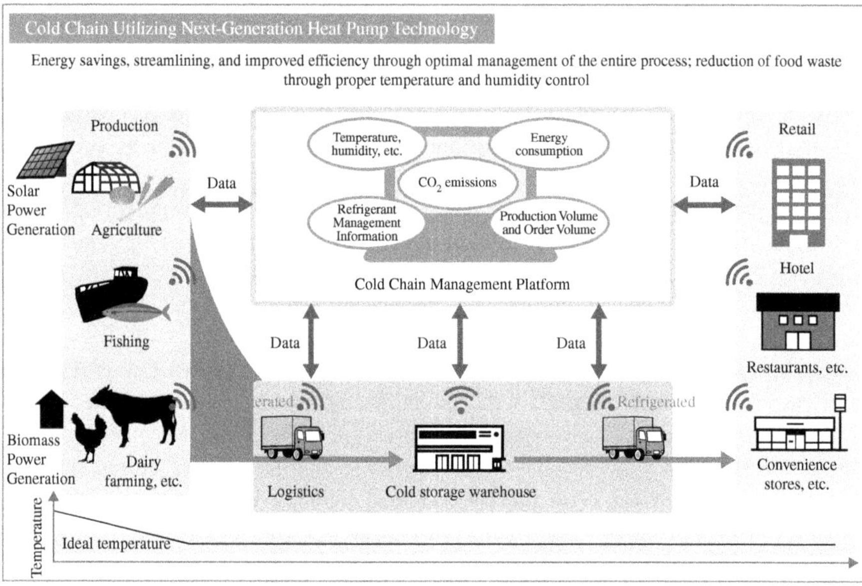

FIGURE 7.19 A cold chain realized by next-generation heat pump technology.

The data to be handled here include temperature and humidity information for foodstuffs, refrigerant management information for freezing and refrigeration equipment, and other related information (e.g. transaction information), not necessarily information on heat pump technology. For this reason, the companies and organizations that build the cold chain are the main actors and are responsible for maintaining it. However, as information on heat pump equipment and refrigerant management is managed on a common platform utilizing AI, etc., it is necessary to create a mechanism to link the two.

7.11.2.2 Establishment and Promotion of a System for Capturing Data on CO_2 Emissions in the Cold Chain

In promoting initiatives in the cold chain, it is also necessary to gain an understanding of consumers. Consideration of CO_2 emissions data from food products [16] is underway, but if CO_2 emissions are simply taken as a total amount, it is difficult to understand the performance of the cold chain, as emissions from the agricultural sector at the production stage are significant.

Therefore, the industry should consider, in cooperation with the Japanese Government, how to include data on the CO_2 emissions throughout the distribution process in the evaluation method and make them visible to consumers; for example, by setting standards for decarbonized cold chains and certifying and labeling them.

7.11.2.3 Subsidize and Publicize the Introduction of Heat Pump Equipment for Horticulture and Other Uses

The use of heat pump technology in agriculture, forestry, and fisheries, such as that used in horticulture facilities, is an essential component of the future food system, but has not yet progressed.

This is partly due to the fact that, in the case of agriculture, forestry, and fishery enterprises, financial capacity is a constraint, making it difficult to introduce expensive equipment. Therefore, the government should actively promote the introduction of such systems by setting up subsidies and publicizing them.

7.12 NEXT-GENERATION HEAT PUMP TECHNOLOGY POLICY FOR INDUSTRY

7.12.1 Decarbonization of Manufacturing Processes and Increase in Productivity and Competitiveness

Decarbonization efforts in the industrial sector (e.g. factories) are essential to achieve carbon neutrality (see Section 6.1). The Sixth Basic Energy Plan and the global warming countermeasures plan also impose strict targets for energy saving and CO_2 reduction (Table 7.6).

In recent years, pressure on companies to decarbonize, from stakeholders across the value chain, has also increased rapidly.

To achieve the targets shown in Table 7.6, decarbonization of the thermal process[12], which accounts for a large proportion of energy consumption in the industrial sector, is essential, and a shift from combustion boilers to industrial heat pumps is required. In addition to this decarbonization, DX must also be promoted if companies and others are to enhance their productivity and competitiveness in the future.

TABLE 7.6
Trends, Indicators, and Targets for the Industrial Sector

Relevant trends	Indicator example	Current status and future outlook	
Decarbonization of the industrial sector	Final energy consumption in the industrial sector	Base year (FY2013)	168 million kL
		2030	Approx. 140 million kL
	Energy-derived CO_2 emissions in the industrial sector	Base year (FY2013)	463 Mt-CO_2
		2030	289 million t-CO_2: 38% reduction compared to base year (FY2013)
		2050	Carbon neutral

Source: Based on the Ministry of Economy, Trade and Industry, and the Agency for Natural Resources and Energy, "The 6th Basic Energy Plan" [10] and "Global warming countermeasures plan" [17].

Where next-generation heat pump technology advances in line with the technology roadmap, it will enable functions that meet the following needs

(1) Expansion of the range of applications for industrial heat pumps (e.g. higher temperatures)
(2) Effective and integrated use of unused heat and waste heat (simultaneous use of hot and cold extraction)
(3) Optimum design and optimum operational control of equipment and systems throughout the entire manufacturing process

As set out in the product technology section of the technology roadmap, the expansion of the range of applications, such as higher temperatures, will lead to the promotion of industrial heat pumps (1). At the same time, the development of the technology required to utilize unused heat (improvement of the performance of heat exchangers, etc.) will also advance the effective utilization of unused and waste heat (2).

In addition, as also mentioned under 'product technology' in the roadmap, the use of digital twin technology, etc., will enable optimum design and operational control of equipment and systems throughout the entire manufacturing process (3).

As described above, next-generation heat pump technology will enable the realization of the above-mentioned functions, which will lead to manufacturing processes with decarbonization and improved productivity and competitiveness, as shown in Figure 7.20 (Figure 6.8 reproduced).

7.12.2 Policy Recommendations for the Industrial Sector

In the industrial sector, air conditioning and hot water supply are also used, but applications that utilize drying and steam generation in high temperature range are important for the future.

Based on this, the following policy directions for the industrial sector are recommended

1. Subsidies for the installation of high temperature heat pump equipment
2. Mandatory and incentivized demand response (DR)

7.12.2.1 Subsidies for the Introduction of High Temperature Heat Pump Equipment

When introduced as a replacement for combustion-based equipment such as boilers, even if operating costs can be reduced, the introduction of such equipment is not promoted because the initial outlay for equipment is high. Therefore, it is necessary for the government to formulate a thermal energy plan and provide financial support, such as subsidies for the introduction of heat pumps, while clearly stating the policy of active introduction of heat pumps.

However, measures should also be taken to curb the sale of conventional equipment once a certain level of uptake has been achieved, so that further uptake does not rely on subsidies.

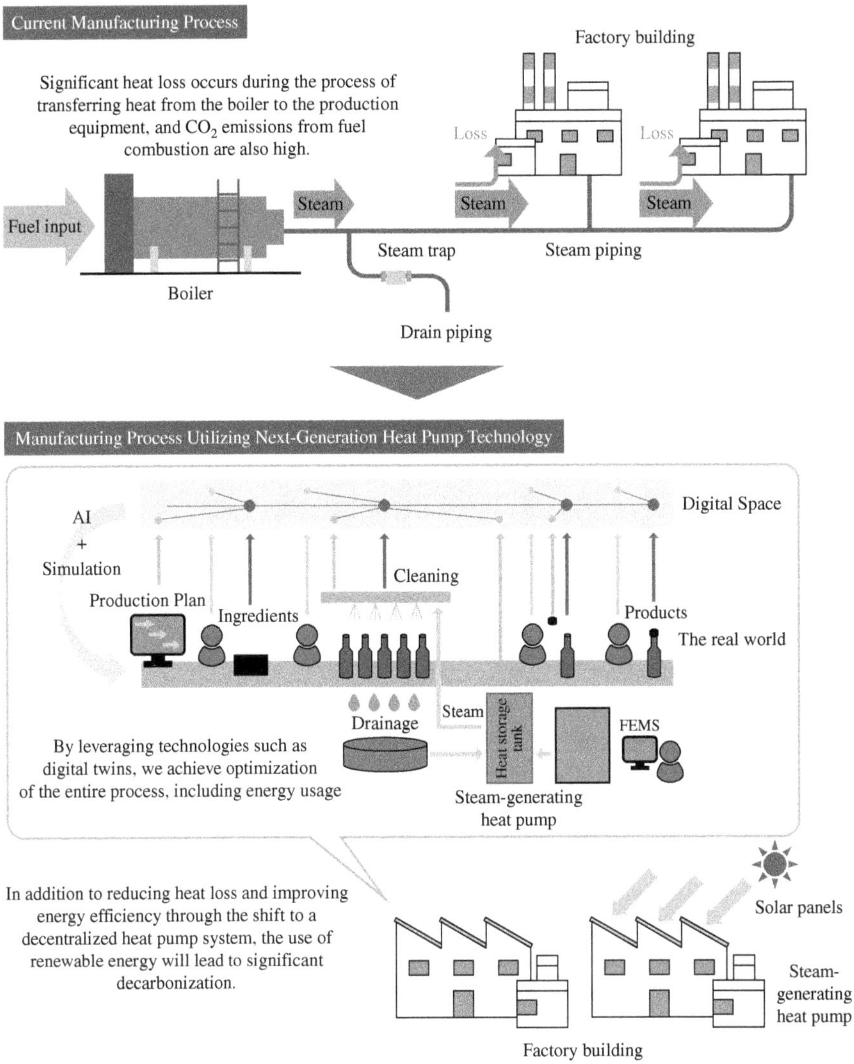

FIGURE 7.20 The current manufacturing process versus one realized by next-generation heat pump technology.

7.12.2.2 Mandating and Incentivizing Demand Response

DR for heat pump technology could be made mandatory or be incentivized.

While it is difficult to adjust the amount of electricity used in factories that use boilers, factories that use heat pumps plus heat storage can make use of heat storage to adjust electricity as far as possible without affecting the production line. In this way, the superior characteristics of next-generation heat pump technology become more apparent, which is expected to boost its introduction in this sector.

7.13 NEXT-GENERATION HEAT PUMP TECHNOLOGY POLICY FOR POWER SUPPLY AND DEMAND

7.13.1 Building a Stable Power System for Decarbonization

The massive introduction of renewable energy is essential for the realization of carbon neutrality. To achieve this, it is important to ensure the necessary regulating power to increase the flexibility of the power system (see Section 6.4).

Against this backdrop, there is a growing need to use decentralized energy resources (DER), which have become increasingly popular in recent years, for DR and virtual power plant (VPP) as a new means of securing the regulating power of the power system. All sectors on the demand side, including not only the consumer sector but also the industrial sector, are required to respond.

The roadmap drawn up by the ten general transmission and distribution companies responsible for the stable supply of electricity also aims to establish technology for controlling supply and demand using DER, with the aim of upgrading technology for adjusting supply and demand and stabilizing the grid (Table 7.7).

Heat pump water heaters with hot water storage tanks and thermal storage air conditioning systems that combine heat pump technology with heat storage systems are expected to be used for DR and VPP as typical DER.

Next-generation heat pump technology will achieve remote and integrated control without lowering the level of user comfort, etc. on the demand side, through optimized operational control technology such as digital twin technology, as described in the product technology section of the technology roadmap, and the construction of a centralized management platform for information, as described in the common platform section.

TABLE 7.7
Trends, Indicators, and Targets for Decentralized Energy Resources

Related trends	Examples of indicators	Current status and targets	
Progress in the diffusion and utilization of DER (supply and demand adjustment and upgrading of grid stabilization technology)	Related technology development and dissemination targets	Mid-2020s onwards	Introduce and utilize next-generation smart meters, etc. as the basis for DER control, maximize the use of existing resources, and promote and utilize DER flexibility
		Late 2040s onwards	Establishment of DER integrated control technology

Source: Based on Electricity Transmission and Distribution Network Council "Towards carbon neutrality in 2050 – Roadmap for the next generation of electricity networks" [18].

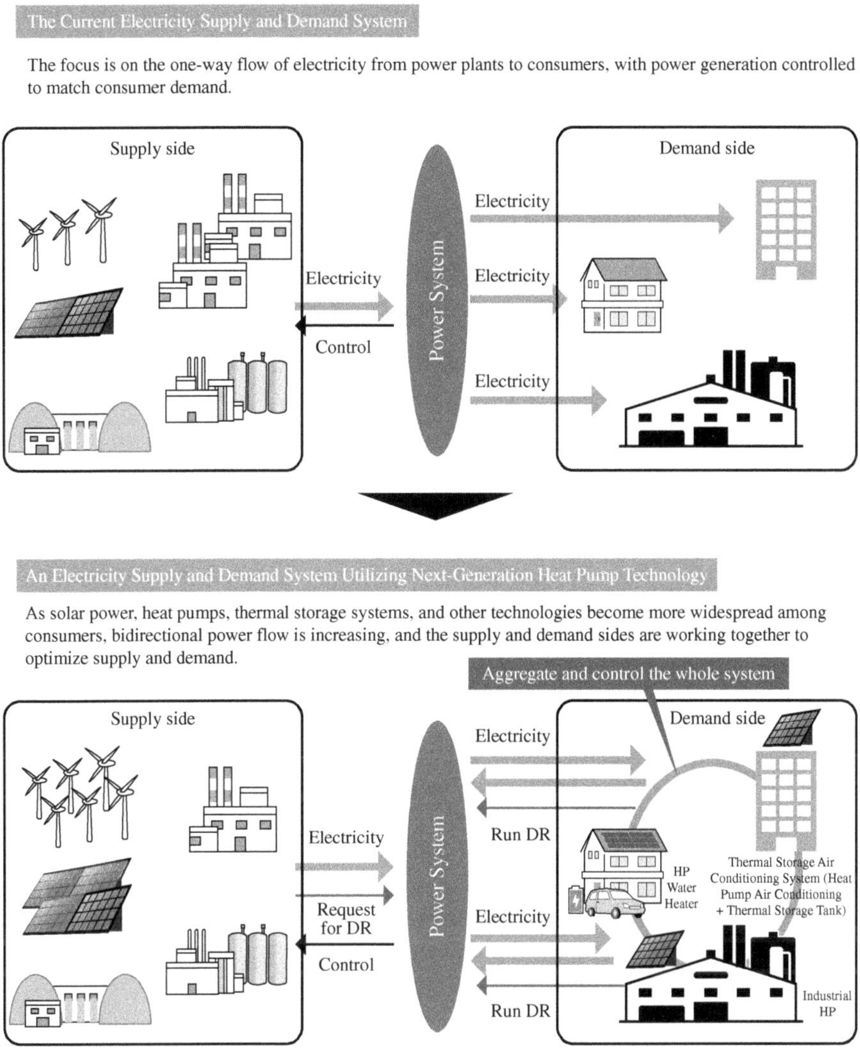

FIGURE 7.21 The current electricity supply and demand system versus a power system realized by next-generation heat pump technology.

This can therefore be utilized for DR and VPP, contributing to the construction of a power system that achieves a stable supply while also achieving decarbonization, as shown in Figure 7.21 (Figure 6.18 repeated).

7.13.2 Policy Recommendations for the Power System

In terms of power supply and demand, it is desirable for systems with DR functions to become more widespread in order to ensure the stability of the electricity supply.

For this reason, the following matters are recommended as policy directions in power supply and demand:

1. Creation of rules for DR
2. Establishment of an optimum tariff system for DR
3. Subsidizing the introduction of thermal storage tanks

7.13.2.1 Creation of Rules for DR

In order to achieve DR, it is necessary to clarify rules such as what kind of interactions are required to implement it, and what kind of time response is required.

7.13.2.2 Establishing an Optimal Tariff Structure for Demand Response

In order to implement DR, it is also necessary to establish a tariff structure that provides incentives.

As for the way DR for thermal energy, which is currently being considered for electricity, the industry dealing with heat pump equipment also needs to engage in dialogue with retail electricity companies to establish a desirable tariff structure.

7.13.2.3 Assistance in Installing Thermal Storage Tanks

To facilitate effective DR functions, it is necessary to promote the installation of thermal storage tanks that can freely change the time of heat production. It would be desirable to have government subsidies or other economic incentives for this as well.

The above are the policy directions that are considered necessary for the realization of next-generation heat pump technology, both in terms of cross-sectoral matters and in terms of sector-specific matters. In order to promote these overall objectives, it is necessary not only to regulate the environmental aspects of heat pump technology, such as refrigerants and equipment performance, but also to create a system in which academia, industry, and government work together to promote the overall optimization and diffusion of heat pump technology.

NOTES

1. Decentralized data management: The digital product passport, which is being considered as an important technology for realizing a circular economy in Europe, involves the construction of a decentralized system as a mechanism for managing product information, and implementation in this direction is being considered in Japan.
2. Total Equivalent Warming Impact (TEWI): An indicator that comprehensively evaluates the direct global warming impact of CO_2 diffused into the atmosphere by equipment refrigerants and the indirect global warming impact of CO_2 generated by the energy consumption of equipment operation. Indicator to assess overall CO_2.
3. Product life climate load: Life Cycle Climate Performance (LCCP) is an overall evaluation index of global warming impact that takes into account the impact of both refrigerants and greenhouse gases generated by electricity consumption, rather than evaluation based on global warming potential (GWP) alone.
4. Support for natural refrigerant equipment: The Ministry of the Environment's natural refrigerant introduction support project (subsidy) only covers refrigerated warehouses, food production plants, and food retail outlets.

5. Tighter regulations on the sale of conventional equipment: Stricter GWP standards for products designated under the Fluorocarbons Law, designation under the Green Purchasing Law, and a ban on sales.
6. Green transformation (GX) is a policy aiming for an industrial and economic society that does not use fossil fuels that generate greenhouse gases, but mainly uses clean energy from renewable energy sources (e.g. solar power, wind power).
7. Carbon pricing: A method of putting a price on the CO_2 emitted by companies. Typical examples include carbon taxes and emissions trading schemes.
8. Matching type: Also known as integral type. A product technology in which an already designed component is combined with other components and several elemental technologies are matched to achieve an overall optimized design.
9. Refrigeration and air conditioning culture: Resources, including personnel and technology, such as equipment and their proper use, maintenance and evaluation methods, testing and research institutes and technicians.
10. Building energy conservation law: Officially known as the Law on the Improvement of Energy Consumption Performance of Buildings, enacted to achieve carbon neutrality by 2050 and a 46% reduction in greenhouse gases by 2030 (compared to FY2013). It aims to reduce greenhouse gas emissions in the building sector, which accounts for approximately 30% of Japan's energy consumption, by improving the energy consumption performance of buildings, and implements measures such as the obligation to comply with energy consumption performance standards for buildings.
11. WELL certification (WELL Building Standard) is a space rating system for buildings developed by the US company Delos in 2014. The system evaluates and certifies buildings from the perspective of human health, focusing not only on ergonomic aspects but also on the wellness of the people who spend time in the space.
12. Thermal processes: Here, the term refers specifically to the demand for steam and hot water in the lower temperature range below 200°C.

REFERENCES

1. Cabinet Office, Government of Japan, Science, Technology and Innovation Promotion Office (CSTP), "Strategic Innovation Promotion Program (SIP) Circular Economy System Construction: Strategy and R&D Plan for Social Implementation," 2023, https://www8.cao.go.jp/cstp/gaiyo/sip/sip_3/keikaku/circulareconomy.pdf
2. The Ministry of the Environment, Document 4-2, "Outline of the Fourth Basic Plan for the Promotion of a Circular Society and Circular Economy Roadmap," Achieving Carbon Neutrality and Circular Economy Simultaneously, p.8, 2023) https://www.env.go.jp/content/000138209.pdf
3. Daikin Industries, Ltd., "Commercial Air Conditioner Subscription ZEAS Connect—Air Conditioner Flat-rate Service," 2023, https://www.ac.daikin.co.jp/shopoffice/zeas_connect
4. The Ministry of Land, Infrastructure, Transport and Tourism, "Level classification of automatic operation," 2023, https://www.mlit.go.jp/common/001226541.pdf
5. The Ministry of the Environment, "Non-fluorocarbons► site for the promotion of decarbonisation and natural refrigerants," 2023, https://www.env.go.jp/earth/ozone/cn_naturalrefrigerant/grant/
6. New Energy and Industrial Technology Development Organization (NEDO), "Development of an industrial heat pump simulator that can quantitatively evaluate the effect of introducing heat pumps," 2023, https://www.nedo.go.jp/content/100943176.pdf
7. Daikin Industries, Ltd./Toyo Keizai Inc., Toyo Keizai Online, Air conditioner 'to increase by about three times by 2050' problem severity: Essential in extreme heat, increasing demand worldwide and environmental concerns," 2020, https://toyokeizai.net/articles/-/363759

8. National Institute of Population and Social Security Research, "Future projections of the number of households in Japan (Nationwide Estimates)," (2018 estimates), (2018), https://www.ipss.go.jp/pp-ajsetai/j/hprj2018/t-page.asp
9. Nippon Kenko Kaigi, "Five Implementation Manifestations for Health Promotion 2025," 2023, https://2025.kenkokaigi.jp/sengen
10. The Ministry of Economy, Trade and Industry, Agency for Natural Resources and Energy, "The 6th Basic Energy Plan," 2021, https://www.enecho.meti.go.jp/category/others/basic_plan/pdf/20211022_01.pdf
11. The Ministry of Land, Infrastructure, Transport and Tourism, Study Group on Energy-Saving Measures for Housing and Buildings Toward a Decarbonized Society, "Outline of energy efficiency and conservation measures, etc. in housing and buildings towards a decarbonised society and how they should be implemented and advanced," 2021, https://www.mlit.go.jp/jutakukentiku/house/jutakukentiku_house_tk4_000188.html
12. The Ministry of Economy, Trade and Industry, Agency for Natural Resources and Energy, Hot Water Heater Energy Saving Project, "Details of Target Equipment," 2023, https://kyutou-shoene.meti.go.jp/materials/ecocute.html
13. The Ministry of Agriculture, Forestry and Fisheries, "Green Food System Strategy," 2021, https://www.maff.go.jp/j/kanbo/kankyo/seisaku/midori/attach/pdf/index-10.pdf
14. The Ministry of Agriculture, Forestry and Fisheries, "'Green Food System Strategy' KPI 2030 Target Setting," (2022), https://www.maff.go.jp/j/council/seisaku/kikaku/goudou/attach/pdf/32-2.pdf
15. The Ministry of Agriculture, Forestry and Fisheries, "Basic Plan for Food, Agriculture and Rural Areas," 2020, https://www.maff.go.jp/j/keikaku/k_aratana/attach/pdf/index-13.pdf
16. The Ministry of Agriculture, Forestry and Fisheries, "'Visualisation' label for greenhouse gases reduction: Cumulative number of demonstration shops reaches 100!," 2023, https://www.maff.go.jp/j/press/kanbo/b_kankyo/230307.html
17. Ministry of Economy, Trade and Industry, Agency for Natural Resources and Energy, "Global warming countermeasures plan," 2021, https://www.env.go.jp/content/900440195.pdf
18. Transmission & Distribution Grid Council, "Toward Carbon Neutrality by 2050—Roadmap for the Next Generation of Electric Power Networks," 2021, https://www.tdgc.jp/information/2021/05/21_1600.html

Appendix 1: Refrigerant Conversion for Next-Generation Heat Pump Technology

OVERVIEW

Refrigerants used in heat pump technology have been converted from chlorofluorocarbons (CFCs), hydrochlorofluorocarbons (HCFCs), and hydrofluorocarbons (HFCs) from the perspective of ozone layer protection, and from the perspective of reducing greenhouse gas (GHG) emissions to achieve carbon neutrality in recent years, conversion to refrigerants with lower global warming potential (GWP) than HFCs with higher GWP is being pursued.

However, low GWP refrigerants may be flammable or toxic, or have lower energy efficiency compared to the use of HFCs. It is therefore important to achieve refrigerant conversion with a view to safety management and reduction of GHG emissions during use and at disposal.

This chapter provides examples of two approaches:

- Measures aimed at carbon neutrality through refrigerant conversion of newly sold equipment only
- Measures to reduce refrigerant leakage at the time of use and to ensure recovery and recycling at the time of disposal in areas where low GWP refrigerants are relatively easy to use (e.g. residential air conditioners, and commercial refrigeration and freezing equipment), as well as for commercial packaged air conditioners and multi-air conditioners for buildings

On this basis, examples of refrigerant use in next-generation heat pumps that both comply with the HFC production and consumption caps based on the Kigali Amendment and reduce total GHG emissions (energy origin and refrigerant emission origin) are given.

A1.1 HISTORY OF REFRIGERANT USE AND REGULATION

A1.1.1 Fluorocarbons Affecting the Global Environment

The refrigerants used in heat pump appliances are mainly fluorocarbons (FCs) with the following characteristics

- Chemically very stable and easy to handle
- Low toxicity to humans

A1.1.1.1 History of Regulations on Global Environment

The development of CFCs in the 1920s led to their widespread use as refrigerants in the USA, and from the 1960s onwards, their consumption increased significantly in industrialized countries.

However, in 1974, Dr Rowland and colleagues at the University of California published a theory that CFC refrigerants emitted into the atmosphere may deplete the ozone layer [1]. Since then, a number of scientific findings have accumulated, showing that because FCs are stable, they reach the stratosphere above and are decomposed by ultraviolet radiation, releasing chlorine radicals, leading to the depletion of the ozone layer in the stratosphere.

A1.1.2 Domestic and International Regulation of Fluorocarbons

A1.1.2.1 Refrigerant Conversion from Fluorocarbons Regulation

In order to protect the ozone layer, the Montreal Protocol (adopted in 1987 and entered into force in 1989) stipulated the phase-out of the production and import of specific FCs[1] with high ozone depletion potential (ODP) and the consumption of HCFCs. A reduction schedule was set out. In Japan, the Act on the Protection of the Ozone Layer Through the Control of Specified Substances and Other Measures (Ozone Layer Protection Law) was enacted in 1988 as a domestic implementing law for the Montreal Protocol.

This necessitated a conversion from CFCs and HCFCs to HFCs, with refrigerant conversion from CFC-12 and HCFC-22 to HFC-404A[2] in low temperature equipment and from HCFC-22 to HFC-410A[3] in air conditioning equipment.

A1.1.2.2 Further Regulatory Legislation Due to Refrigerant Conversion

Although HFC refrigerants were developed as a measure to protect the ozone layer, they still have the drawback that they have a large GWP.

For example, the GWP values of the aforementioned HFC-404A and HFC-410A are 3,920 and 2,090 respectively, which are reversed from the 1,810 of HCFC-22.[4] Therefore, the Kyoto Protocol, adopted in 1997 and entered into force in 2005, stipulates the control of emissions of three gases—HFCs, perfluorocarbons (PFCs), and sulphur hexafluoride (SF6)—which are alternative FCs with large GWPs, and requires a refrigerant conversion to a smaller GWP. However, the Kyoto Protocol did not go far enough, with the USA withdrawing from the Kyoto Protocol on the grounds that it was unfair that developing countries such as China were not obliged to reduce emissions, and that global warming measures would have a negative impact on the economy, so an alternative framework was sought to replace the Kyoto Protocol.

At the 28th Meeting of the Parties to the Montreal Protocol held in Kigali, Rwanda, on October 15, 2016, an amendment proposal was adopted to provide for reductions in HFC production and consumption in each country (Kigali Amendment[5]).

A1.2 REFRIGERANT NUMBERS AND PROPERTIES OF REFRIGERANTS

A1.2.1 Provisions for Refrigerant Numbers

Refrigerant numbers are defined by Standard 34 of the American Society of Refrigeration and Air Conditioning Engineers (ASHRAE)[6] or by the international standard ISO 817.

The refrigerant number for a single refrigerant is a number based on the refrigerant composition with an R appended, expressed as, for example, R32 for HFC-32 and R1234yf for HFO-1234yf. HFO refrigerant is the designation for HFC refrigerants, which are compounds composed of hydrogen, fluorine, and carbon with carbon-to-carbon double bonds in the molecule, easily decomposed in the atmosphere and characterized by low GWP.

The number based on the refrigerant composition is formulated as follows: The digit in the thousands place represents the number of double bonds between carbons, the digit in the hundreds place represents the number of atoms of carbon –1, the digit in the tens place represents the number of atoms of hydrogen +1, the digit in the ones place represents the number of atoms of fluorine, and isomers are denoted by a, b, c or another lower-case letter.

If the components of the mixed refrigerant are the same but have different composition ratios, the number is followed by a capital letter A, B, or C to indicate the difference in composition ratios. Figure A1.1 shows how the refrigerant numbers are assigned.

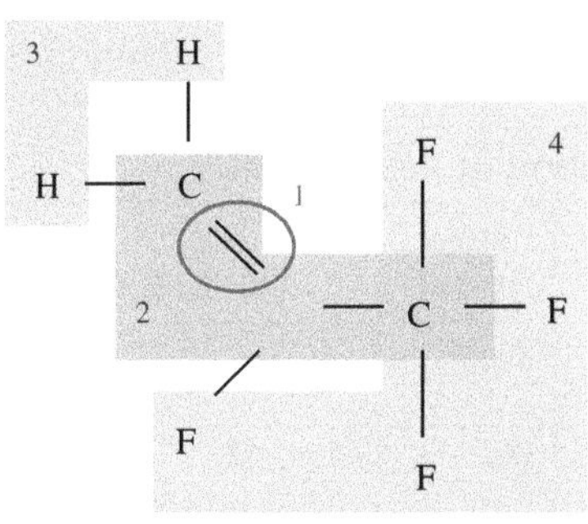

	1	2	3	4	
	Thousands place	Hundreds place	Tens place	Ones place	Adjacent characters
R-	1	2	3	4	yf
	Number of carbon-carbon double bonds	Number of carbon atoms -1	Number of hydrogen atoms + 1	Number of fluorine atoms	Distinguishing between structural isomers

FIGURE A1.1 How to number single refrigerants (example of R1234yf).

As mixed refrigerants, non-azeotropic refrigerant mixture is assigned the number 400 and azeotropic mixed refrigerants are assigned the number 500[7]. The numbers are assigned sequentially in the order in which they are applied, with the same number if the constituent refrigerant species are the same, and are distinguished by capital letters A, B, C, etc. in different mixing ratios[8].

In addition, other organic compounds are given numbers in the 600s (last two digits are the number of atoms of carbon −4) and other inorganic compounds are given numbers in the 700s (last two digits are the molecular weight)[9].

A1.2.2 Properties of Refrigerants and Considerations when Selecting Them

The properties of refrigerants are classified according to toxicity and flammability. Toxicity is determined as A (lower toxicity) or B (higher toxicity) depending on whether the operational exposure limit (OEL)[10], an indicator of chronic toxicity, is above or below 400 ppm. In the case of refrigerant leakage, since refrigerant does not remain in the room for a long time, a method of determining A/B by adding an acute toxicity exposure limit (ATEL)[11], an indicator of acute toxicity, to those with chronic toxicity is being considered.

Flammability is classified according to four criteria: Presence or absence of flame spread, lower flammability limit (LFL), heat of combustion (HOC), and burning velocity (Table A1.1).

Non-flammable (A1) refrigerants (R22, R404A, R410A, etc.) have been used in the past, but as shorter atmospheric lifetimes to reduce GWP often lead to higher reactivity (toxicity and flammability) in the atmosphere [2], in order to meet the demand for low GWP, mildly-flammable (A2L) refrigerant (R32, R1234yf, etc.) conversions have been progressing. And, in the final year of the Kigali Amendment (2036) and moving toward carbon neutrality in 2050, even lower GWP is required,

TABLE A1.1
Flammability Classification of Refrigerants

Classification	Notation	Flame Spread	LFL (kJ/kg)	HOC (kJ/kg)	Combustion Rate
1	No Flame Propagation	None		–	–
2L	mildly flammability	Yes	>3.5%(ISO) >0.10kg/m³(ASHRAE)	and <19000	and ≤10 cm/s
2	Flammable	Yes	>3.5%(ISO) >0.10 kg/m³(ASHRAE)	and <19000	–
3	Higher flammability	Yes	≦3.5%(ISO) ≦0.10 kg/m³(ASHRAE)	or ≧19000	–

Source: Based on Japan Refrigeration and Air Conditioning Industry Association, ISO817:2014—Safety Classification of Refrigerants [3].

and the possibility of using strong flammability (A3) refrigerants (e.g. R290) is being studied and safety risk assessments are being carried out.

The maximum amount of refrigerant that can be used in equipment is limited by standards for safety reasons. One of the criteria for the refrigerant charge limit is defined as the refrigerant concentration limit (RCL).

This is defined as the lowest of the following

1. ATEL (Acute Toxicity Exposure Limit): the lower concentration limit causing acute toxicity
2. ODL (Oxygen Deprivation Limit): the lower concentration limit at which oxygen deprivation occurs
3. One-fourth or one-fifth of the LFL(Lower Flammability Limit), which is the lower concentration limit for ignition

For most combustible refrigerants (A2L, A2 and A3), one-quarter of the LFL is the lowest.

For A1 refrigerants, the maximum refrigerant volume is limited by the ODL for most refrigerants, but there are also refrigerants where the ATEL is smaller than the ODL and the maximum refrigerant volume is limited by the ATEL (Figure A1.2).

Performance is also important in the selection of refrigerants. The boiling point (Bubble point) is the temperature at which the saturation vapor pressure of a liquid is equal to the external pressure, and a refrigerant with a low boiling point has a higher saturation pressure at the same temperature than a refrigerant with a high boiling point. As a result, their condensation and evaporation pressures in the refrigeration cycle increase, leading to a tendency for higher refrigeration capacity and a higher coefficient of performance (COP) (Figure A1.3).

In general, high performance can also be obtained by selecting a refrigerant with a large compressor suction density and a refrigerant with a large difference between saturated gas enthalpy and saturated liquid enthalpy.

The following factors should then also be taken into account when selecting the appropriate refrigerant for each piece of equipment

- Equipment size: For example, in centrifugal and screw type equipment, refrigerants with lower pressure are selected as they cannot withstand high pressure
- Temperature range used: Select a refrigerant that can provide high performance in the operating temperature range

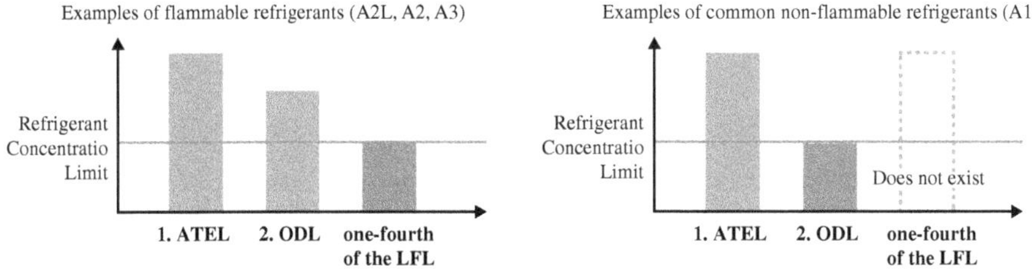

FIGURE A1.2 Standards for the maximum amount of refrigerant that can be used in equipment. *Note:* Definitions according to ASHRAE Standard 34-2022 are illustrated.

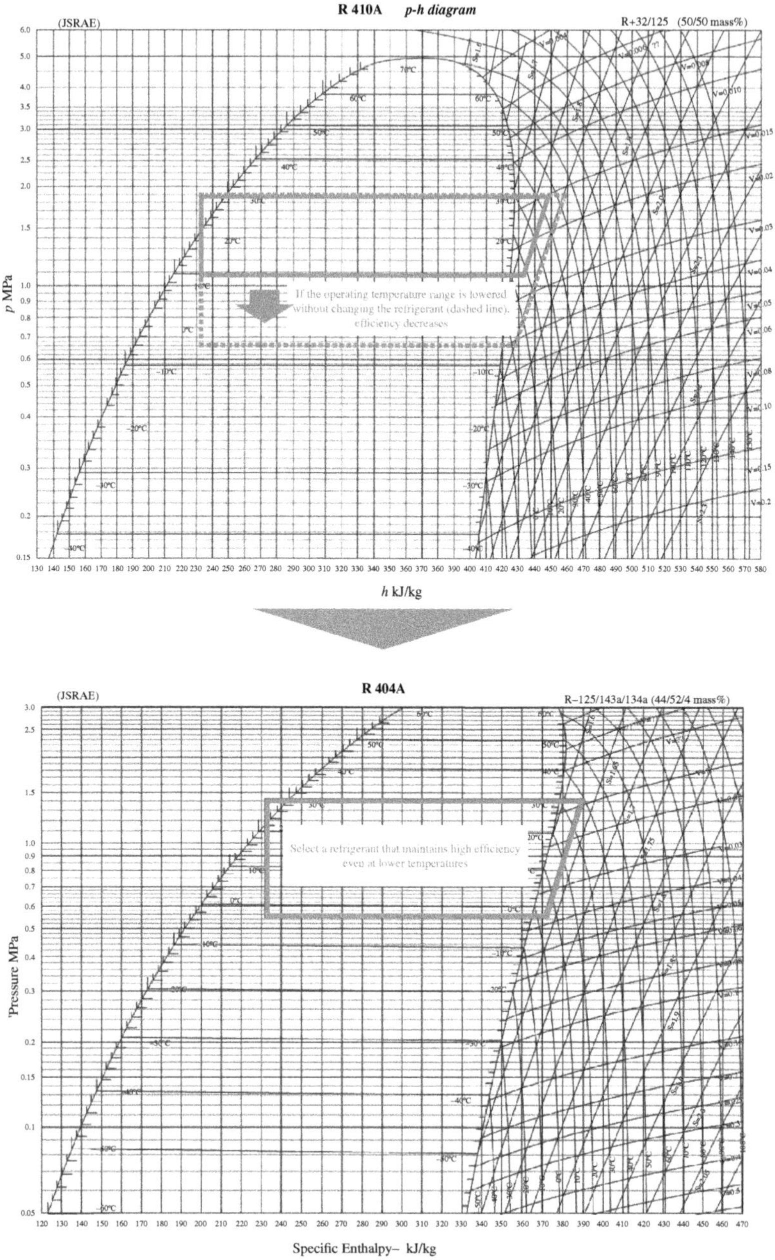

FIGURE A1.3 The difference in performance depending on the temperature range used and the type of refrigerant (e.g. the difference in performance when used in a lower temperature range). *Note:* The p-h diagram was created using the refrigeration cycle calculation program software Ver. 2, and text boxes and other information were added. (From Japan Society of Refrigeration and Air Conditioning Engineers, "Refrigeration Cycle Calculation Program Software Ver. 2" [4]).

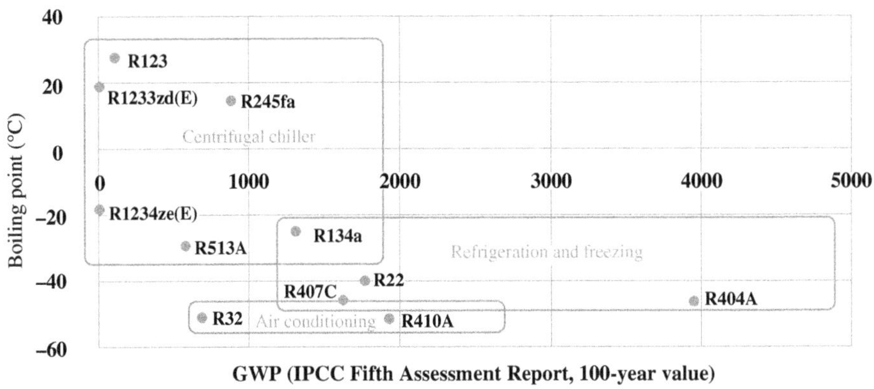

FIGURE A1.4 Map of refrigerant applications (arranged by boiling point and GWP). (From Japan Fluorocarbon Manufacturers Association, "List of environmental and safety data for specific fluorocarbons (CFCs/HCFCs) and fluorocarbons" [5]).

In order to select the refrigerant to be used for each appliance, not only must the ODP and GWP of the refrigerant be taken into account but also its toxicity and flammability along with an understanding of the maximum refrigerant quantities specified in the standard.

A safety risk assessment should be conducted if necessary, and the equipment performance when the refrigerant is used should be determined by refrigeration cycle simulation and drop-in evaluation. For reference, the relationship between the boiling point and GWP of each refrigerant is shown in Figure A1.4.

A1.2.3 Application Status of Low GWP Refrigerants by Heat Pump Equipment

Figure A1.5 summarizes the relationship between low GWP candidate refrigerants and GWP values for each heat pump equipment application.

Candidates for refrigerants with GWP below 100 are mildly-flammable or strongly flammable and highly toxic ammonia (R717). However, for commercial packaged air conditioners and multi-air conditioners for buildings, no specific candidate refrigerant with a GWP below 100 has been listed.

A1.3 SCENARIOS FOR LOW GWP REFRIGERANT CONVERSION

A1.3.1 Scenarios for Domestic Conversion to Low GWP

Assumptions need to be made in order to predict future refrigerant conversions. The Ministry of Economy, Trade, and Industry (METI) and the Ministry of the Environment have developed a scenario for future conversion to low GWP refrigerants (Figure A1.6).

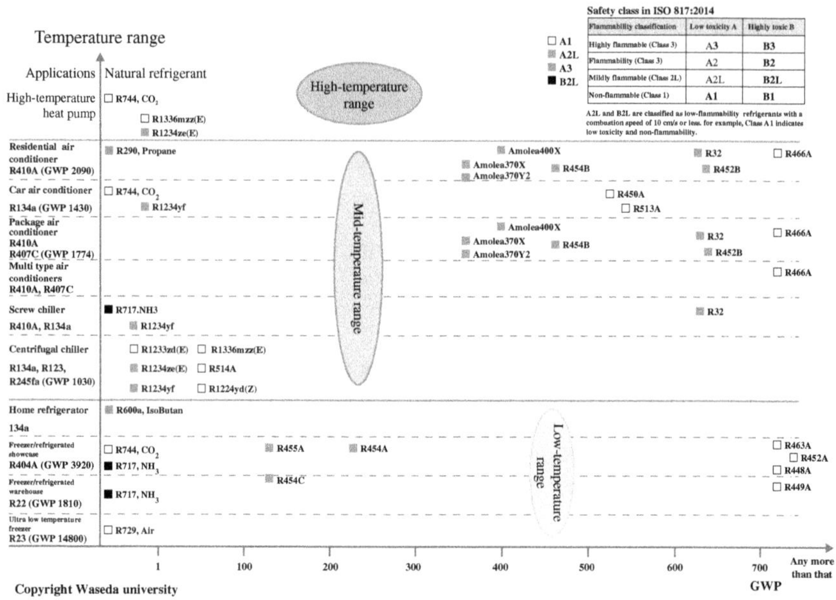

FIGURE A1.5 Map of refrigerant applications (arranged by temperature range and GWP). (From Waseda University, w-refrigerant.com, "Map of refrigerant applications" [6]).

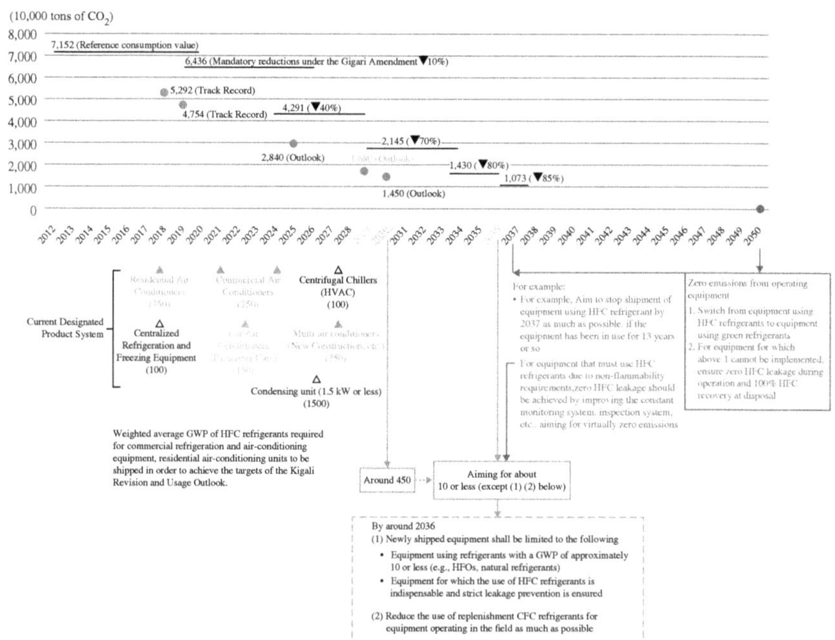

FIGURE A1.6 Low GWP scenarios of the Ministry of Economy, Trade and Industry (METI) and the Ministry of the Environment. (From METI, "Report on the evaluation and study of the enforcement status of the revised Fluorocarbons Emissions Control Law of 2013" [7]).

This scenario requires the following to be achieved

- Kigali Amendment (the Ozone Layer Protection Act in Japan): Average GWP of new models to be below about 450 by 2030 in order to meet the target.
- Achieving carbon neutrality by 2050: Average GWP of new models to be below about 10 by 2036

Next, in line with this estimation/assumption, the forecast for future refrigerant conversion for each type of equipment is described. Here, residential air conditioners, commercial packaged air conditioners, multi-air conditioners for buildings, built-in refrigeration and freezing equipment, condensing units, and vehicle air conditioners are direct expansion type equipment[12] and chillers are inter-expansion type appliances[13]. Table A1.2 shows a list of projected refrigerant conversions by appliance (water heaters are excluded as they already use natural refrigerant as a direct expansion type).

TABLE A1.2
List of Projected Refrigerant Conversions by Appliance

Type of Equipment	Current Main Refrigerant	Candidate Refrigerants in Transition	Candidate Refrigerants with GWP below 10 other than Natural Refrigerants	Final Refrigerant Candidate if only Natural Refrigerants are Allowed
Residential air conditioners	R32	Hydrofluoroolefin (HFO) mixed refrigerant (e.g. R454C)	HFO mixed refrigerant (under development)	A3 refrigerant (e.g. R290)
Commercial packaged air conditioners	R32 (partly R410A)	HFO mixed refrigerant (e.g. R454C)	HFO mixed refrigerant (under development)	Conversion to chiller with A3 refrigerant (e.g. R290)*
Multi-air conditioner for buildings	R410A	R32, HFO mixed refrigerant (e.g. R454B)	HFO mixed refrigerant (under development)	CO_2, Conversion to chiller with A3 refrigerant (e.g. R290)*
Built-in refrigeration and freezing equipment	R134a, R404A, R410A (partially CO_2)	HFO mixed refrigerant (e.g. R454C)	HFO mixed refrigerant (under development)	CO_2, A3 refrigerant (e.g. R290)

(Continued)

TABLE A1.2 (*Continued*)
List of Projected Refrigerant Conversions by Appliance

Type of Equipment		Current Main Refrigerant	Candidate Refrigerants in Transition	Candidate Refrigerants with GWP below 10 other than Natural Refrigerants	Final Refrigerant Candidate if only Natural Refrigerants are Allowed
Condensing units					
	Refrigeration and freezing units	R404A, R410A	HFO mixed refrigerant (e.g. R454C)	HFO mixed refrigerant (under development)	CO_2, A3 refrigerant (e.g. R290)
	Medium size condensing unit	R404A, R410A (partly CO_2)	– Possible A2L HFO mixed refrigerant conversion, but with challenges	– HFO mixed refrigerant (under development)	CO_2, separate type showcases to be converted to built-in type and conversion to chiller with A3 refrigerant (e.g. R290)*
	Large size condensing units	R404A, R134a, R407C	– Possible A2L HFO mixed refrigerant conversion, but with challenges	– HFO mixed refrigerant (under development)	Conversion to chiller with A3 refrigerant (e.g. R290)*
Chiller					
	Scroll chillers	R410A	R32, HFO mixed refrigerant (e.g. R454C)	HFO mixed refrigerant (under development)	A3 refrigerant (e.g. R290)
	Screw chillers	R404A, R134a	HFO mixed refrigerant (e.g. R454C)	HFO mixed refrigerant (under development)	A3 refrigerant (e.g. R290)

(*Continued*)

TABLE A1.2 (*Continued*)
List of Projected Refrigerant Conversions by Appliance

Type of Equipment		Current Main Refrigerant	Candidate Refrigerants in Transition	Candidate Refrigerants with GWP below 10 other than Natural Refrigerants	Final Refrigerant Candidate if only Natural Refrigerants are Allowed
	Centrifugal chillers	Below 1 MPa: R123, R245fa. 1 MPa and above: R134a, R513A	–	Less than 1 MPa: R514A R1233zd(E) R1224yd(Z). 1 MPa or more: R1234yf R1234ze(E)	€ –
Vehicle air conditioner		R134a	–	R1234yf	CO_2, R290
Transport refrigeration and freezing units		R404A, R134a	R452a	–	
High-temperature heat pumps		Below 100°C: CO_2 Above 100°C R1234ze(E), R1233zd(E), R1336mzz(E)	–	Less than 100°C:– 100°C or more R1234ze(E) R1233zd(E), R1336mzz(E)	Below 100°C: CO_2 Above 100°C: A3 refrigerant (e.g. hydrocarbon)

When converting to chillers, it should be noted that there may be negative effects such as increased equipment installation costs and increased power consumption.

HFO mixed refrigerant or natural refrigerant is considered as a candidate refrigerant with a GWP of 10 or less for many appliances. However, HFO mixed refrigerant is still under development and is not yet ready for the market, while natural refrigerant (including the strong A3 refrigerant as an option) has many issues at present, such as lower energy performance and increased electricity consumption.

Hydrocarbons have been considered and partially introduced in built-in refrigeration and freezing equipment and residential air conditioners, which are small and have a low refrigerant charge, but it would be quite difficult in reality to reduce the GWP of all equipment to below 10.

A1.3.2 Refrigerant Conversion Scenarios for Residential Air Conditioners

The refrigerant currently used in residential air conditioners for domestic use is R32, a mildly flammable (A2L) refrigerant, and the parts of the world where equipment is sold by domestic manufacturers include Japan, Europe, and North America. A possible future scenario for low GWP refrigerant conversion is shown in Figure A1.7.

A1.3.2.1 Transitional Refrigerant Candidate: HFO Mixed Refrigerant

If HFO refrigerants such as HFO-1234yf are used on their own, an outdoor unit approximately 1.5 times larger may be required to achieve the same coefficient of performance (COP) of the equipment. Due to cost and performance issues, it is unlikely that HFO stand-alone refrigerant will be selected as the final refrigerant (a refrigerant that can achieve carbon neutrality).

There is currently no prospect for an HFO mixed refrigerant[14] to replace R32 with a GWP of 10 or less, although the challenge is how high pressure refrigerants can be developed for HFO mixed refrigerants. If the pressure can be increased to about R22 level, it may be possible to use the refrigerant depending on ingenuity. Note that, if the refrigerant has to be changed to achieve the 2029 and 2030 reduction targets, A2L refrigerants with a GWP of 150 or less (e.g. R454C) could be adopted in the interim, before the final refrigerant is agreed.

A1.3.2.2 Final Refrigerant Candidate: Natural Refrigerant

If the situation arises where only natural refrigerants are acceptable as final refrigerants, the only option are A3 refrigerants such as R290. Note that North America has its own restrictions on A3 refrigerants and it may be difficult to use A3 refrigerants in direct expansion type appliances.

R290 has a lower pressure than R32, making it difficult to direct drop-in replacement in current equipment, but ingenuity is likely to eliminate any related

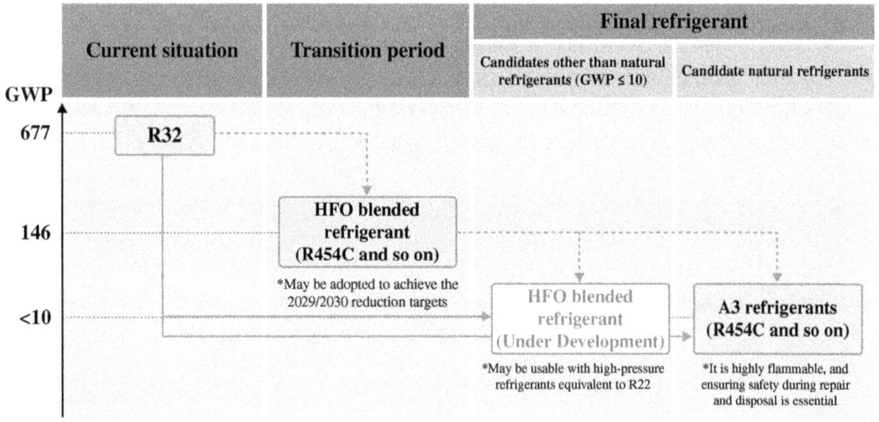

FIGURE A1.7 Scenario for residential air conditioner conversion to low GWP refrigerants.

performance problems. However, R290 is a highly flammable refrigerant, so the challenge is how to ensure safety during operations such as repairs and disposal.

Safety at work can be ensured by a qualification system, but without a national qualification system, enforcement is weak. At the time of disposal, many items flow outside the legal channels of the Home Appliance Recycling Law, and safety cannot be ensured unless this is ensured to flow through legal channels.

Note that more than 270,000 residential air conditioners using R290 have already been sold in China [8], and in Japan, Panasonic is selling air conditioners equipped with the R290, which does not have an outdoor unit, in the European market [9].

A1.3.3 Refrigerant Conversion Scenarios for Commercial Packaged Air Conditioners

The refrigerant currently used in commercial packaged air conditioners is R32 (partially R410A), and the parts of the world where equipment is sold by Japanese manufacturers are Japan, Europe, and North America. Figure A1.8 shows an image of future scenarios for low GWP refrigerant conversion.

A1.3.3.1 Transitional Refrigerant Candidate: HFO Mixed Refrigerant

Commercial packaged air conditioners have long piping connecting outdoor and indoor units, which is not easy to replace when embedded in a building. It would be difficult to use HFO single refrigerant without changing the connective piping, as the pressure loss would be too high.

At present, there is no prospect for R32 or R410A substitute HFO mixed refrigerant with GWP below 10, but if one such mixed refrigerant were developed with pressure equivalent to R22, it may be possible to use it with a little ingenuity.

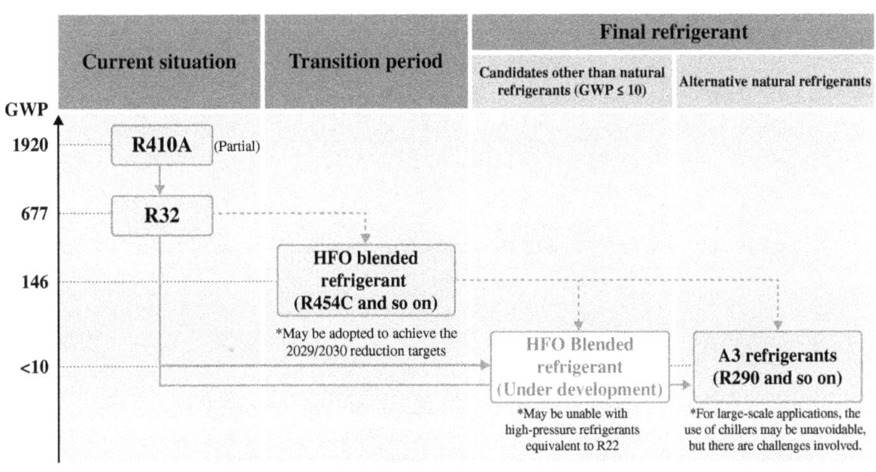

FIGURE A1.8 Scenario for conversion to low GWP refrigerants for commercial packaged air conditioners.

In the case of commercial packaged air conditioners, it is not advisable to change to another A2L refrigerant with a higher GWP than that before the final refrigerant. However, it may have to be adopted depending on the balance with other equipment in order to achieve the 2029 and 2030 reduction targets. In such cases, A2L refrigerants with GWP below 150, such as R454C, could be adopted.

A1.3.3.2 Final Refrigerant Candidate: Natural Refrigerant

If the situation arises where only natural refrigerants are acceptable as final refrigerants, the only option are A3 refrigerants such as R290. The only direct expansion type equipment that can use A3 refrigerant are the small appliances[15] in terms of maximum refrigerant quantity, and for large commercial packaged air conditioners, there may be no other alternative but to use A3 refrigerant chillers. In that case, the following issues are to be addressed.

A1.3.4 Refrigerant Conversion Scenarios for Multi-Air Conditioners for Buildings

Domestic manufacturers currently sell multi-air conditioners for buildings to Japan, Europe, and North America. The refrigerant currently used in these appliances is R410A and the most recent challenge for multi-air conditioners for buildings is to convert this to R32 (A2L). In order to install indoor units in small rooms, appliances must be equipped with a refrigerant leak detector, a safety measure to prevent ignition accidents in the event of refrigerant leakage, and a shut-off valve.

An image of a future scenario for low GWP refrigerant conversion is shown in Figure A1.9.

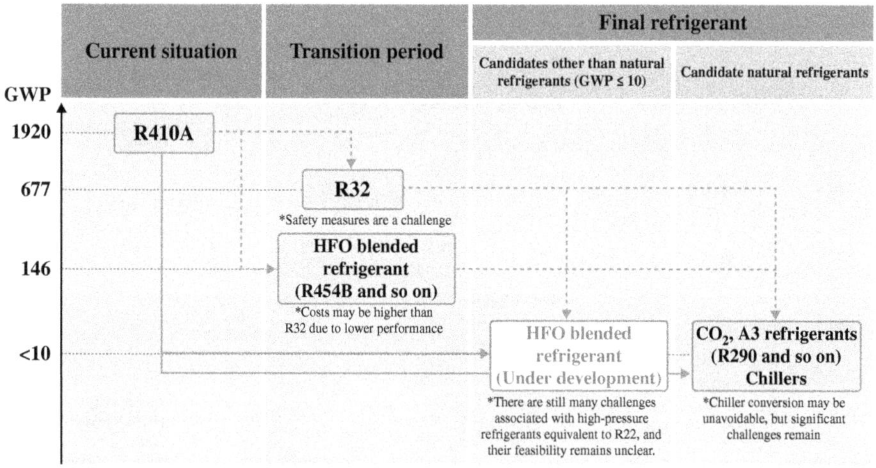

FIGURE A1.9 Scenario for conversion to low GWP refrigerants for multi-air conditioner for buildings.

A1.3.4.1 Transitional Refrigerant Candidate: HFO-Mixed Refrigerant

The use of HFO stand-alone refrigerant in multi-air conditioners for buildings is likely to be difficult due to too high pressure losses in the connecting pipes, but an HFO-mixed refrigerant as an alternative to R410A with a GWP below 10 is currently not on the cards. The use of HFO-mixed refrigerant in multi-air conditioners for buildings is likely to be difficult due to high pressure losses in the connecting pipes.

Even if an HFO mixed refrigerant with a pressure equivalent to R22 and a GWP of 10 or less is developed for multi-air conditioners for buildings, it will require considerable ingenuity and effort to use it.

Furthermore, it is unclear whether it can be used, as it would make it difficult to use existing pipework.

If a suitable replacement is not developed, it might be that an existing refrigerant such as R454B (GWP 466), which has a lower GWP than the currently used R410A, is used in the interim. However, it can only be used for a short period of time from a GWP perspective, and its performance will be lower than R32, so energy-saving technology will be required to restore the performance, which may lead to higher costs.

A1.3.4.2 Candidate Final Refrigerant: Natural Refrigerant

If the situation arises where only natural refrigerants are acceptable as final refrigerants, the only option may be A3 refrigerants such as R290, although there is also the possibility of CO_2 being chosen as an alternative refrigerant. However, in such a case, direct expansion type equipment would be difficult to apply with A3 refrigerant in terms of maximum refrigerant quantity, and there may be no other alternative but to substitute chillers with A3 refrigerant. In that case, the following issues are to be addressed

1. In Japan, there is a culture of disliking chillers that require water piping due to the risk of water leaks (similar to commercial packaged air conditioners)
2. Inter expansion type equipment like chillers performs worse than direct expansion type equipment such as commercial packaged air conditioners
3. Poor workability in Japan, as steel pipes and threaded joints are used for water piping. In order to make it an easy-to-install alternative to multi-air conditioners for buildings, plastic pipes (e.g. polybutene) and one-touch couplings need to be made available for water piping, as it is in Europe
4. The whole process needs to be systematized so that equipment design and construction can be carried out easily and inexpensively
5. Under the High Pressure Gas Safety Law, notification is required for A3 refrigerant (flammable gas) with a refrigeration capacity of 3 tonnes or more but fewer than 20 tonnes, and this needs to be relaxed in order to promote the widespread use of equipment

A1.3.5 Refrigerant Conversion Scenarios for Built-In Refrigeration and Freezing Equipment

The refrigerants currently used in built-in refrigeration and freezing equipment such as built-in showcases are R134a, R404A, R410A, etc. The target country for sales of equipment by domestic manufacturers is Japan. An image of future scenarios for low GWP refrigerant conversion is shown in Figure A1.10.

A1.3.5.1 Transitional Refrigerant Candidate: HFO Mixed Refrigerant

For large built-in refrigeration and freezing equipment[16], the refrigerant circuit needs to be divided into several parts to meet the maximum amount of A3 refrigerant that can be filled as specified in the standard. The potential for cost issues necessitates the use of A2L refrigerant.

HFO-1234yf has poor performance in the refrigeration temperature range, so HFO mixed refrigerant with a GWP of 10 or less is needed, which has good performance in the refrigeration temperature range. HFO mixed refrigerant could be the final refrigerant, as a refrigerant with a pressure equivalent to HFO-1234yf is sufficient. Before the final refrigerant is agreed upon, A2L refrigerants with GWP below 150, such as R454C, could be adopted if the refrigerant has to be changed to meet the 2029 and 2030 reduction targets.

A1.3.5.2 Final Refrigerant Candidate: Natural Refrigerants

If a situation arises where only natural refrigerants are acceptable, some CO_2 may remain as alternative refrigerant, but most likely A3 refrigerants due to poor performance with CO_2. Certain customers who prefer natural refrigerants are also changing their adoption policy from CO_2 to A3 refrigerants due to performance considerations. Many built-in showcases using A3 refrigerant from foreign manufacturers are now being imported.

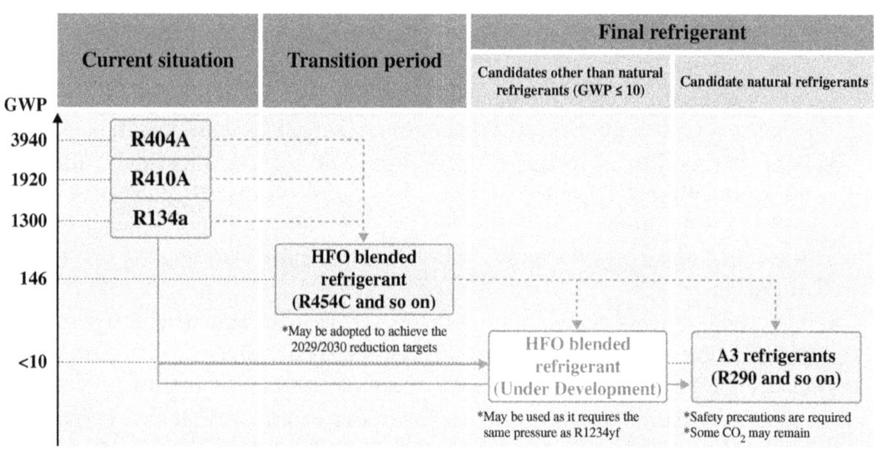

FIGURE A1.10 Scenario for conversion to low GWP refrigerants for built-in refrigeration and freezing equipment.

When A3 refrigerant is used, there are no major problems when the equipment is operating, provided that the necessary safety measures are taken based on the results of the risk assessment. However, during transportation, storage, installation, repair, and removal, the High Pressure Gas Safety Law and the Japan Refrigeration and Air Conditioning Industry Association standard (JRAIA) must be followed. It is also necessary to inform industrial waste contractors about disposal methods after equipment has been removed.

In addition, when repairs are carried out on site, it is necessary to repair the refrigerant circuit in the shop; for example, when the showcase cannot be pulled out due to interference with a hanging wall. In such cases, safety guarantees and good knowledge of work methods are required. And from a safety perspective, it is preferable to integrate (cassette) refrigerant circuits, including evaporators, so that the refrigerant circuits can be pulled out together.

A1.3.6 Condensing Units: Refrigerant Conversion Scenarios for Refrigeration and Freezing Units

The target country for sales of refrigeration and freezing equipment by domestic manufacturers is Japan and the application is refrigeration refrigerators (e.g. prefabricated refrigerators). The refrigerants currently used in refrigeration and freezing units are R404A or R410A. Figure A1.11 shows an image of future scenarios for low GWP refrigerant conversion for this type of appliance.

A1.3.6.1 Transitional Refrigerant Candidate: HFO Mixed Refrigerant

An HFO mixed refrigerant with a GWP of 10 or less is currently not in sight as an alternative to R404A or R410A. HFO-1234yf has a poor performance in the refrigeration temperature range, so an HFO mixed refrigerant with a GWP of 10 or less

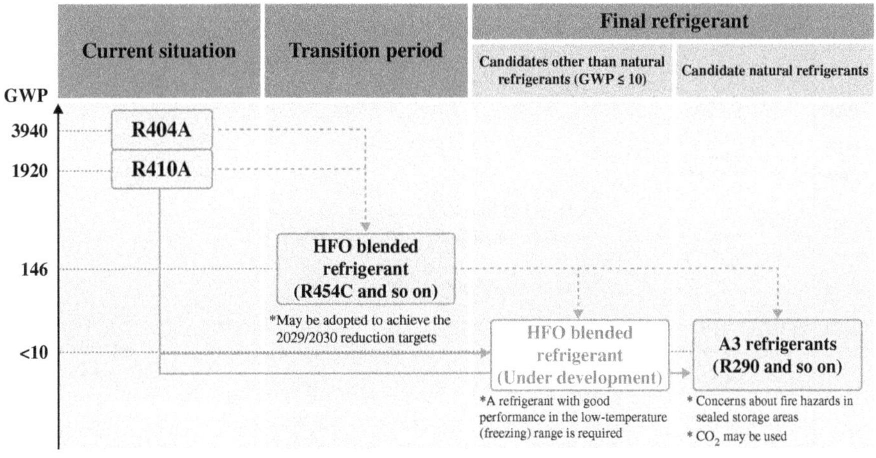

FIGURE A1.11 Scenario for conversion to low GWP refrigerants for refrigeration and freezing units.

is needed for good performance in the refrigeration temperature range. Before the final refrigerant is agreed, A2L refrigerants with a GWP below 150, such as R454C, would be considered acceptable, if the refrigerant needs to be changed to achieve the 2029 and 2030 reduction targets.

A1.3.6.2 Final Refrigerant Candidate: Natural Refrigerants

If only natural refrigerants are allowed, A3 refrigerant with good performance will be adopted in many cases. For A3 refrigerant, safety assurance in accordance with the High Pressure Gas Safety Act and JRAIA standards will be required during operation, and industrial waste contractors will need to be informed about the disposal method of the equipment.

As the unit is installed in a closed cabinet, if the A3 refrigerant were to accidentally ignite, it would be much more hazardous than in a showcase unit. Refrigeration and freezing units are relatively easy to convert to A3 refrigerant, as they are mainly integrated types and the refrigerant circuit is not opened on site except for repair work. Nevertheless, CO_2 may still be adopted due to concerns about the degree of ignition-related hazard in sealed units.

A1.3.7 Refrigerant Conversion Scenarios for Medium-Sized Condensing Units

The refrigerants currently used in medium-sized condensing units are R404A, R410A, and A1 with GWP below 1500 (R404A and R410A alternatives) and CO_2, with equipment sold by domestic manufacturers in Japan and with applications in shops (separate type showcases), refrigerated warehouses, etc. Figure A1.12 shows an image of future scenarios for low GWP refrigerant conversion.

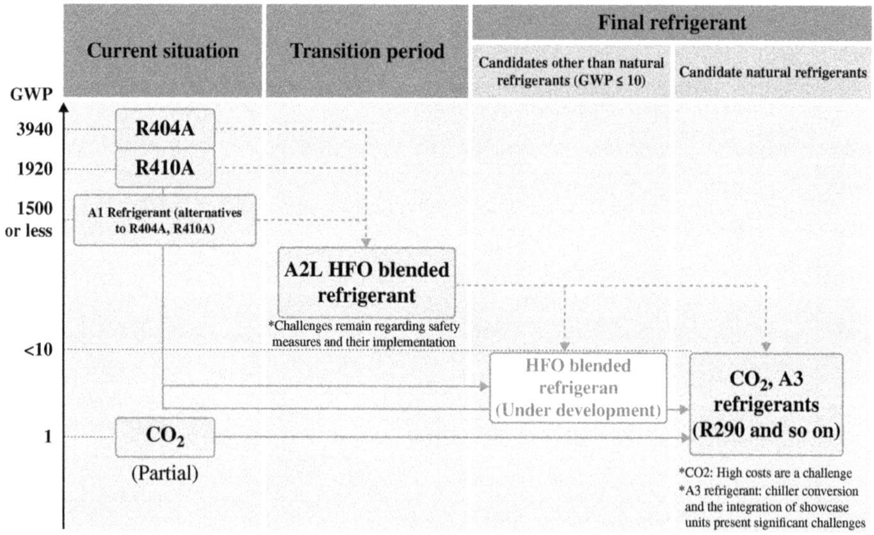

FIGURE A1.12 Scenario for medium-sized condensing unit conversion to low GWP refrigerants.

A1.3.7.1 Transitional Refrigerant Candidate: HFO Mixed Refrigerant

HFO mixed refrigerant (A2L) as an alternative to R404A and R410A with a GWP of 10 or less is not in sight at present. In addition, when A2L refrigerant is used, it is highly likely that equipment will have refrigerant leak detectors and shut-off valves as a safety measure, as is the case with multi-air conditioners for buildings. On top of this, it is necessary to conduct promotional activities among equipment design and installation companies to ensure safety when A2L refrigerant is used. In low temperature equipment, design and installation are often carried out by small, local companies, and these activities are not easy.

A1.3.7.2 Final Refrigerant Candidate: Natural Refrigerants

If the situation arises where only natural refrigerants are acceptable, the following options are possible

- Use CO_2
- Use a brine chiller (brine cooler) with A3 refrigerant
- Use a built-in showcase instead of a separate type appliance

If CO_2 is used as the final refrigerant, different refrigerants will be used for air conditioners and condensing units, which means that parts cannot be standardized and the piping and parts will be wider, which will inevitably lead to higher costs.

In addition, the performance will be worse than with HFC refrigerant, and it will cost even more to achieve a performance comparable to that of HFC refrigerant. CO_2 refrigerant is covered by the Ministry of the Environment's introduction support program[17] as of 2022, but may no longer be economically viable if subsidies are no longer offered.

The conversion to brine chillers with A3 refrigerant is plagued by deteriorating performance as well as the same issues as those described for the conversion to chillers in multi-air conditioners for buildings. In addition, the use of flammable refrigerants (A2L, A2, and A3) in separate type showcases, which are the direct expansion type equipment found in shops, is a hurdle in terms of informing relevant personnel about safety guarantee methods.

Therefore, there is a possibility that some separate type showcases could be replaced by built-in showcases using A3 refrigerants like R290. If some of the separate type showcases are replaced by built-in showcases in a shop, the air conditioning equipment has a higher COP, which may conversely lead to energy saving. If all separate type showcases are replaced by built-in showcases, the amount of heat generated in the shop may increase too much and the power system performance may deteriorate due to the increased power of the air conditioning equipment.

A1.3.8 Refrigerant Conversion Scenarios for Large Condensing Units

The refrigerants currently used in large condensing units include R404A, R134a, and R407C, with Japan being the target country for equipment sales by domestic manufacturers. An image of future scenarios for conversion to low GWP refrigerants is shown in Figure A1.13.

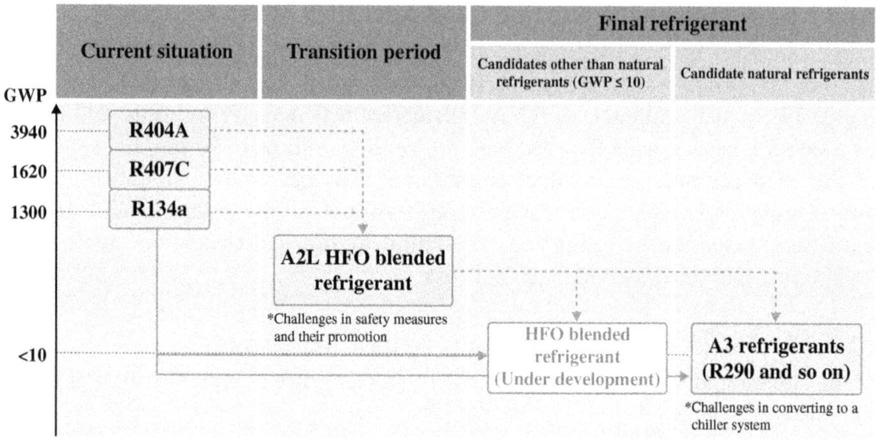

FIGURE A1.13 Scenario for conversion to low GWP refrigerants for large-sized condensing units.

A1.3.8.1 Transitional Refrigerant Candidate: HFO Mixed Refrigerant

Due to the pressure resistance aspects of screw compressors, it is difficult to use CO_2 refrigerant.

HFO-1234yf has poor performance in the refrigeration temperature range, so HFO mixed refrigerant with a GWP of less than 10 with good performance in the refrigeration temperature range is needed to cover the requisite temperature range. The use of A2L refrigerant in direct expansion type units presents the same challenges as in medium-sized condensing units.

A1.3.8.2 Final Refrigerant Candidate: Natural Refrigerants

If the situation arises where only natural refrigerants are acceptable, the only option may be brine chiller with A3 refrigerant.

A1.3.9 Scroll Chiller Refrigerant Conversion Scenario

The refrigerant currently used in scroll chillers[18] is R410A and the parts of the world where the equipment is sold by domestic manufacturers include Japan and Europe. Figure A1.14 shows an image of future low GWP refrigerant conversion scenarios.

A1.3.9.1 Transitional Refrigerant Candidate: HFO Mixed Refrigerant

HFO mixed refrigerant to replace R410A with a GWP of 10 or less is not in sight at present. As using different refrigerants for air conditioning and chillers would increase costs due to the lack of component commonality, the refrigerant will be converted to R32 refrigerant for a time, but there is a need for a lower GWP refrigerant conversion.

In Japan, steel pipes and threaded joints are used for water piping, but to expand the area of use of chillers, plastic (e.g. polybutene) and one-touch couplings need to be made available, as they are in Europe.

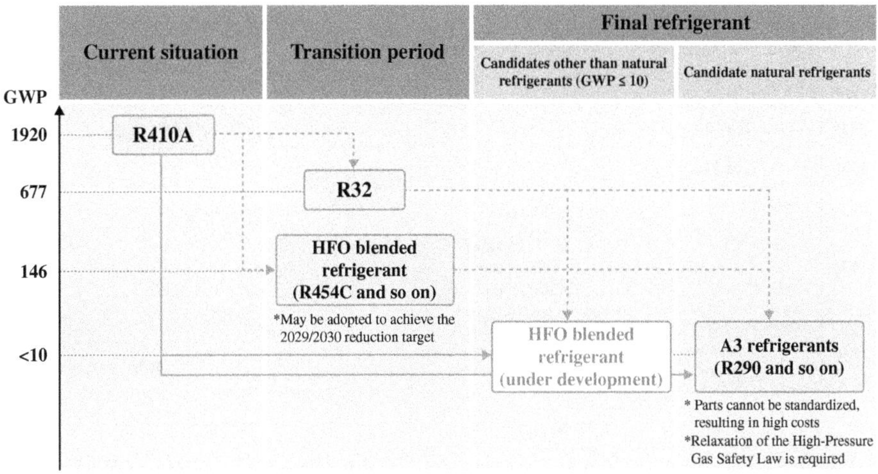

FIGURE A1.14 Scenario for conversion to low GWP refrigerants for a scroll chiller.

A1.3.9.2 Final Refrigerant Candidate: Natural Refrigerants

If a situation arises where only natural refrigerants are allowed, A3 refrigerants such as R290 may be the only option. In Japan, the High Pressure Gas Safety Act requires notification for flammable gases with a refrigeration capacity of 3 tonnes or more but fewer than 20 tonnes, and A3 refrigerants would be difficult to convert to A3 refrigerants unless this is relaxed.

A1.3.10 Refrigerant Conversion Scenarios for Screw Chillers

The refrigerants currently used in screw chillers (large chillers using screw compressors) are R404A and R134a, and the parts of the world where equipment is sold by domestic manufacturers include Japan and Europe. An image of future scenarios for low GWP refrigerant conversion is shown in Figure A1.15.

A1.3.10.1 Transitional Refrigerant Candidate: HFO Mixed Refrigerant

An HFO mixed refrigerant to replace R404A with a GWP of 10 or less is not in sight at present. If the refrigerant has to be changed before the final refrigerant in order to achieve the 2029 and 2030 reduction targets, an A2L refrigerant with a GWP of 150 or less, such as R454C, could be adopted.

A1.3.10.2 Final Refrigerant Candidate: Natural Refrigerants

If a situation arises where only natural refrigerants are acceptable, A3 refrigerants may be the only option.

A1.3.11 Refrigerant Conversion Scenarios for Centrifugal Chillers

The refrigerants in centrifugal chillers[19] are broadly classified into refrigerants above 1 MPa[20] and below 1 MPa, which are regulated by the High Pressure Gas Safety Act.

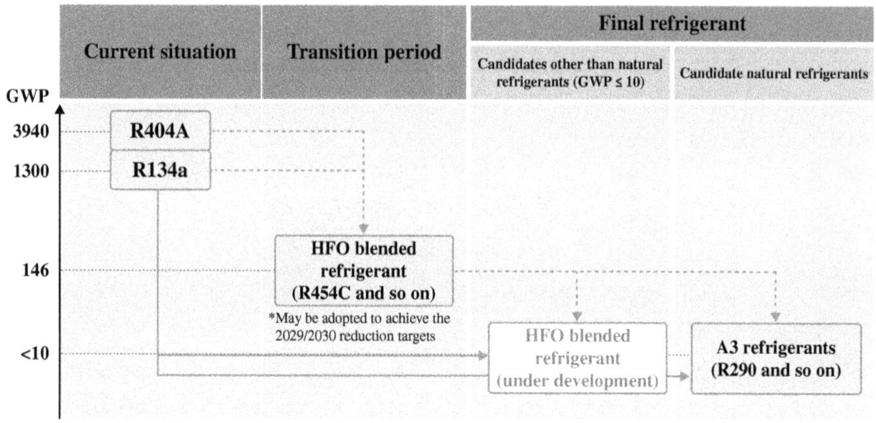

FIGURE A1.15 Scenario for conversion to low GWP refrigerants for a screw chiller.

Low pressure refrigerants are generally considered suitable for high capacity, low temperature applications, whereas low pressure refrigerants are for small capacity, high temperature applications [10].

An image of future scenarios for low GWP refrigerant conversion is shown in Figure A1.16.

Conventionally, R134a and R513A were used as refrigerants above 1 MPa, while R123 and R245fa were used as refrigerants below 1 MPa. In order to promote low

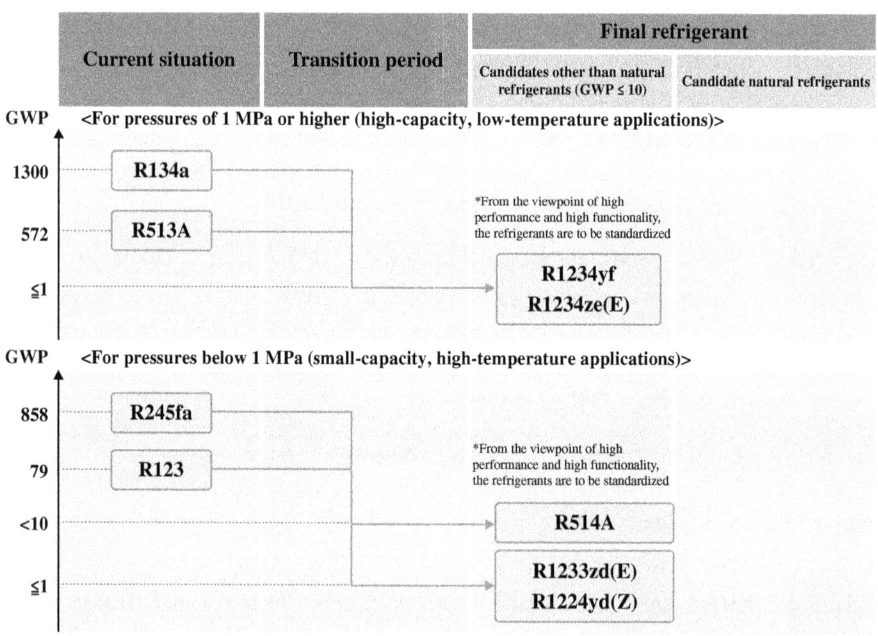

FIGURE A1.16 Scenario for conversion to low GWP refrigerants for a centrifugal chiller.

GWP, various refrigerants are beginning to be adopted as next-generation refrigerants, as follows

- Refrigerants above 1 MPa: R1234yf, R1234ze(E), etc.
- Refrigerants below 1 MPa: R514A, R1233zd(E), R1224yd(Z), etc.

The GWP of these next-generation refrigerants, which are highly energy-efficient for centrifugal chillers, is below 10. From the perspective of aiming for higher performance and functionality, the refrigerants used are being standardized.

Refrigerants below 1 MPa, which are specific to centrifugal chillers, have a high theoretical efficiency, become liquid at atmospheric pressure depending on the temperature, and have a low risk of leakage. It is expected that the transition will continue to be centered on refrigerants below 1 MPa, while very large machines, for which the use of refrigerants below 1 MPa is difficult, will be consolidated among refrigerants above 1 MPa.

A1.3.12 Refrigerant Conversion Scenarios for Vehicle Air Conditioners

The refrigerant currently used in vehicle air conditioners is R134a, and the target markets for sales of equipment by domestic manufacturers are worldwide: Japan, Europe, North America, and China. An image of future scenarios for low GWP refrigerant conversion is shown in Figure A1.17.

The European car air conditioner directive (Directive 2006/40/EC) is leading the way in the conversion to R1234yf as an alternative refrigerant to R134a. Japan and North America are also gradually converting.

R1234yf has a low GWP of 4. There is no need to further reduce low GWP, but vehicle electrification[21] needs to be addressed and air conditioning systems are being electrified. Electric vehicles (EVs), especially battery EVs (BEVs), do not have a heat source equivalent to the combustion engine of a conventional vehicle and were initially heated by auxiliary heating equipment using PTC heaters (electric heaters). In this case, however, the power consumption due to heating is higher, especially in winter, and it is important to reduce air conditioning power consumption in order

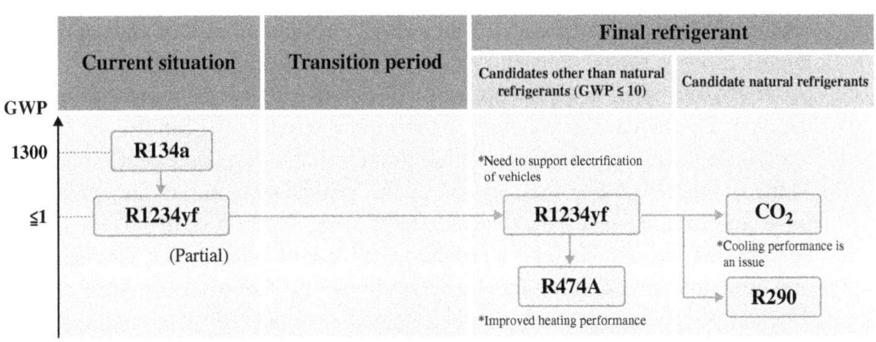

FIGURE A1.17 Scenario for conversion to low GWP refrigerants for vehicle air conditioning.

to extend the cruising range [11]. For this reason, heat pumps in combination with auxiliary heating equipment or full-scale heat pumps without auxiliary heating have been promoted.

In this case, R1234yf is not suitable for heat pump operation at low ambient temperatures due to its thermodynamic properties. Some consider that systems that emphasize heating performance and use CO_2 (R744) as refrigerant are more desirable [12]. CO_2 has poor cooling performance compared to R1234yf in high outdoor temperature conditions[22] and the market adoption of such systems is limited at present.

In recent years, there has been an attempt to extend the range of vehicles while ensuring air conditioning performance in the entire range from low to high ambient temperatures. To this end, research and development of new, low GWP mixed refrigerants, including R474A and systems compatible with them [13], as well as development is underway on systems that use the flammable refrigerant R290, which could not be used under conventional legal standards [14].

A1.3.13 Refrigerant Conversion Scenarios for Transport Refrigeration and Freezing Units

R404A is the main refrigerant used in transport refrigeration and freezing units installed in trucks for constant-temperature transportation to cool (and in some cases heat) the cargo interior (storage), while R134a is used in some medium-temperature, refrigerated-only products. R134a is mainly sold by domestic manufacturers in Japan, while R404A is also available for overseas markets.

Matters to consider for potential refrigerants include the following

(1) Suitability for refrigeration cycles with high temperature differences and high pressure ratios in order to achieve heat transfer between the low temperature inside the sub-zero chamber and directly above the asphalt road surface, which is hot due to the retention of waste heat
(2) Difficulty for ventilation systems to avoid the explosion range of flammable fluids due to temperature control in airtight boxes. Non-flammable refrigerants are preferred.
(3) Polymer materials are extensively used to absorb displacement and provide damping, as they are subjected to vibrations from the road surface and engine. Highly reactive refrigerants and those with low molecular weight cause rubber hoses and other components made of polymeric materials to swell and degrade. Therefore, it is necessary to carry out various verifications, such as material resistance evaluations, to determine suitable candidate refrigerants.
(4) Many compressors use the rotation of the vehicle's engine as the driving force and have a shaft seal section to expose the rotating shaft outside the compressor casing. Under the influence of engine speed, the discharge pressure and temperature increase rapidly while the inlet pressure and temperature decrease rapidly, depending on the vehicle running conditions. To prevent atmospheric suction from the seal section, the refrigerant pressure generally needs to be operated at atmospheric pressure or higher.

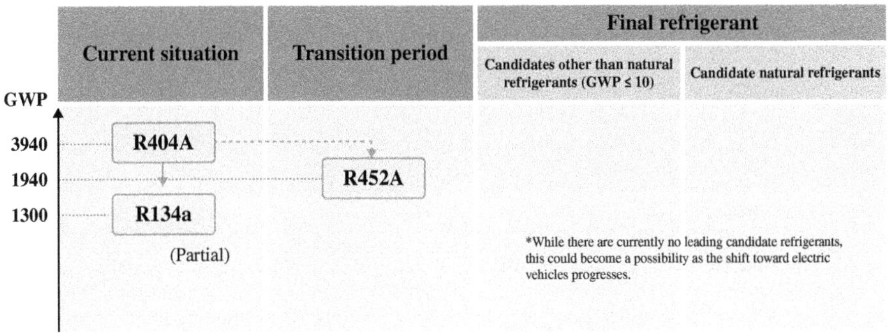

FIGURE A1.18 Scenario for conversion of transportation refrigeration units to low GWP refrigerants.

Although there are currently no promising candidate refrigerants with low GWP and non-flammability, the following national and international trends regarding refrigerant conversion are taking place.

According to the United Nations Environment Programme's (UNEP) Assessment Report of the Refrigeration, Air Conditioning and Heat Pumps Technical Options Committee(RTOC) [15], products using R452A (GWP 2140) were launched on the European market by two major companies until 2017, and in Japan, products using R452A (GWP 2140) was launched in 2019.

However, when further attempts are made to adopt low GWP refrigerants, there is skepticism about their development and uptake due to the obstacles posed by issues in points (1)–(4) above. It is therefore unlikely to be treated in the same way as other models in regulations and promotion measures for low GWP. It should be noted that if the switch to electric drive vehicles progresses in the future, issues (3) and (4) above will become less of a challenge and the development of new low GWP refrigerant products may be accelerated.

Figure A1.18 shows an image of a future scenario for low GWP refrigerant conversion.

A1.3.14 Refrigerant Conversion Scenario for High Temperature Heat Pumps

High temperature heat pumps are broadly classified into two temperature ranges: for water heating and heating where the operating temperature range is below 100°C, and for heating above 100°C.

For the former, for hot water supply and heating, heat pump water heaters using CO_2 refrigerant are the mainstream, but some equipment using HFC refrigerants such as R134a and R245fa, which were manufactured before the Kigali Amendment to the Montreal Protocol, are still in use.

In recent years, equipment with an operating temperature range of 100°C or higher has also been developed. In Japan, HFO refrigerants such as R1234ze(E), R1233zd(E), and R1336mzz(E) are often used, but the market is still in the

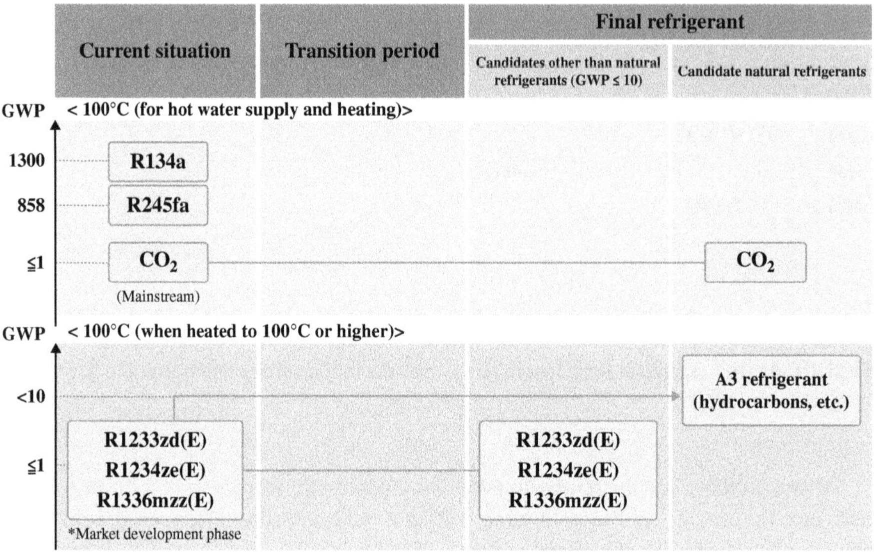

FIGURE A1.19 Scenario for conversion to low GWP refrigerants for high temperature heat pumps.

development stage. The main target country for equipment sales by domestic manufacturers is Japan.

Therefore, in both temperature ranges below and above 100°C, refrigerants with a GWP of less than 10 are already the mainstream, and these low GWP refrigerants are likely to be the main refrigerant of choice when heat pump equipment becomes more widely used in high temperature applications in the future.

Figure A1.19 shows an image of future scenarios for low GWP refrigerant conversion.

A1.4 SCENARIO FOR ENHANCED REFRIGERANT RECOVERY AND REGENERATION

As described in Section A1.3, the METI and the Ministry of the Environment have developed a scenario to promote low GWP in order to comply with the Kigali Amendment and to achieve carbon neutrality.

However, it is practically difficult to achieve the GHG emission reduction targets in the Paris Agreement and, beyond that, to achieve carbon neutrality by simply using low GWP refrigerants in equipment. It is important to achieve this through a comprehensive approach that includes reducing the amount of refrigerant used, reducing refrigerant leakage, and improving refrigerant recovery and regeneration rate (Figure A1.20).

This section describes the current status of refrigerant recovery and regeneration/destruction, its impact on HFC reduction, and future possibilities.

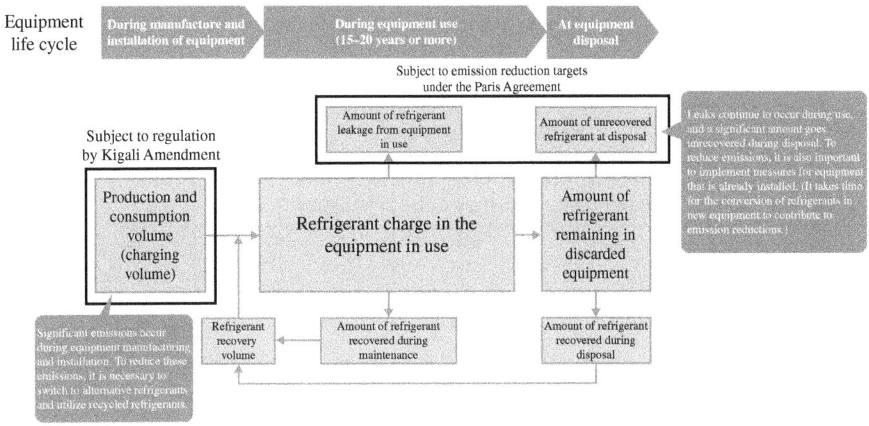

FIGURE A1.20 The need for measures according to the stage of the material flow of refrigerants in heat pumps.

A1.4.1 Current Status and Challenges of Refrigerant Recovery and Recycling

Manufacturers and importers of refrigerants are obliged by the Ozone Layer Protection Act to reduce refrigerant supply volume in line with the Kigali Amendment, which limits the supply of fluorocarbons. Meanwhile, the management of refrigerants is regulated by three laws: the Home Appliance Recycling Law, the Act on Rational Use and Proper Management of Fluorocarbons (Fluorocarbon Emissions Control Act) and the Law for the Recycling of End-of-Life Automobiles (Automobile Recycling Law). This section discusses The Home Appliance Recycling Law and the Fluorocarbons Emissions and Proper Management Act.

A1.4.1.1 Refrigerant Recovery Rates

Table A1.3 consolidates the data on the recovery and recycling/destruction of refrigerants published by the two aforementioned laws in 2020. The data shows that a total of 7,939 tonnes of fluorocarbons had been recovered. Household equipment totaled 2,705 tonnes in 2020 and commercial heat pump equipment totaled 5,234 tonnes. The recovered fluorocarbons include a mixture of CFCs, HCFCs, and HFCs, but the amount of HFC refrigerants recovered will also increase in the future, as more HFCs will be disposed of.

Recovery rates have remained stagnant at around 30% for domestic appliances and 41% for commercial appliances, which is less than half of the total.

A1.4.1.2 Recovery Rates for Residential Air Conditioners

The recycling rate of residential air conditioners for domestic use after recovery of refrigerant is as high as 87% (Table A1.3).

TABLE A1.3
Amount of Refrigerant Recovered and Recycled (Actual Tonnes)

2020	Household Appliances			Commercial Equipment		
	Air conditioning	Refrigerators	Washing machines	CFCs	HCFC	HFC
Amount recovered (tonnes)	2,505	161	38	124	2,408	2,702
Recovery rate	Approx. 30			41%		
Amount recycled (tonnes)	2,224	91	30	43	1,101	1,060
Renewable rate	87%			35%	46%	39%

Note: "Recycled amount" is the total of the amount handed over to Type 1 fluorocarbons recyclers or those recognized by prefectural governors as those who reliably hand over to fluorocarbons destroyers and the amount recycled by the fluorocarbon collectors themselves. R410A and R22 are the main recovered refrigerants in residential air conditioning, but their ratios are not disclosed.

Source: Based on the METI, "Results of the total amount of fluorocarbons filled and recovered from commercial refrigeration and air conditioning equipment/Based on the fluorocarbons Emissions Control Law in FY2020" [16]; the Ministry of the Environment, "Home appliance recycling results in FY2020" [17].

This is probably due to the fact that recycling of resources is a prerequisite for home appliances under the Home Appliance Recycling Law, and that refrigerant types are fewer and easier to recycle than those for commercial use. However, the equipment recovery rate for discarded residential air conditioning is low at 37.6% (Figure A1.21), so the overall refrigerant regeneration rate is estimated to be around 30–35%. In order to increase the amount recycled in the future, it will be important to improve the recovery rate of the appliances themselves under the Home Appliance Recycling Law.

A1.4.1.3 Recovery Rate of Commercial Heat Pump Equipment

The refrigerant recovery rate for commercial heat pump equipment is 41%, which is higher than for residential air conditioning (Figure A1.22). The refrigerant recovery rate for recovered refrigerants is low at 42% in total for CFCs and only 39% for HFCs (Table A1.3).

This may be due to the fact that the Act on Ensuring the Recovery and Destruction of Fluorocarbons from Specified Products (Fluorocarbons Recovery and Destruction Law), which was the predecessor to the Fluorocarbons Emissions Control Law, was based on the assumption that refrigerants would be destroyed, and this may have influenced the continuation of conventional destruction processes after the implementation of the Fluorocarbons Emissions Control Law.

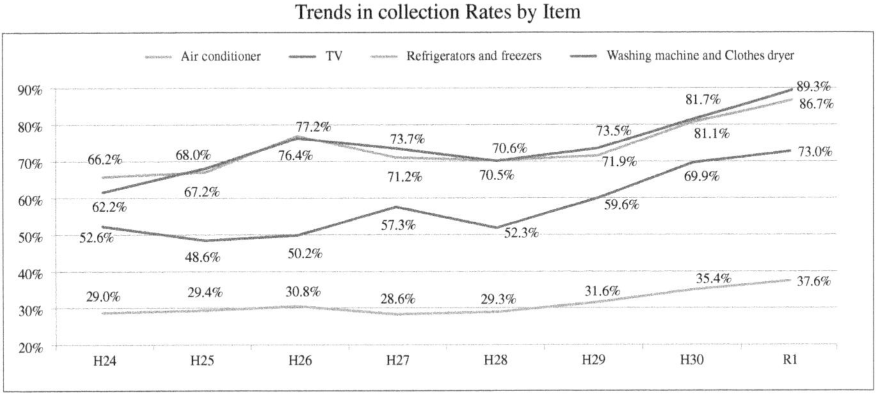

FIGURE A1.21 Recovery rate of equipment for residential air conditioners for home use. (From METI, "Report on the Evaluation and Examination of the Implementation of the Home Appliance Recycling Scheme [18]).

In addition, unlike residential air conditioners, commercial heat pump equipment uses a large number of refrigerant types, so that multiple refrigerants are mixed in cylinders and recovery machines. Furthermore, it cannot be ignored that the difficulty in identifying feasible handover routes may be a factor in the lack of progress in reclamation.

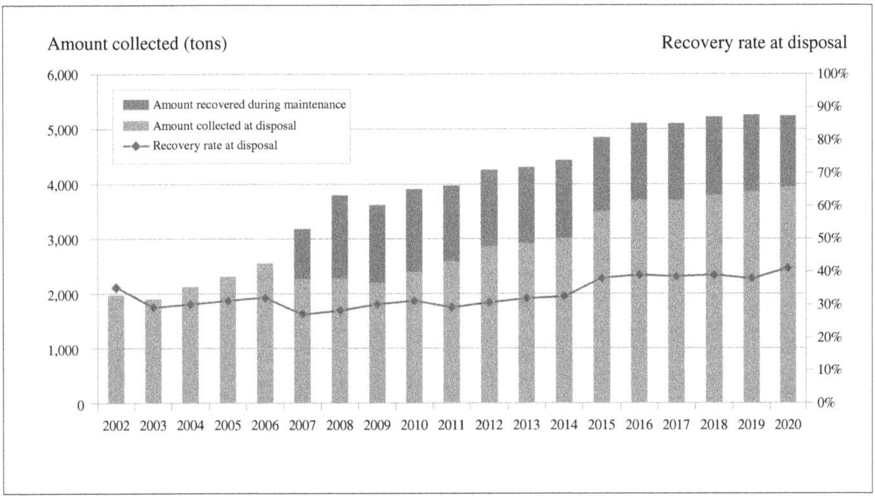

FIGURE A1.22 Changes in the recovery of fluorocarbons and other materials based on the Fluorocarbons Emissions Control Law. *Note:* "Recovery rate at the time of disposal" is broken down into the number of units delivered × refrigerant recovery rate per unit of equipment. (Based on the METI, "Results of the total amount of fluorocarbons filled and recovered from commercial refrigeration and air conditioning equipment/Based on the Fluorocarbons Emissions Control Law in FY2020" [19].)

Since the refrigerant recovery rate is 41% and the renewable rate is 42%, the overall regeneration rate is estimated to be around 17%.

A1.4.2 Challenges in Improving Refrigerant Recovery Rates

As mentioned above, refrigerant recovery and subsequent treatment are regulated by the Home Appliance Recycling Law and Fluorocarbons Emissions Control Law, but this is still insufficient and further improvements are required.

Efforts to improve the recovery rate of commercial refrigeration and air conditioning equipment have been reported in a detailed study by a joint council of the METI and the Ministry of the Environment. In this report, the factors that lead to unrecovered fluorocarbons at the time of disposal are classified (Figure A1.23), and it is reported that the causes can be classified into the following three main factors

- Factor A: Uncollected quantity due to the fact that collection operations were not carried out
- Factor B: Uncollected quantities that are considered to be uncollected because collection operations were not carried out or, if they were carried out, were left uncollected (combined factor of Factors A and C)
- Factor C: Uncollected quantity due to the fact that collection operations were carried out, but the quantity remained uncollected

The results of the survey on the percentage of factors A to C for each piece of equipment (Table A1.4) show that the failure to carry out recovery operations accounts for a large percentage of uncollected items. The factors that led to uncollected equipment can be summarized in the following four points, in order of volume size (Table A1.5).

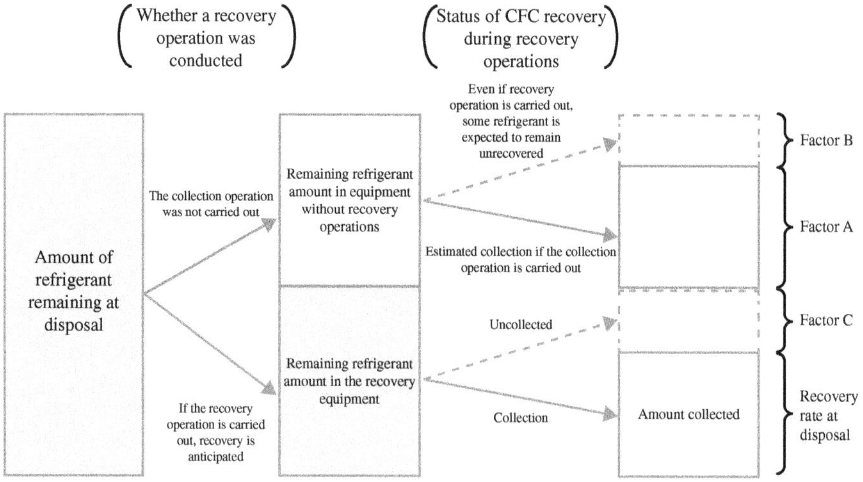

FIGURE A1.23 Factors leading to non-recovery. (From Ministry of the Environment, "Direction of measures to improve the recovery rate of fluorocarbons at disposal" [20]).

TABLE A1.4
Breakdown of Uncollected Fluorocarbons at the Time of Equipment Disposal by Model, Size, and Factor

Model		Reference) Breakdown of Remaining Refrigerants at Disposal*	Factor Analysis		
			Factor A	Factor B	Factor C
Air conditioners/ refrigeration & freezing equipment	Large centrifugal	4%	0%	0%	2–3%
	Large screw	0.3%	0%	0%	0%
Air conditioners/ refrigeration & freezing equipment	Medium	41%	6–7%	9–10%	14–16%
	Small size	23%	10%	0%	0%
Refrigeration & freezing equipment	Medium	28%	12–19%	3–10%	1–3%
	Small size	3%	1%	0%	0%
Total		100%	29–37%	13–20%	17–22%

* Percentage of total refrigerant remaining at disposal as 100%.
Estimated value of total refrigerant remaining at disposal in 2017: 9,890t.
Source: Based on the Ministry of the Environment, "Direction of measures to improve the recovery rate of fluorocarbons at disposal" [20, 21].

TABLE A1.5
Factors Leading to Uncollected Fluorocarbons at the Time of Equipment Disposal (in Order of Size of Uncollected Quantities)

Order	Factors	Supplementary Information
(1)	Medium-sized refrigeration and freezing equipment not collected	e.g. during shop refurbishment or demolition
(2)	Collection is carried out on medium-sized air conditioning such as multi-air conditioners for buildings, but the collection rate did not increase and the remaining parts were left uncollected	Combined factors with factors A and C
(3)	Small air conditioning equipment is not collected	e.g. during shop refurbishment or demolition
(4)	Medium-sized air conditioners, such as multi-air conditioners for buildings, which have not been implemented but will remain to be collected even if they are implemented due to equipment failure, etc.	Technical issues, man-hour factors, etc.

In this context, there are cases where equipment is disposed of without recovery operations being carried out (e.g. at the time of building demolition or shop closure), where the operator (e.g. equipment supplier) who carries out fluorocarbon recovery operations on a daily basis does not intervene. This has been pointed out as a disincentive to recovery due to the existence of multiple subcontracting structures in construction.

In view of this situation, the Fluorocarbons Emissions Control Act was amended in April 2020 as a measure against (1) and (3) in particular, and a fine will be imposed if equipment is disposed of without recovering the fluorocarbons.

The other two reasons for nonrecovery, (2) and (4), are often found to be the case when refrigerant is almost completely drained from the equipment at the time of disposal, making it impossible to recover refrigerant, or when the power supply at the disposal site is turned off, thus making it impossible to recover refrigerant or pump down the equipment. It is therefore important to prevent refrigerant leakage during use and to ensure that all parties involved in the disposal are aware of what is required for the proper recovery of fluorocarbons.

Another factor preventing proper recovery is the fact that it is unclear who is responsible for the disposal of waste. Furthermore, in the recovery of refrigerant from medium-sized air conditioning equipment, operators may lack the skills to adequately recover refrigerant from the refrigerant circuit, and this skill needs to be improved. In addition, as medium-sized air conditioning unit has many valves, it is more difficult to recover refrigerant from than small air conditioning equipment. Product manufacturers need to develop products with refrigerant recovery in mind.

A1.5 MEASURES TO INCREASE REFRIGERANT REGENERATION RATES

A1.5.1 Refrigerant Regeneration for Emission Regulations

As the Ozone Layer Protection Act is a "tap regulation" (regulation of refrigerant supply) and the Fluorocarbons Emissions Control Act is an "emission regulation" (regulation of refrigerant emissions), it is important to distinguish between the two.

For emission regulations, methods to increase recovery rates are effective, but if the recovered refrigerant is destroyed and new refrigerant is produced for the required supply, the consumption of refrigerant resources will continue.

It is therefore important to increase the refrigerant regeneration rate in order to curb the new production of refrigerant in real terms (Figure A1.24).

As a measure to increase the refrigerant regeneration rate, it is necessary to first increase the recovery rate and then encourage people to choose regeneration.

To increase the recovery rate, it is necessary to increase the residual volume at the time of disposal to be recovered. To encourage people to choose regeneration, barriers need to be lifted and incentives need to be offered. Therefore, as measures to increase the refrigerant recovery rate, it is also necessary to computerize refrigerant management and provide incentives for recovery and regeneration, which are effective in both increasing the recovery rate and the regeneration selection rate.

In addition, the use of recycled refrigerants is highly effective during the transitional period of refrigerant conversion, as described in Section A1.3. Even if refrigerants for new equipment are converted to low GWP refrigerants, high GWP

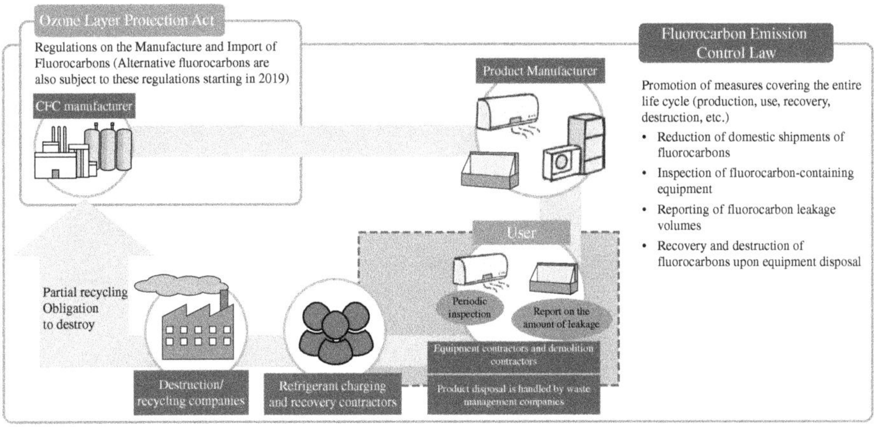

FIGURE A1.24 Relationship between the Ozone Layer Protection Act and the Fluorocarbons Emissions Control Law. (From Ministry of the Environment, "Briefing on the Revised Fluorocarbon Emission Control Act in FY2021: Trends Surrounding Fluorocarbons (Common)," [22]).

refrigerants will still be needed for service refueling of equipment still in use (service use) for the life of the product, which is 15–25 years.

As service refrigerants have a high GWP and account for a large proportion of the overall production and consumption quota, production and consumption can be reduced by ensuring refrigerant recovery and regeneration from existing, high GWP refrigerant equipment. This section describes the results of quantitative estimates of the effects of this approach.

A1.5.2 Increase in Residual Volume at Disposal

A1.5.2.1 Prevention of Leakage during Use

Efforts to prevent leakage at use are necessary because the amount of refrigerant recovered and recycled cannot be increased if there is little refrigerant at the time of recovery in the first place.

The following details are now obligatory under the Fluorocarbons Emissions Control Act of 2013 (Figure A1.25).

- Maintenance and preservation of the installation and use environment by the Class 1 specified product management (owner)
- Carrying out simple and periodic inspections
- Prohibit in principle the filling of fluorocarbons until repairs are carried out in case of confirmed leakage, etc.
- Preparation and maintenance of inspection and maintenance records

Simple inspections must be carried out once every 3 months, and periodic inspections must be carried out at least once a year for products above a certain scale[23].

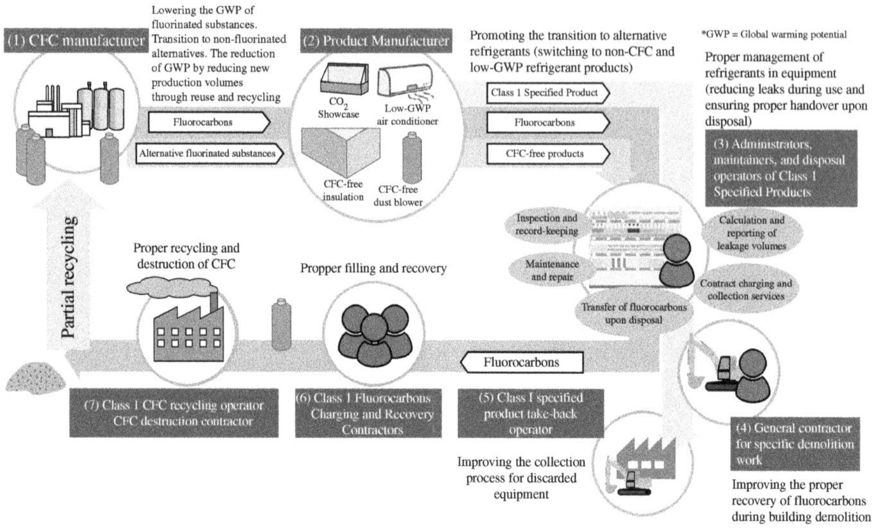

FIGURE A1.25 Overall picture of the fluorocarbons emissions and proper management act. (From Ministry of the Environment, "Overview of the Fluorocarbon Emissions Control Act" [23]).

In addition, regulations requiring reporting of calculated leakage quantities were enacted when leakage of a considerable amount or more occurs at a certain business operator. The product management of fluorocarbons is required to report the amount of calculated leakage of fluorocarbons to the minister in charge of the project in accordance with the ministerial ordinance, and the minister in charge of the project is to compile and publish this information. The difference between the amount recovered at the time of servicing and the amount subsequently filled is defined as the "amount of leakage."

A1.5.2.2 Causes of Refrigerant Leakage

The causes of refrigerant leakage are described in the "Guidelines for prevention of Refrigerant Leakage from fluorocarbons-based refrigeration and air conditioning equipment" by the Japan Refrigeration and Air Conditioning Industry Association (JRAIA) [24], and according to an analysis conducted on information on 200 leakage incidents from the High Pressure Gas Safety Institute of Japan from 2008 to 2011, more than 80% were caused by fatigue and corrosion [25].

The causes of fatigue are inadequate fixing of valves and insufficient attention to resonance and fatigue strength. Causes of corrosion include poor installation of pipe covers and cold insulation, inadequate and unsuitable material selection and design for harsh environments, and insufficient consideration of corrosion inhibitors. Leakage due to age-related deterioration is another cause, as 65% of all leaks occur more than 10 years after installation, which means that proper maintenance has not been carried out.

In light of this, JRAIA has established guidelines for the prevention of refrigerant leakage from refrigeration equipment, based on information on accidents from equipment manufacturers and the High Pressure Gas Safety Institute of Japan. The main points of the guidelines are the expansion of the scope of application to a wide range of parties involved (from design and manufacture to disposal), the implementation of periodic inspections, and the management of inspection records, which are similar to the ideas behind the current Fluorocarbons Emissions Control Act. In addition, the "Guidelines for the prevention of refrigerant leakage from fluorocarbons-based refrigeration and air conditioning equipment" [26] and others have developed and promoted equipment management technologies such as abnormal leakage detection and remote monitoring using Internet of Things (IoT), AI, and other digital transformations (DXs). These technologies are expected to simplify inspection work as determined by the Fluorocarbons Emissions Control Act.

A1.5.3 Eliminating Barriers to Choosing Regeneration

Although the current policy direction is to actively promote regeneration, until the 2013 amendments, it was mandatory to destroy refrigerants as a post-recovery process. Due in part to this, even today, the internal regulations of equipment managers who hand over refrigerants still assume destruction, which means that recovered refrigerants are often not sent for regeneration.

In the future, it will be necessary to make the public aware of the need to switch from destruction to regeneration as far as possible, and to design systems such as cylinder management to prevent the mixing of multiple refrigerant types.

A1.5.4 Digitalization of Refrigerant Management

The Fluorocarbons Emissions Control Act was amended in 2019, requiring the issuance and storage of fluorocarbons recovered certificates to waste and recycling companies, and the storage of documents confirming the presence or absence of equipment by building demolition companies, etc. [27]. Stricter refrigerant management of those used on the market is expected to have the effect of reducing unintentional, related emissions.

As shown in Figure A1.26, the confirmation document is digitalized and operated as a software as a service (SaaS) in the form of a refrigerant management system (RaMS) by the Japan Refrigerant and Environmental Protection Agency [28].

The use of such a digital management system not only frees the user from cumbersome document handling, but also prevents the occurrence of unclear refrigerant quantities that can occur during admin work, such as errors in the description. Furthermore, the system is considered to be effective for refrigerant management, as it enables the aggregation of calculated leakage quantities, etc.

A1.5.5 Providing Incentives for Recovery and Regeneration

In order to improve future refrigerant recovery and renewable rates, it is considered necessary to provide incentives for recovery. As refrigerants are a depleting resource (see Appendix 3), future shortages of refrigerants are essential.

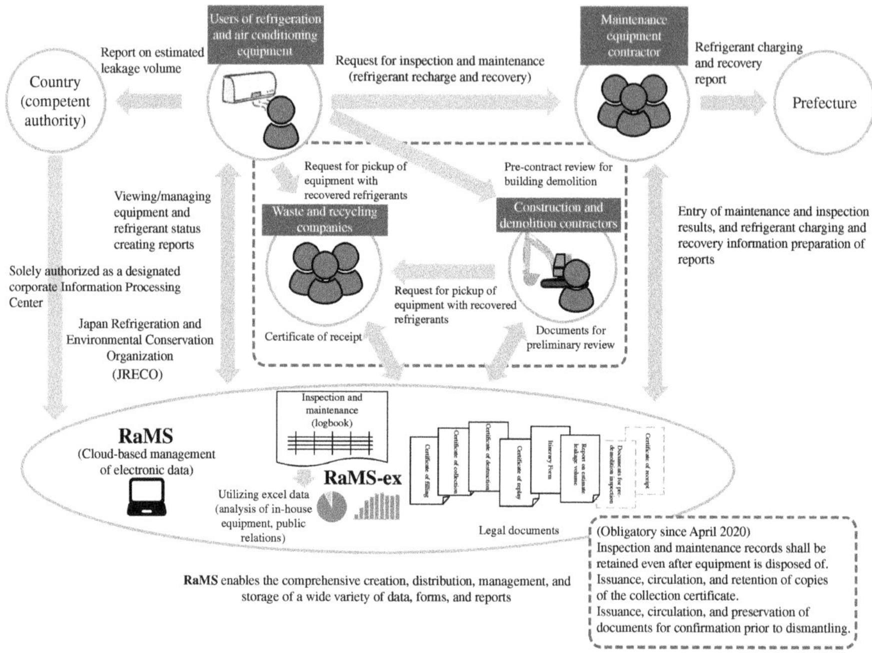

FIGURE A1.26 Overview of the RaMS. (From Based on Japan Refrigerant and Environmental Protection Organisation "What is RaMS (Refrigerant Management System)" [28]).

Considering that prices will eventually soar, a system should be put in place to ensure that all concerned are aware that it is a valuable resource and, if possible, that it is purchased for a fee.

The Kigali Amendment allows not only the use of recycled refrigerants separately from the production and consumption quota, but also the addition of the amount destroyed to the national production and consumption quota. However, the way in which this is to be operated is not clear and has not been addressed [29].

An easy-to-operate system design should be considered, and the use of the amount recycled and destroyed to achieve production and consumption targets should also be considered. However, this requires component analysis of refrigerants, which is expensive. To promote this, support schemes such as a buy-back scheme for recycled refrigerants and destroyed quantities that contribute to increasing national production and consumption quotas are also needed.

A1.6 PROSPECTS FOR REFRIGERANT CONVERSION AND EMISSION REDUCTIONS

A1.6.1 Low GWP Refrigerant Conversion and Limitations

As discussed in Section A1.3, according to the METI, in order to achieve carbon neutrality in 2050 through refrigerant conversion of newly installed equipment alone, the average GWP of newly installed equipment by 2036 needs to be below 10.

For this reason, the development of low GWP refrigerants and equipment using them is still ongoing.

In equipment where the amount of refrigerant is low and safety measures are easy to implement, and in equipment where HFO refrigerants with low GWP are being developed, it is expected that low GWP will be achieved through refrigerant conversion.

However, if a refrigerant that can safely replace the refrigerant currently used in each appliance without loss of performance has not been developed, such as in the case of air conditioning equipment with large amounts of refrigerant, it may not always be possible to convert to low GWP refrigerants while maintaining the form factor of each appliance.

For example, it may be necessary to convert the equipment configuration itself, such as converting a direct expansion type multi-air conditioner for buildings to an inter-expansion type chiller. When converting the equipment configuration, such as from a direct expansion type to a chiller, the power consumption is expected to be higher than that of conventional equipment due to the need for heat exchange from refrigerant to water, and the cost is also expected to be higher in order to improve performance. In addition, installation and operating conditions will also change, for example, water piping will be required and water leakage measures will be necessary, making it difficult to introduce the system across the board.

A1.6.2 Compliance with the Kigali Amendment and Promotion of Comprehensive Measures to Reduce Refrigerant Emissions

Instead of promoting only low GWP refrigerants as mentioned above, it is necessary to look at the entire life cycle of heat pumps.

It has been shown that by combining leakage prevention during use, thorough recovery at disposal, and promotion of recycling, GHG emission reductions equivalent to those achieved by promoting only low GWP refrigerants can be expected and compliance with the Kigali Amendment can also be achieved (see Section A1.8). Therefore, comprehensive measures such as the recycling of refrigerants will be promoted alongside low GWP through refrigerant conversion.

It should be noted that, in Japan, the refrigerant management standards for preventing refrigerant leakage are very strict, even by global standards, and the refrigerant recovery rate at the time of disposal is less than 40%. While this may not be sufficient to combat global warming, it is a prominent measure in other major industrialized countries, which do not have a system for reporting the amount recovered or calculating recovery rates [30]. It is also important for Japan to utilize this strength and spread its initiatives worldwide.

A1.7 ESTIMATION OF FUTURE BENEFITS FROM ENHANCED REFRIGERANT RECOVERY AND REGENERATION

In order to ascertain the effects of the widespread use of next-generation heat pumps that would solve the problems of refrigerant recovery and regeneration, estimates were made of the future effects of reduced energy usage (energy savings) and GHG emission reductions, including the effects of refrigerant emission control.

A1.7.1 Approach to Estimating the Effects of Emission and Energy Usage Controls

The assumptions for the estimation of the effects of emission and energy usage controls are shown in Table A1.6. In addition to the effects of improvements to the heat pump equipment itself and the expansion of the scope of application, the effects of overall optimization of refrigerant measures and energy savings were also estimated by comparing several scenarios.

The calculation flow in the estimation of effects of controls is shown in Figure A1.27.

A1.7.1.1 Calculation Method for Number of Units Shipped

The number of future shipments is estimated based on the number of past shipments of each heat pump model and other economic indicators. In doing so, the amount of refrigerant initially filled per unit for each model is set.

Once the equipment has been shipped, the number of units stocked each year[24] shall be calculated, assuming that disposal takes place according to the waste rate distribution curve[25] to be set for each model.

A1.7.1.2 Calculation of Refrigerant Amount

Based on the refrigerant leakage rate at the time of use set for each model, it is assumed that leakage will occur every year, and it is taken into account that refrigerant recovery and filling will be carried out due to regular servicing. Set the annual leakage rate, recovery rate from servicing, and filling rate from servicing

TABLE A1.6
Assumptions for Estimating the Effects of Emission and Energy Usage Controls

Item	Setting
Estimation period	2020–2040 (in 1-year increments)
Fields covered by the estimation	Air conditioning (residential, commercial, industrial) Refrigeration (residential, industrial)
Parameters covered by estimation	Amount of refrigerant leakage during use Amount of refrigerant recovered/unrecovered at disposal Energy usage and GHG emissions associated with usage (heat pumps and competing equipment*

* Competing appliances for heat pump equipment are assumed to be gas and oil-based heating for residential air conditioning and absorption chillers for commercial/industrial central air conditioning.

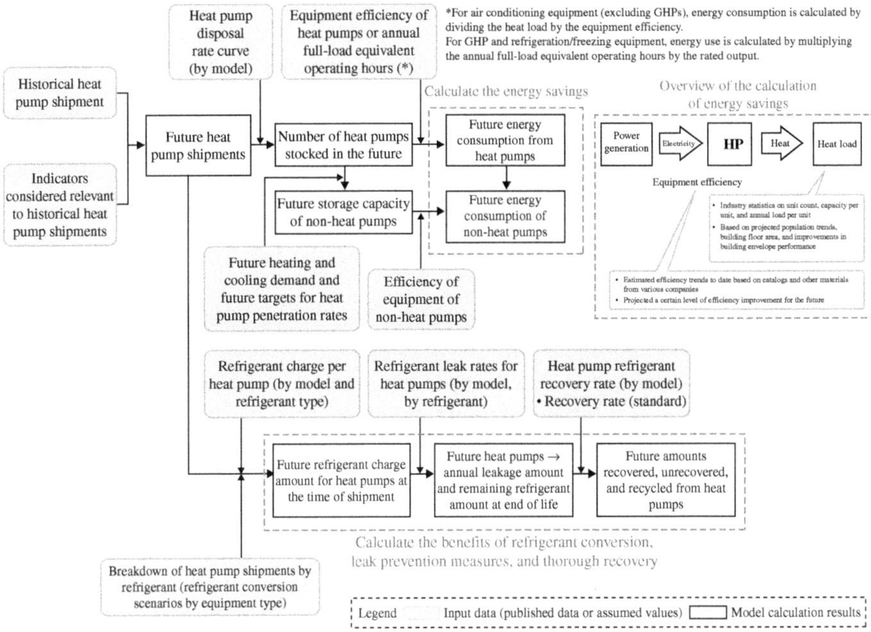

FIGURE A1.27 Flowchart to show how to estimate the effects of controls.

to be consistent with the actual values of the most recent recovery and filling rate from servicing[26]. For a given year's refrigerant stock amount[27], the refrigerant stock amount for the following year is estimated by calculating leakage, recovery at servicing, and filling at servicing.

A1.7.1.3 Calculation of Equipment to Be Disposed of

Multiply the refrigerant stock amount at the time of disposal by the waste recovery rate to calculate the amount recovered and the emissions at the time of disposal, which is the difference between the refrigerant stock amount and the amount recovered at the time of disposal.

A1.7.1.4 Method of Calculating Energy Usage

Considering the trend of improvements in energy efficiency over time of the equipment to be shipped, the energy usage at that point in time is calculated using the expected future heat load and the average energy efficiency at that moment[28].

However, for Gas Heat Pumps (GHP) and refrigeration and freezing equipment, energy usage is calculated by multiplying the rated output by the same number of hours for all models, as information on changes in heat load over time was not available and only the rated output of the equipment was known. In addition, as heat pumps use electricity, the electricity emissions factor[29] for each point in time is used to calculate the GHG emissions associated with energy usage.

A1.7.2 Approach to Setting Various Parameters

Depending on the model, one of two methods was selected for the estimation

(1) Taking into account the trend of improved energy efficiency over time for the equipment being shipped, the energy consumption at a given point in time is calculated using the projected future heat load and the average energy efficiency at each point in time
(2) Energy usage was calculated by multiplying the rated output by a fixed number of hours

A1.7.2.1 Method of Calculating Energy Usage

For residential air conditioners, package air conditioners[30], and central air conditioning[31], energy usage was estimated by multiplying energy efficiency against the estimated past to future heat load from the report by the Heat Pump & Thermal Storage Technology Center of Japan and others [31].

A1.7.2.2 Method of Calculating Heat Load

It is assumed that the heat load is shared between the heat pump equipment and other equipment such as combustion systems (competing equipment), referring to the assumption in the same report [31] that the share of heat pump equipment increases over time.

A1.7.2.3 Methodology for Calculating Energy Efficiency

Energy efficiency figures are based on Figures A1.28 and A1.29, which were calculated using the methodology described in the report [31].

A1.7.3 Method for Calculating the Heat Load of GHP and Refrigeration Equipment

For GHP and refrigeration and freezing equipment, information on changes in heat load over time was not available and only information on the rated output of the equipment could be ascertained, so energy usage was calculated by multiplying the representative rated output for each capacity category (described below) by a set number of hours (3,500 hours) for all models.

A1.7.3.1 Method of Calculating the Number of Products Shipped

For residential air conditioners, package air conditioners, and central air conditioning, which correspond to point (1) above, the number of future shipments was set so that the future trends in heat load in the existing literature and the future trends in the number of stock units resulting from the estimation generally coincide.

For GHP and refrigeration and freezing equipment, which correspond to point (2) above, the actual past trends in the number of units shipped for each model and the Japan Refrigeration and Air Conditioning Industry Association's "Refrigeration and air conditioning equipment: Voluntary statistical data file" were referenced [33], the economic indicators with the highest correlation were

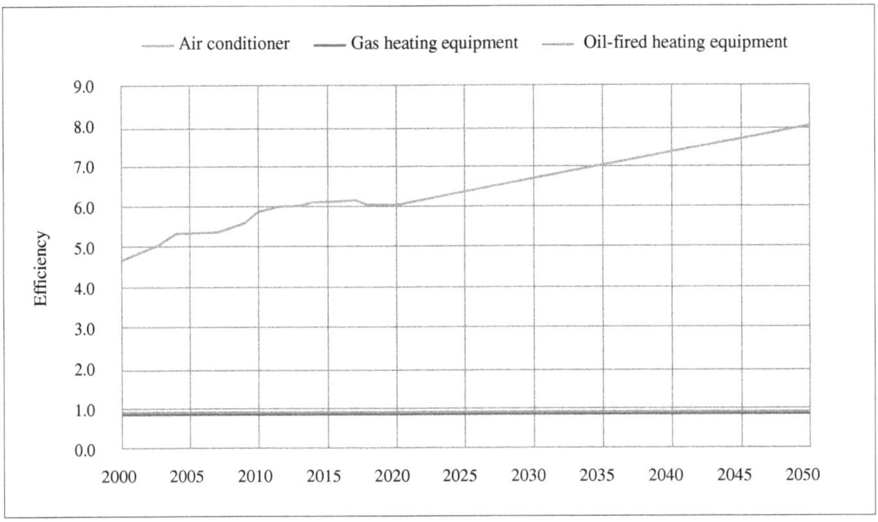

FIGURE A1.28 Assumed energy efficiency of residential air conditioners and gas/oil heaters, etc. (From Heat Pump and Thermal Storage Technology Centre of Japan/Japan Electro-Heat Centre, based on "FY2022 Electrification Dissemination Outlook Survey Report" [32]).

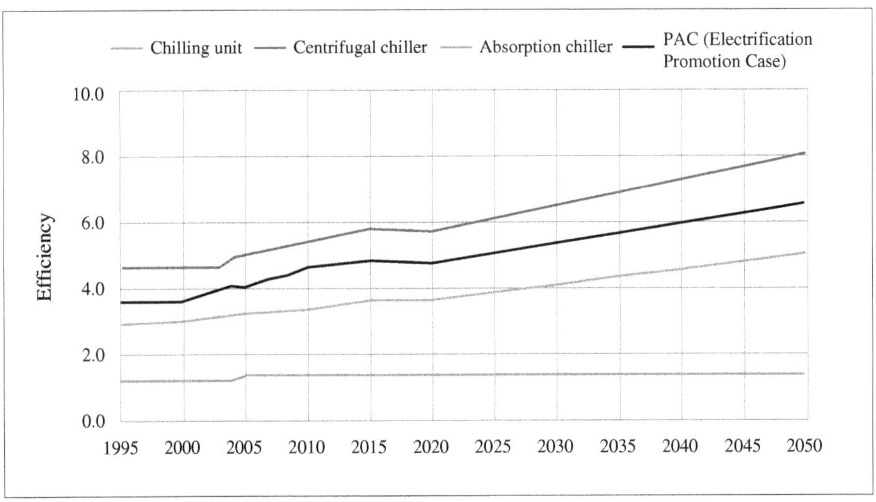

FIGURE A1.29 Assumed energy efficiency of commercial air conditioning, etc. (From Heat Pump & Thermal Storage Center of Japan/Japan Electro-Heat Center, based on "FY2022 Electrification Dissemination Outlook Survey Report" [32]).

selected, and the future estimated values of those economic indicators were used as the basis (Mitsubishi Research Institute, Inc.). The future trends in the number of units shipped were estimated based on the estimated values of the economic indicators (see the forecasted values of real gross domestic product, private-sector capital investment, and private-sector final consumption expenditure by the Mitsubishi Research Institute Inc.). For semi-hermetic condensing units, the future change in total capacity of shipped equipment was estimated, as the correlation was greater with the total capacity of shipped equipment than with the number of units shipped.

As the status of refrigerant conversion and the transition of improvement in the in-use leakage rate are considered to differ depending on the equipment capacity of the same model, it was decided to estimate the future number of units shipped for each of two or three capacity (compressor output) categories, depending on the model.

The percentage of future shipments by capacity category was estimated on the assumption that the trend in the percentage of past shipments will remain unchanged in the future.

The approach to estimating the number of shipments by model is shown in Table A1.7.

A1.7.3.2 Composition Ratio of Refrigerant Used of Shipped Equipment

Trends in the composition ratio of refrigerant used in shipped equipment were established because the refrigerant used in shipped equipment has changed over time, even for the same model.

The past equipment shipped was set based on various literature and catalogues, while the future was set under the two scenarios shown in Table A1.8.

A1.7.3.3 Amount of Refrigerant Filled Per Unit

The amount of refrigerant to be filled per unit of shipped equipment[32] was set for each model, capacity category (see Table A1.7 for details of categories) and refrigerant type.

First, it was assumed that the initial filling amount is roughly proportional to the capacity, and the HFC stock amount by model in 2012 was set to match the value published by the METI [34], taking into account differences in the filling amount by refrigerant type.

Then, based on the factory production filling volume from the published value by METI and the published value of the on-site installation filling volume based on the Fluorocarbons Emissions Control Act, a correction was made by multiplying a uniform coefficient for the model[33] for which on-site installation filling is considered to exist.

A1.7.3.4 Distribution Curve of Waste Rates

It is known that the proportion of equipment disposed of increases with the number of years the equipment has been in use, and that the transition follows a curve similar to that shown in Figure A1.30.

Therefore, a proportion of the equipment shipped in each year of use was disposed of in proportion to the number of years of use, and was subject to refrigerant

TABLE A1.7
Approach to Estimating the Number of Units Shipped by Model

Model	Concept of Estimation of Number of Shipments	Capacity Category Setting
Residential air conditioners	The number of units to be shipped in the future is set so that the future trends in heat load in the existing literature and the future trends in the number of stock units obtained as a result of the estimation are roughly in line with each other	–
Commercial packaged air conditioners		Less than 4.5 kW, 4.5 kW and above – less than 7.5 kW, 7.5 kW and above
Multi-air conditioner for buildings Air conditioning for facilities		Less than 7.5 kW – 7.5 kW or more
Centrifugal chillers		–
Chilling unit		–
GHP	Estimated based on future estimates of economic indicators (private-sector capital investment) that have a high correlation with historical shipment figures	–
Semi-hermetic condensing units	Estimated based on future estimates of economic indicators (private-sector capital investment) that have a large correlation with the total capacity of equipment shipped in the past	Less than 7.5 kW, 7.5 kW and above
Hermetic condensing units	※	–
Separate type showcase	Estimated based on future estimates of economic indicators (real gross domestic product) that have a high correlation with past shipment figures	Less than 7.5 kW, 7.5 kW and above
Built-in showcase		–
Refrigeration and freezing units	Estimated based on future estimates of economic indicators (private final consumption expenditure) that have a large correlation with the number of units shipped in the past	Less than 1.5 kW, 1.5 kW or more – less than 7.5 kW, 7.5 kW or more

Note: ※Hermetic condensing units are calculated by extending the historical trend of shipments, as it was not possible to select an economic indicator with a high correlation.

recovery at disposal. For the following models not listed in Figure A1.30, the curves for the other models listed were applied.

- Commercial packaged air conditioners, facility air conditioners, and GHP: A curve for commercial packaged air conditioners was applied as a representative example.
- Condensing units, showcases, refrigeration and freezing units: Industrial refrigeration units (commercial refrigerators)
- Chilling units: Centrifugal chillers

TABLE A1.8
Concept of Composition Ratio of Refrigerant Used for Shipped Equipment

Parameters	[Scenario 1] Pursuit of Refrigerant Conversion	[Scenario 2] Enhanced Refrigerant Recovery and Recycling
Composition ratio of refrigerant used in equipment to be shipped in the future	Achieve the target values of the designated product system based on the Fluorocarbons Emissions Control Law; switch to natural refrigerant to achieve an average GWP of 450 in 2030 and 10 in 2036 (in the case of packaged air conditioners, which are direct expansion type equipment, instead of using R290, an A3 refrigerant, use R290 or R717). Assumed conversion to chilling units	Residential air conditioning: R454C will become the mainstream in the future, but it is assumed that it will not switch to R290. Commercial and industrial air conditioning: Target values for the designated product system based on the Fluorocarbons Emissions Control Act have been achieved. (Direct expansion type equipment such as scenario 1 will not be converted to chillers). Industrial refrigeration/freezing: same as scenario 1.

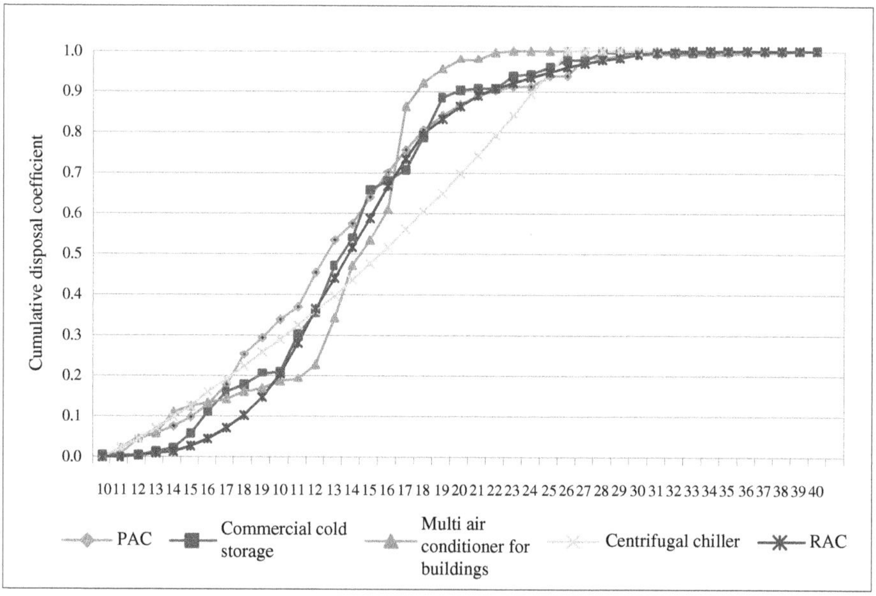

FIGURE A1.30 Cumulative waste rate distribution by model over time. (From METI, "Survey on Commercial Refrigeration Equipment and Residential Air Conditioners (3) Other Changes (Review of Disposal Factors)" [35]).

A1.7.3.5 Refrigerant Leakage Rate in Use

The values shown in Table A1.9 are known for the in-use refrigerant leakage rate by model. However, this value includes the portion recovered during equipment servicing, so the net leakage rate was estimated excluding this figure.

TABLE A1.9
Emission Factors in Use by Model

Classification of Equipment		Previous Coefficient (Calculated Value Applicable to 2007 Stock)	Revised Coefficient*	[Reference Values] 2006 IPCC Guidelines		[Reference Value] Germany	[Reference Value] Canada
Large refrigeration and freezing equipment	Centrifugal chillers	2.3%	7%	$2\% \leq x \leq 15\%$	Chillers	7% (1)	17% (3)
	Screw chillers	2.8%	12%	$10\% \leq x \leq 35\%$	Industrial Refrigeration including food processing and cold storage	7% (1)	
Medium-sized refrigeration and freezing equipment	Transport refrigeration and freezing units	9.0%	15%	$15\% \leq x \leq 50\%$	Transport refrigeration	15–25%	
	Refrigeration and freezing units	1.1%	17%	$10\% \leq x \leq 35\%$	Medium and large commercial refrigeration	1.5–15% (2)	
	Condensing units	–	13%	$10\% \leq x \leq 35\%$	Medium and large commercial refrigeration	1.5–15% (2)	
	Separate type refrigerated showcase	0.7%	16%	$7\% \leq x \leq 25\%$	Medium and large commercial refrigeration	1.5–15% (2)	
Commercial air conditioning	Commercial packaged air conditioners	0.9%	3%	$1\% \leq x \leq 10\%$	Residential and commercial air conditioning, including heat pumps	6.0%	17% (3)
	Packaged air conditioners for buildings	0.9%	3.5%	$1\% \leq x \leq 10\%$	Residential and commercial air conditioning, including heat pumps	6.0%	
	Industrial packaged air conditioners (PAC)	0.3%	4.5%)	$1\% \leq x \leq 10\%$	Residential and commercial air conditioning, including heat pumps	6.0%	
	GHP	4.4%	5.0%	$1\% \leq x \leq 10\%$	Residential and commercial air conditioning, including heat pumps	6.0%	

(Continued)

TABLE A1.9 (*Continued*)
Emission Factors in Use by Model

Classification of Equipment			Previous Coefficient (Calculated Value Applicable to 2007 Stock)	Revised Coefficient*	[Reference Values] 2006 IPCC Guidelines		[Reference Value] Germany	[Reference Value] Canada
Residential air conditioning (RAC)			0.2%	2%	1%≤x≤10%	Residential and commercial air conditioning, including heat pumps	2.5%	–
Small refrigeration and freezing equipment	Integrated type equipment			2%	1%≤x≤15%	Stand-alone commercial application	1.5–15% (2)	17% (4)
		Built-in refrigerated showcases	0.02%					
		Ice makers	0.02%					
		Water coolers	0.02%					
		Commercial refrigerators	0.01%					
Chilling units	Chilling units			6%	2%≤x≤15%	Chiller	–	–
		Chilling units for refrigeration and freezing	2.0%					
		Chilling units for air conditioning	2.0%					
Car air conditioners (MAC)			5.2%	5.2% (as usual)	10%≤x≤20%	Mobile air conditioning	10%	15%

(1) Industrial Refrigeration, (2) Commercial Refrigeration, (3) Stationary Air Conditioning, (4) Commercial Refrigeration.

The emission factors for Japan include the emissions of refrigerants recovered during equipment maintenance, so simple international comparisons are not possible. The emission factor from equipment in normal operation is considerably lower than this, as it also includes emissions due to accidents and breakdowns.

Source: Based on the METI, "The situation and future issues surrounding refrigerant HFCs in Japan" [36].

A1.7.3.6 Recovery Rate during Servicing

In practice, it is generally considered that after several years of leakage, recovery, and filling at servicing in a given year, the amount of refrigerant contained returns to the same amount as at the time of shipment. For the sake of simplicity, it was assumed that a certain percentage of refrigerant leakage in use and recovery/filling during servicing occurs each year, and that the amount of refrigerant contained decreases monotonically until the year of disposal of the equipment.

A1.7.3.7 Refrigerant Recovery Rate at Disposal

The values shown in Table A1.10 are representative published values for refrigerant recovery rates at disposal for different models. The concept of the assumed models is also shown in Figure A1.31. However, the value in Table A1.10 is a ratio based on the residual amount of refrigerant at disposal, which was calculated based on a different concept from that of this estimation model and, therefore, if it is used in this estimation model to calculate the recovery at disposal, it will not be consistent with the actual recovery at disposal values based on the most recent Fluorocarbons Emissions Control Act. Therefore, a correction was made by multiplying a standard coefficient for all models to match the actual amount

TABLE A1.10
Refrigerant Recovery Rate at Disposal by Model

Model		Number of Units	Volume	Per Unit
Air conditioning/refrigeration and freezing	Large Centrifugal chiller	92–116%	42–58%	45–60%
	Large screw chiller	152–175%	82–153%	52–101%
Air conditioning/refrigeration and freezing	Medium-sized	60–63%	25–27%	39–42%
	Small-sized	58%	66–69%	113–119%
Refrigeration and freezing	Medium-sized*	21–29%	12–18%	55–85%
	Small-sized	49–51%	68–138%	133–271%
Total		53%	33–37%	63–70%

When expanding the survey results to prefectural reported values (number of units and volume), estimates were made using multiple estimation methods to ensure the reliability of the analysis. Table A1.10 shows the results of these multiple methods as a range of estimation results.

Note: "Number of units" represents the ratio of the number of units recovered at disposal to the number of units disposed of, while "per unit" represents the ratio of the amount of refrigerant recovered at disposal to the amount of refrigerant remaining in the disposed equipment. The product of the two is therefore the waste refrigerant recovery rate (the ratio of the quantity of refrigerant recovered at the time of disposal to the quantity of refrigerant remaining in the waste equipment), which corresponds to the "volume" value.

Source: Based on Ministry of the Environment, "Direction of measures to improve the recovery rate of fluorocarbons at time of disposal" [37].

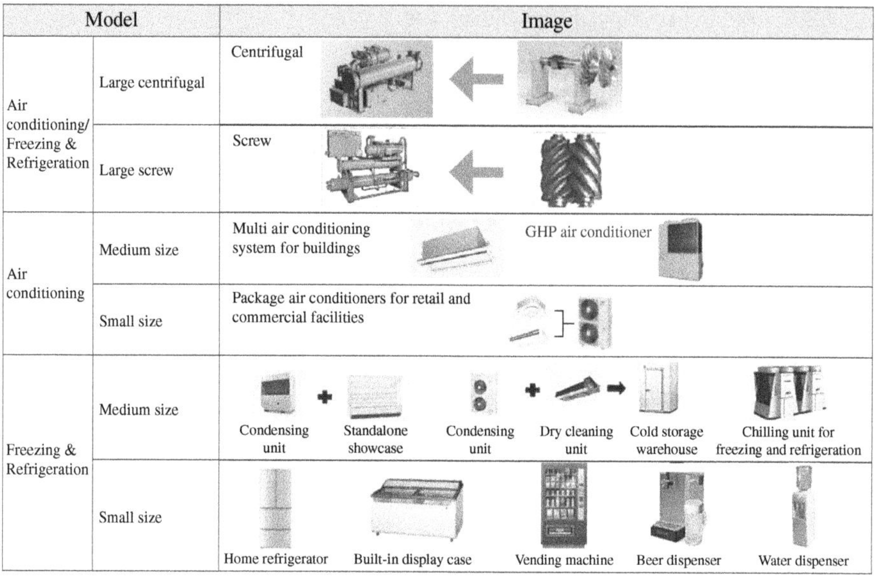

FIGURE A1.31 Concept of models. (From Ministry of the Environment, "Direction of measures to improve the recovery rate of fluorocarbons at disposal" [37]). Classification method for equipment tailored to the purpose of organizing the recovery rate at the time of disposal.)

recovered at the time of disposal in FY2020 (1,712 tonnes for HFCs) [36] in accordance with the Fluorocarbons Emissions Control Act, even if it is used in this estimation model.

For residential air conditioners, the refrigerant recovery rate at the time of disposal in this estimation model was also set to be consistent with the actual number of units for recycling in FY2020 (3.82 million units) and the actual amount of fluorocarbons recovered per unit (656 g/unit) [37].

The recovery rate at disposal in 2020 used in this estimation model is shown in Table A1.11. Note that the estimation scenarios also assume a case where the recovery rate at the time of disposal in the future increases, which is discussed below in Table A1.12.

A1.7.3.8 Refrigerant Regeneration Rate

Refrigerants recovered from equipment are either destroyed or regenerated. Currently, the renewable rate for residential air conditioners is around 90%[34] and for commercial refrigeration and air conditioning equipment, the average is around 30%[35].

Note that for commercial refrigeration and air conditioning equipment, information on the renewable rate by model was not available, so a standard value was used for all models.

TABLE A1.11
Establishment of Recovery Rates at Time of Disposal (FY2020)

Model	Classification in Tables A1.9–A1.10	(1) Recovery Rate at Time of Disposal (Number of Units Recovered/ Number of Units Disposed of)	(2) Refrigerant Recovery Rate per Unit at Time of Disposal (Amount Recovered per Unit/Amount Remaining per Unit)	(3) Recovery Rate at Time of Disposal ((1) × (2))
Centrifugal chillers	Large centrifugal chillers	100%	60%	60%
Multi-air conditioner for buildings, GHP	Medium-sized air conditioning equipment	66%	43%	28%
Air conditioning equipment for shops and facilities	Small-sized air conditioning equipment	70%	95%	66%
Chilling units, semi-hermetic condensing units, separate type showcases, refrigeration and freezing units	Medium-sized refrigeration and freezing equipment	27%	75%	20%
Hermetic condensing units, built-in showcases	Small-sized refrigeration and freezing equipment	60%	95%	57%
Residential air conditioners	–	53%	95%	51%

Note: As simple calculations may result in (1) and (2) exceeding 100%, the upper limit for (1) was set at 100% and the upper limit for (2) was set at 95%, assuming that there is a technically difficult amount of 5% to recover, but adjusted so that the value of (3), the product of the two, is maintained in such cases.

A1.8 SETTING FUTURE PROJECTION SCENARIOS FOR NEXT-GENERATION HEAT PUMP TECHNOLOGY

A major crossroads in the roadmap to next-generation heat pumps is how to proceed with refrigerant conversion. For this reason, the following scenarios were set up as shown in Table A1.12 to compare the predicted results. The set-up of the past and future changes in the composition of specification refrigerants in shipped equipment is shown in Figures A1.32 to A1.35.

[Scenario (1)] Refrigerant conversion pursuit scenario
[Scenario (2)] Enhanced refrigerant recovery and regeneration scenario

TABLE A1.12
Scenarios for Roadmap to Refrigerant Conversion

Main Parameters	[Scenario (1)] Pursuit of Refrigerant Conversion Scenario	[Scenario (2)] Enhanced Refrigerant Recovery and Regeneration Scenario
Composition ratio of refrigerant used in equipment to be shipped in the future (Table A1.8 repeated)	Achieve the target values of the designated product system based on the Fluorocarbons Emissions Control Law; switch to natural refrigerant to achieve an average GWP of 450 in 2030 and 10 in 2036 (in the case of packaged air conditioners, which are direct expansion type equipment, instead of using R290, an A3 refrigerant, use R290 or R717). Assumed conversion to chilling units	Residential air conditioners: R454C will become the mainstream in the future, but it is assumed that it will not switch to R290 Commercial and industrial air conditioning: Target values for the designated product system under the Fluorocarbons Emissions Control Act were achieved Scenario (1). (Avoid conversion of direct expansion type equipment to chillers as in Scenario (1)) Industrial refrigeration and freezing equipment: Same as Scenario (1)
Leakage rate of refrigerants in use	The current leakage rate will remain constant in the future	Toward achieving the reduction rate targets (2025, 2030) set by the Global Warming Countermeasures Plan [38]
Refrigerant recovery rate at disposal	Constant in the future at the current recovery rate (approx. 40%)	Achieve the recovery rate targets in the global warming countermeasures plan (60% in 2025 and 75% in 2030) and reach 95% in 2040
Refrigerant regeneration rate from waste recovery	Maintain the current renewable rate (residential air conditioners: Approx. 90% / other: Approx. 30%) and keep it constant in the future	Recovery rate of refrigerants other than for residential air conditioners: 90% by 2040

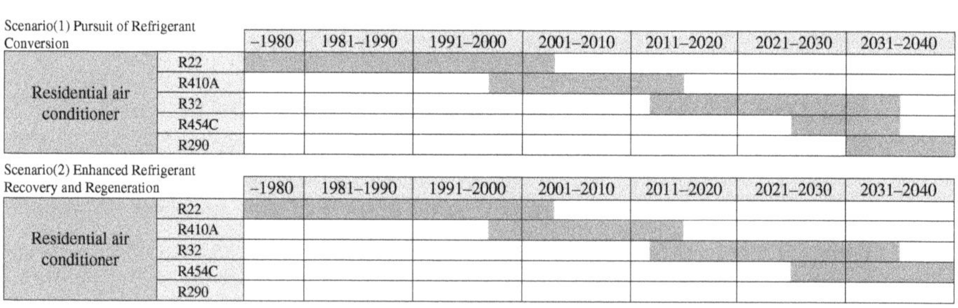

FIGURE A1.32 Refrigerant composition of shipped equipment (residential air conditioners).

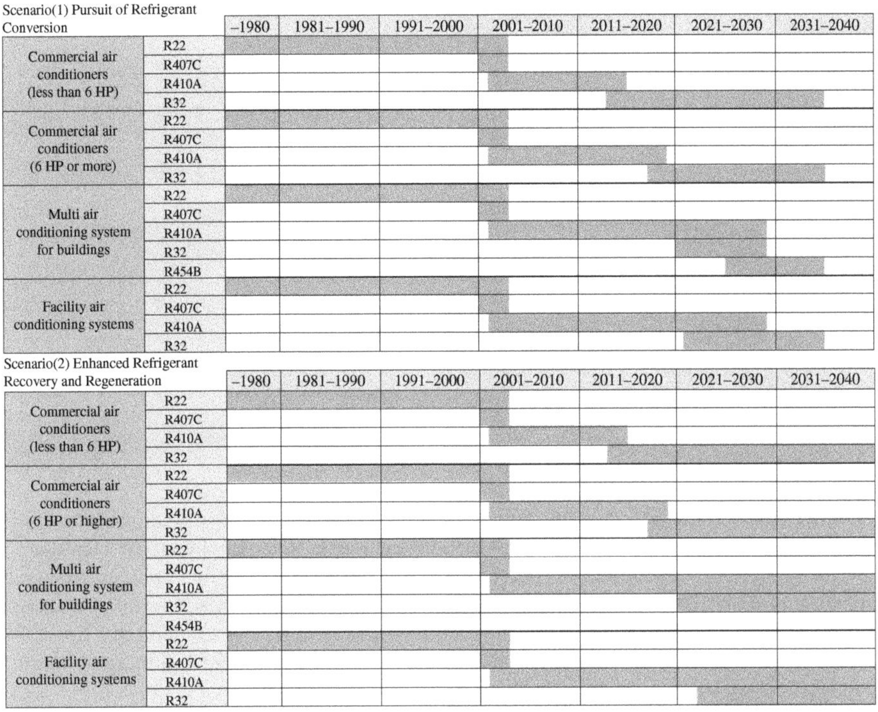

FIGURE A1.33 Refrigerant composition of shipped equipment (packaged air conditioners). *Note:* (1) In the scenario for pursuing refrigerant conversion, it is assumed that conversion to chilling units using R290 or R717 will progress from 2026 onwards, and that after 2036 there will be no more shipments as packaged air conditioners and all units will be chilling units using R290.

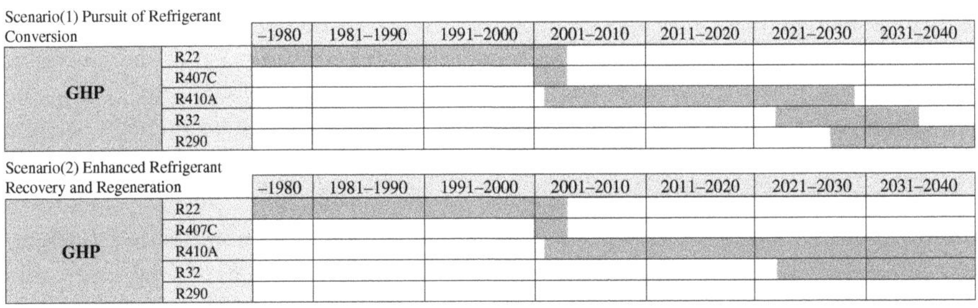

FIGURE A1.34 Refrigerant composition of shipped equipment (GHP).

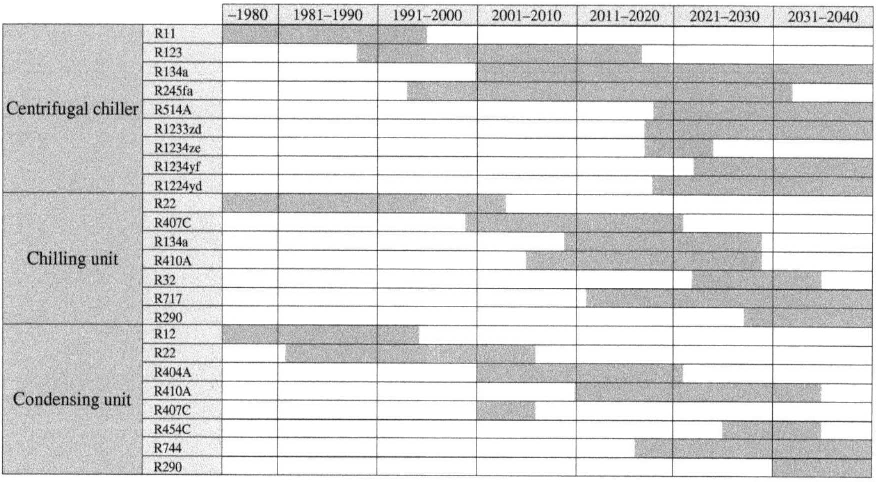

FIGURE A1.35 Refrigerant composition of shipped equipment (centrifugal chillers, chilling units, and condensing units). *Note:* These three models were assumed to be common to the two scenarios. In addition, the "condensing unit" ratio was common for semi-hermetic condensing units, hermetic condensing units, separate type showcases, built-in showcases, and refrigeration and freezing units. However, for chilling units it is assumed that (1) in the pursuit of refrigerant conversion scenario, shipments of R717 and R290 will increase significantly as the substitution from package air conditioners progresses after 2026.

When looking at the effects of next-generation heat pumps in terms of reduced energy usage (energy savings) and GHG emissions, an important question is whether compliance with the Kigali Amendment, which requires refrigerant measures, will be achieved. It is also necessary to consider the extent to which the project can contribute to carbon neutrality, including refrigerants. The benefits of next-generation heat pumps are now presented from these two perspectives.

A1.8.1 Compliance with the Kigali Amendment

The Kigali Amendment regulates the production and consumption of new refrigerants, but here we will focus on the demand side of the equipment and look at the consumption of new refrigerants. In this case, the target for 2036 is 10.73 million t-CO_2 [39].

The consumption of refrigerant consists of the initial charge for new equipment and the charge during servicing for installed equipment. In the pursuit of refrigerant conversion scenario, low GWP of refrigerants for all equipment progresses, and after 2036 the initial amount of refrigerants is almost zero in terms of CO_2 equivalent, but there remains around 100~ 3 million tCO_2 of maintenance fill for installed equipment (Figures A1.36 and A1.37).

If this is added to the approximately 4 million t-CO_2[36] used in applications other than heat pump equipment, the target of the Kigali Amendment will be fully met.

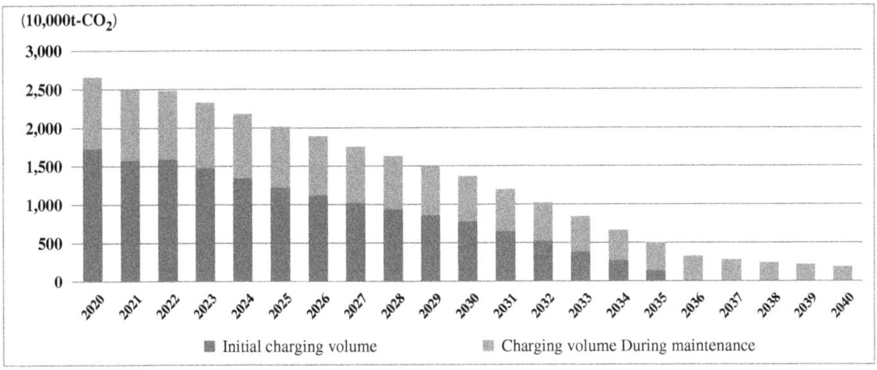

FIGURE A1.36 Amount of filling of heat pump appliances converted to CO_2. *Note:* All appliances: Pursuit of refrigerant conversion scenario.

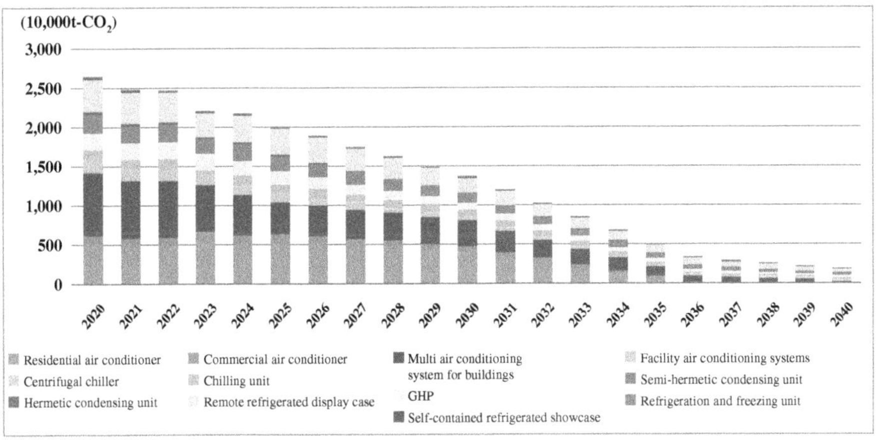

FIGURE A1.37 Amount of CO_2 converted from filling heat pump appliances by model. *Note:* All appliances: Pursuit of refrigerant conversion scenario.

On the other hand, in the enhanced refrigerant recovery and regeneration scenario, HFCs will continue to be used in residential air conditioners and multi-air conditioners for buildings, where refrigerant conversion is difficult, so that an initial amount of 4 million t-CO_2 will remain in 2036 (Figures A1.38 and A1.39). However, with enhanced regeneration from servicing and recovery at disposal, much of the required filling volume could be covered by recycled refrigerant in 2036. However, as R454C will continue to be used in residential air conditioners, recycled refrigerant alone will not be sufficient, resulting in a certain amount of virgin refrigerant input being required (Figure A1.40).

Taking into account the demand for applications other than heat pump equipment, it is seen as possible to achieve the Kigali Amendment.

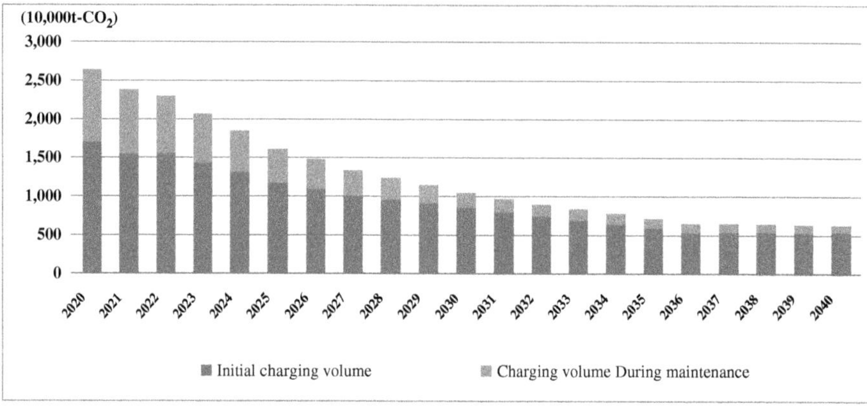

FIGURE A1.38 Filling of heat pump equipment. *Note:* All appliances: Enhanced refrigerant recovery and regeneration scenario.

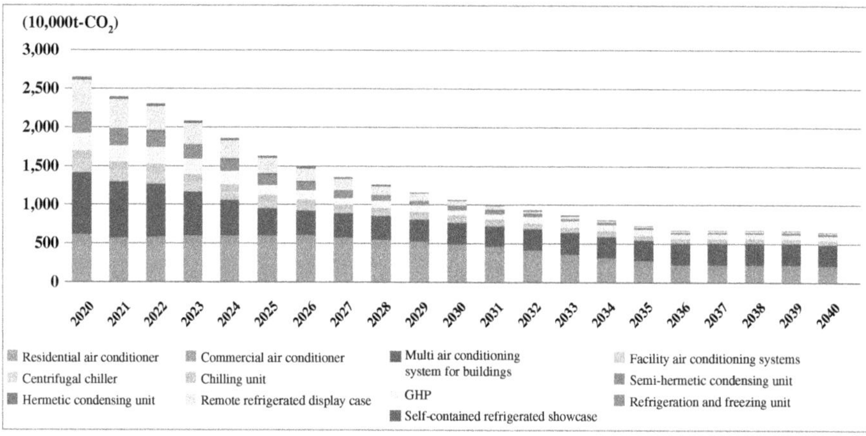

FIGURE A1.39 Amount of heat pump equipment filled, by model. *Note:* All appliances: Enhanced refrigerant recovery and regeneration scenario.

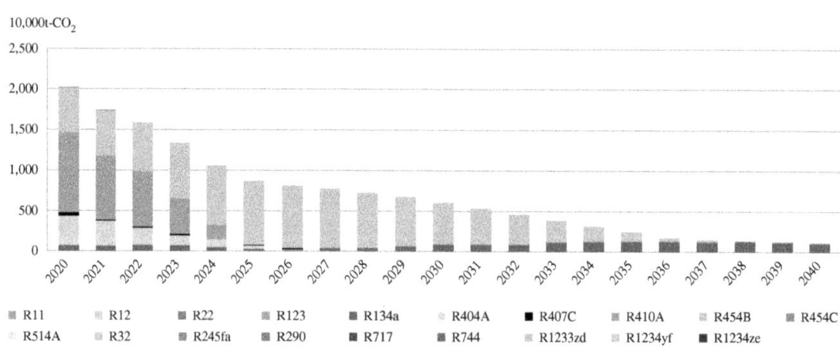

FIGURE A1.40 Virgin refrigerant requirements, by refrigerant. *Notes:* All appliances: enhanced refrigerant recovery and regeneration scenario. "Virgin refrigerant requirement" is calculated as initial fill + fill at servicing – regenerated volume.

A1.8.2 Contribution to Carbon Neutrality

A1.8.2.1 Reduction of GHG Emissions

Before moving on to comparisons between scenarios, it is worth confirming that heat pumps are highly energy efficient to begin with, and that there are significant savings to be made by replacing competing equipment as the performance of heat pumps improves.

Figure A1.41 shows the total load of household air conditioning. The total load itself is assumed to decrease slightly, taking into account the future trend of population decline. The breakdown shows that, although competing equipment accounts for a certain proportion of the total load, as combustion-based equipment is used for heating, this proportion is expected to decrease, as it is assumed to be increasingly replaced by residential air conditioners.

Figure A1.42 shows how GHG emissions would then change.

Despite heat pumps (residential air conditioners) increasing load sharing, emissions are significantly reduced, partly due to the low carbonization of electricity, and

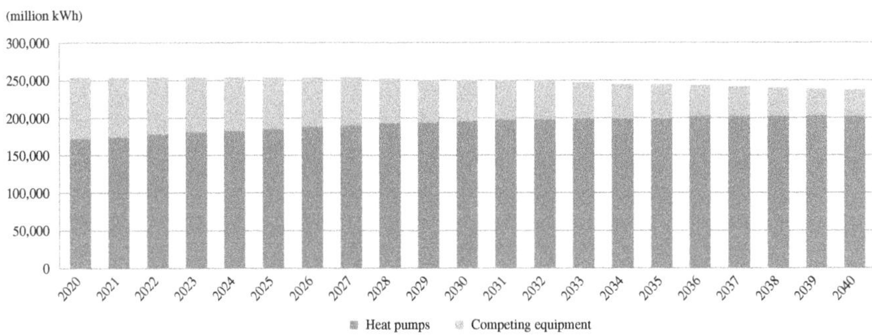

FIGURE A1.41 Load of residential air conditioners by type.

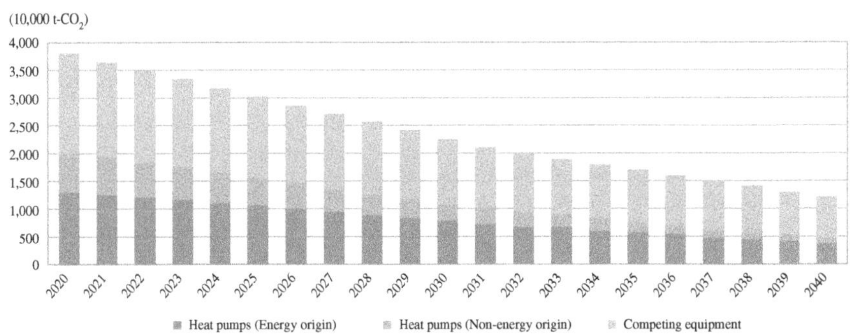

FIGURE A1.42 GHG emissions from residential air conditioners (enhanced refrigerant recovery and regeneration scenario). *Note:* "Energy origin" is GHG emissions from electricity consumption during residential air conditioners usage (assuming that the emissions factor is close to zero toward 2050). "Non-energy origin" are GHG emissions associated with uncollected refrigerants at disposal. Same from here onwards.

emissions are reduced to one-third of the peak level in 2040 compared to the peak level in the 2020s.

A1.8.2.2 GHG Emissions in the Pursuit of Refrigerant Conversion and Enhanced Refrigerant Recovery, and Regeneration Scenarios

So how do these results differ depending on the refrigerant conversion scenario?

When looking at equipment units, the pursuit of refrigerant conversion scenario is difficult to compare as some equipment will be converted to chillers, so the comparison is made on an equipment-wide basis.

Figures A1.43 and A1.44 show the evolution of GHG emissions under the pursuit of refrigerant conversion scenario and the enhanced refrigerant recovery and regeneration scenario respectively.

No significant differences occur between the two scenarios. This means that GHG emissions can be reduced by enhancing refrigerant recovery and regeneration, even if refrigerant conversion is not required uniformly for all appliances.

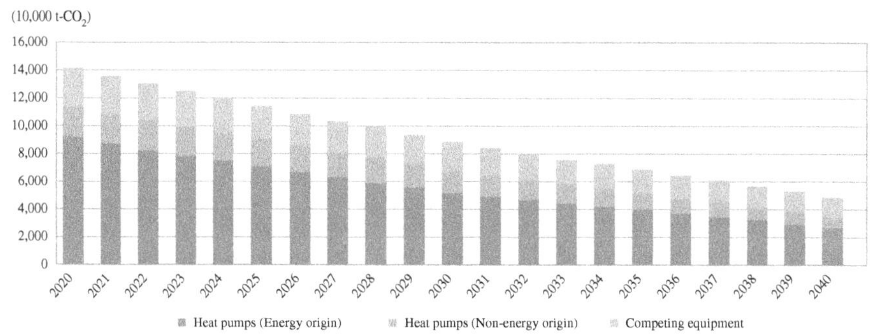

FIGURE A1.43 All appliances: GHG emissions in the pursuit of refrigerant conversion scenario. *Note:* "Non-energy origin" are GHG emissions associated with uncollected refrigerants at disposal

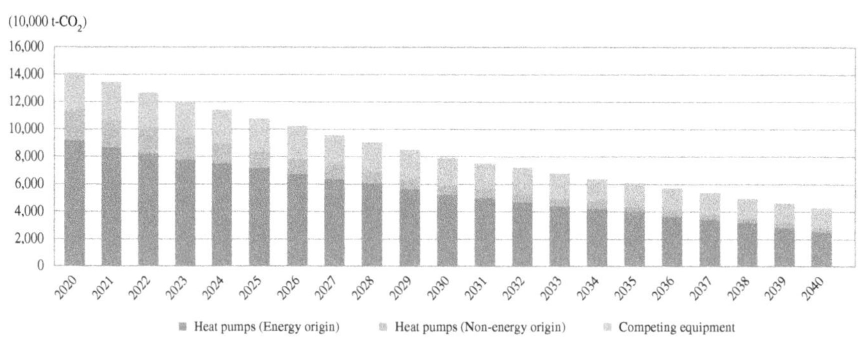

FIGURE A1.44 All appliances: GHG emissions in refrigerant recovery and regeneration enhancement scenario. *Note:* "Non-energy origin" are GHG emissions associated with uncollected refrigerants at disposal

NOTES

1. Specific fluorocarbons: Trichlorofluoromethane (CFC-11), dichlorodifluoromethane (CFC-12), trichlorotrifluoroethane (CFC113), dichlorotetrafluoroethane (CFC-114), and chloropentafluoroethane (CFC-115).
2. HFC-404A: A mixed refrigerant consisting of HFC-125/HFC-143a/HFC-134a = 44/52/4% by weight.
3. HFC-410A: HFC-32/HFC-125 = 50/50% by weight mixed refrigerant.
4. GWP figures: Values from the Fourth Assessment Report (2007) of the Intergovernmental Panel on Climate Change (IPCC) are used.
5. Kigali Amendment: see Chapter 2, Section 4 for details on the HFC reduction schedule under the Kigali Amendment and on fluorocarbons measures under national legislation.
6. American Society of Heating, Refrigerating and Air-Conditioning Engineers (ASHRAE): The ASHRAE is an international association for individuals and organizations involved in air conditioning, including heating, ventilation, and refrigeration. It also develops industrial standards and guidelines for air conditioning systems.
7. Mixed refrigerants: The 400s are non-azeotropic mixed refrigerants, i.e. refrigerants that exhibit properties as a mixture with separated dewpoint and boiling point over the entire composition range; the 500s are azeotropic mixed refrigerants, i.e. refrigerants with a constant boiling point and phase change as if they were a single component of a multi-component refrigerant;
8. A, B, C: Order of application, e.g. A454A, R454B, R454C.
9. Number in the 700 range: As an example, ammonia with a molecular weight of 17 would be R717. If the molecular weight is above 100, the number is in the 7000 range, with the last three digits indicating the molecular weight.
10. OEL: Operational exposure limit (workplace exposure limit). Time-weighted average exposure concentrations at which almost all workers will not be adversely affected in a normal 8-hour day, 40-hour week.
11. ATEL: Acute-toxicity exposure limit. The lower concentration limit for acute toxicity.
12. Direct expansion type equipment: Equipment with refrigerant circulating to the room.
13. Inter-expansion type appliances: Appliances in which refrigerant circulates only outdoors and the secondary medium is circulated indoors.
14. HFO mixed refrigerant: A mixture of HFO refrigerant and other refrigerants.
15. Small appliances: Equipment designed for use with HFC refrigerants containing approximately 2 kg or less of refrigerant.
16. Large built-in refrigeration and freezing equipment: Equipment with more than 1 kg of HFC refrigerant.
17. Support project for introduction: The Ministry of the Environment's subsidy programme "Project for promoting the de-fluorination and decarbonisation of refrigeration and freezing equipment that supports the cold chain." To reduce emissions of fluorocarbons, the project subsidizes part of the costs for the introduction of natural refrigerant equipment capable of decarbonizing in refrigerated warehouses, food manufacturing plants, and food retail outlets.
18. Scroll chillers: Medium-sized chillers using scroll compressors.
19. Centrifugal chillers: Very large chillers using centrifugal compressors.
20. 1 MPa or higher: Article 2 of the High Pressure Gas Safety Act defines high-pressure gas as "compressed gas whose pressure is 1 MPa or higher at normal service temperature or whose pressure is 1 MPa or higher at a temperature of 35°C."
21. Electrification of vehicles: Hybrid electric vehicles (HEV), battery electric vehicles (BEV), and fuel cell vehicles (FCV) are covered.
22. CO_2 refrigerant: CO_2 has a critical temperature of approximately 31°C. If the outside temperature is higher than this, the latent heat cannot be utilized, resulting in poor performance.
23. Above a certain scale: The rated output of the electric motor or internal combustion engine used for the compressor of the equipment concerned is 7.5 kW or more.

24. Number of units stocked each year: Number of units of equipment in use.
25. Waste rate distribution curve: Curve specifying how many years in use and what percentage of the equipment is disposed of.
26. Servicing fill amount: The amount of fill other than at installation.
27. Refrigerant stock amount: The amount of refrigerant remaining in equipment in use.
28. Average energy efficiency: Weighted average value according to the number of stocked units.
29. Electricity emissions factor: Calculated on the assumption that the emissions will be zero in 2050.
30. Packaged air conditioners: Commercial packaged air conditioners, multi-air conditioners for buildings, air conditioners for facilities, etc.
31. Central air conditioning: Centrifugal chillers, chilling units (chillers), etc.
32. Refrigerant charge quantity: sum of shipping charge amount and on-site installation charge amount.
33. On-site installation refrigeration equipment: Packaged air conditioners for commercial packaged air conditioners, multi-air conditioners for buildings, air conditioners for equipment, GHP, centrifugal refrigeration units, semi-hermetic condensing units, separate type showcases, refrigeration and freezing units. However, it is assumed that all centrifugal chillers are filled at the time of on-site installation, while for other models, both factory-filled and on-site-filled amounts are assumed to exist.
34. About 90% recycling rate: Calculated as the ratio of the weight of recycled or reused fluorocarbons to the "recovered weight of fluorocarbons" of air conditioners from the "Annual Report on Home Appliance Recycling (FY2022 edition)" of the Japan Electrical Appliance Manufacturers' Association.
35. Around 30% renewable rate: Calculated as the ratio of the amount recycled by recyclers to the total amount of HFCs handed over to destructors and reused etc. from the amount of fluorocarbons recovered and destruction law and the amount of fluorocarbons recycled and destroyed under Fluorocarbons Emissions Control Law of the Ministry of the Environment. Note that the breakdown between the amount derived from recovery at the time of maintenance and the amount derived from recovery at the time of disposal is not known.
36. Amount used of around 4 million t-CO_2: Assumed from the National Institute for Environmental Studies' Greenhouse Gases Inventory Office, Data on Greenhouse Gases Emissions in Japan (FY1990–2021) (preliminary figures).

REFERENCES

1. Mario. J. Molina and F. Sherwood Rowland, "Stratospheric sink for chlorofluoromethanes: Chlorine atom-catalysed destruction of ozone," Nature Vol.249, p.810–12, 1974.
2. Hideo Kajiwara, "Selection of next-generation low GWP refrigerants considering risk trade-offs," Synthеology Vol.6, No.4, p.209–18, 2013, https://www.jstage.jst.go.jp/article/synth/6/4/6_209/_pdf/-char/ja
3. Japan Refrigeration and Air Conditioning Industry Association, "Figure 6—ISO 817: 2014 – Safety class of refrigerants," 2021, https://www.jraia.or.jp/outline/file/pdf_11.pdf
4. Japan Society of Refrigerating and Air Conditioning Engineers, "Refrigeration cycle calculation program software Ver. 2," 2006, https://www.jsrae.or.jp/books/books_soft.php
5. Japan Fluorocarbon Manufacturers Association, "List of environmental and safety data of specified fluorocarbons (CFCs/HCFCs) and fluorocarbons," 2023, http://www.jfma.org/data.html
6. Waseda University, w-refrigerant.com, "Map of refrigerant applications," 2019, https://www.w-refrigerant.com/tech_documents/冷媒用途マップ/

7. Ministry of Economy, Trade and Industry/Industrial Structure Council Manufacturing Industry Subcommittee Chemical Substance Policy Committee • Working Group on Measures against Fluorocarbons and Related Substances • Central Environmental Council, Global Environment Committee • Subcommittee on Measures against Fluorocarbons and Related Substances, "Report on the evaluation and study of the enforcement status of the revised Fluorocarbons Emissions Control Law of 2013," p.14, 2022, https://www.meti.go.jp/shingikai/sankoshin/seizo_sangyo/kagaku_busshitsu/flon_godo/pdf/20220628_1.pdf
8. Acceleratejapan, "R290 RAC sales in China exceed 278,000 units," 2021, https://acceleratejapan.com/chinese_rac_makersandrsquo_domestic_r290_sales_exceed_270_000_units/
9. Sankei Shimbun Co. Ltd, "Panasonic to develop air conditioners using 'natural refrigerant' propane," 2023, https://www.sankei.com/article/20230622-BJSZSJ7LEZJABD7F36NF3G3LGY/
10. Japan Society of Refrigerating and Air Conditioning Engineers, "Refrigeration cycle control," (Technical Book Series of the Japan Society of Refrigerating and Air Conditioning Engineers), p.178, 2018.
11. Hidetoshi Chisaka, Shinsuke Kobayashi, Yoshinori Sakatani, and Futahiko Nakagawa, "Evaluation of transportation of renewable power by electric vehicles to various houses," Journal of Automotive Technology, Vol. 52, No. 3, p.728, 2021, https://www.jstage.jst.go.jp/article/jsaeronbun/52/3/52_20214370/_pdf
12. Holger Großmann, Comparing the refrigerant R1234yf and CO_2: (ATZ worldwide Volume 118), p.70, 2016, https://link.springer.com/article/10.1007/s38311-016-0119-0
13. Daikin Industries, Ltd., "Introduction to developments automotive refrigerant R-474A (Development)," 2021.
14. ZF Friedrichshafen, AG Press Release Press Release, "At the top of the world in torque density: ZF presents the most compact electric drive for passenger cars," 2023, https://press.zf.com/press/ja/releases/release_57601.html
15. United Nations Environment Programme (UNEP), "Refrigeration, Air Conditioning and Heat Pumps Technical Options Committee Evaluation Report," p.121, 2019. Available from: https://ozone.unep.org/sites/default/files/2019-04/RTOC-assessment-report-2018_0.pdf
16. Ministry of Economy, Trade and Industry, Reference Material 3: "Results of the total amount of fluorocarbons filled and recovered from commercial refrigeration and air conditioning equipment based on the Fluorocarbons Emissions Control Law in FY 2020," p.4, 2021, https://www.meti.go.jp/shingikai/sankoshin/seizo_sangyo/kagaku_busshitsu/flon_godo/pdf/012_s03_00.pdf
17. Ministry of the Environment, "Home appliance recycling results in FY2020," "Appendix 3: Status of the total amount of recommodification, etc. carried out by manufacturers, etc. and designated corporations during one year (1 April 2020 – 31 March 2021) based on the home appliance recycling law," 2021, https://www.env.go.jp/content/900517648.pdf
18. Ministry of Economy, Trade and Industry/Working Group on Electric and Electronic Equipment Recycling, Subcommittee on Waste and Recycling, Industrial Science and Technology Policy and Environment Bureau, Industrial Structure Council, Subcommittee on Evaluation of Home Appliance Recycling System, Recycling-based Society Committee, Central Environmental Council, "Report on Evaluation and Examination of the Implementation Status of the Home Appliance Recycling System," p.5, 2022, https://www.meti.go.jp/press/2022/06/20220623002/20220623002-b.pdf
19. Ministry of Economy, Trade and Industry, Reference Material 3: "Results of the total amount of fluorocarbons filled and recovered from commercial refrigeration and air conditioning equipment based on the Fluorocarbons Emissions Control Law in FY 2020," p.2, 2021, https://www.meti.go.jp/shingikai/sankoshin/seizo_sangyo/kagaku_busshitsu/flon_godo/pdf/012_s03_00.pdf

20. Ministry of the Environment/Working Group on Measures against Fluorocarbons and Related Substances, Chemical Substance Policy Subcommittee, Manufacturing Industry Subcommittee, Industrial Structure Council • Subcommittee on Measures against Fluorocarbons and Related Substances, Global Environment Committee, Central Environmental Council, "Direction of measures to improve the recovery rate of fluorocarbons at disposal," p.6, 2019, https://www.env.go.jp/content/900512924.pdf
21. Ministry of the Environment/Working Group on Measures against Fluorocarbons and Related Substances, Chemical Substance Policy Subcommittee, Manufacturing Industry Subcommittee, Industrial Structure Council • Subcommittee on Measures against Fluorocarbons and Related Substances, Global Environment Committee, Central Environmental Council, "Direction of measures to improve the recovery rate of fluorocarbons at disposal," p.8, 2019, https://www.env.go.jp/content/900512924.pdf
22. Ministry of the Environment, Global Environment Bureau, Global Warming Prevention Division, Fluorocarbon Measures Office, "Briefing on the Revised Fluorocarbon Emission Control Act in FY2021: Trends Surrounding Fluorocarbons (Common)," "Overview of fluorocarbon countermeasures," 2022, https://www.env.go.jp/content/900512924.pdf
23. Ministry of the Environment, Portal Site of the Fluorocarbon Emission Control Act, "Overview of the Fluorocarbon Emission Control Act," 2023, https://www.env.go.jp/earth/furon/gaiyo/gaiyo.html
24. Japan Refrigeration and Air Conditioning Industry Association, "Guidelines for the prevention of refrigerant leakage from fluorocarbons-based refrigeration and air conditioning equipment" (JRA GL-14), 2016.
25. Ministry of Economy, Trade and Industry (METI)/Japan Refrigeration and Air Conditioning Industries Association, 4th High Pressure Gas Subcommittee, Safety Subcommittee, Industrial Structure Council, Document 4-2, "Response to refrigerant leakage accidents," p.1, 2013, https://www.meti.go.jp/shingikai/.sankoshin/hoan_shohi/koatsu_gas/pdf/004_04_02.pdf
26. Japan Refrigeration and Air Conditioning Industry Association, "Guidelines for the prevention of refrigerant leakage from fluorocarbons-based refrigeration and air conditioning equipment" (JRA GL-17), 2021, https://www.jraia.or.jp/download/e-book/GL-17_2021/#target/page_no=1
27. Ministry of the Environment, "Briefing on the Revised Fluorocarbon Emission Control Act in FY2022 (Trends Surrounding Fluorocarbons)," p.18, 2013, https://www.env.go.jp/earth/furon/files/r04_gaiyou_all_rev.pdf
28. Japan Refrigerant and Environment Conservation Organization(JRECO), "What is RaMS (Refrigerant Management System)?," 2023, https://jreco-rams.jp/rams.html
29. Ministry of the Environment/Working Group on Measures against Fluorocarbons and Related Substances, Chemical Substance Policy Subcommittee, Manufacturing Industry Subcommittee, Industrial Structure Council • Subcommittee on Measures against Fluorocarbons and Related Substances, Global Environment Committee, Central Environmental Council, "Future regulation of HFCs based on the amendment of the Montreal Protocol Kigali," p.8, 2017, https://www.env.go.jp/content/900422743.pdf
30. Ministry of the Environment, "Leading the World in Fluorocarbon Countermeasures ~Following the Enactment of the Revised Fluorocarbon Emission Control Act" (Press Conference by the Minister of the Environment), p.6, 2019, https://www.env.go.jp/content/900405956.pdf
31. Heat Pump & Thermal Storage Technology Center of Japan/Japan Electro-Heat Center, "FY2022 survey report on the outlook for electrification dissemination," 2022, https://www.hptcj.or.jp/Portals/0/data0/press_topics/R4TyousaHoukoku/R4DenkaFukyuMitoshi.pdf

32. Heat Pump & Thermal Storage Technology Center of Japan /Japan Electro-Heat Center, "FY2022 Electrification Dissemination Outlook Survey Report," 2022, (in Japanese), https://www.hptcj.or.jp/Portals/0/data0/press_topics/R4TyousaHoukoku/R4DenkaFukyuMitoshi.pdf
33. Japan Refrigeration and Air Conditioning Industry Association, "Refrigeration and air conditioning equipment: Voluntary statistical data file," 2023, https://www.jraia.or.jp/statistic/006.html
34. Ministry of Economy, Trade and Industry (METI), Manufacturing Industries Bureau, Chemical Substance Management Division, Ozone Layer Protection Promotion Office, "On refrigeration and freezing equipment," p.6, 2014, https://www.meti.go.jp/shingikai/sankoshin/seizo_sangyo/kagaku_busshitsu/flon_taisaku/pdf/003_02_00.pdf
35. Ministry of Economy, Trade and Industry (METI), Manufacturing Industries Bureau, Industrial Structure Council Chemical and Bio Subcommittee on Global Warming Prevention Measures, Industrial Structure Council Chemical and Bio Subcommittee, 21st Meeting on Global Warming Prevention Measures Subcommittee, Document 1-1 "Revision of Use-Time Emission Factors for Refrigeration and Air Conditioning Equipment," p.13 "Survey on Commercial Refrigeration and Air Conditioning Equipment and Residential Air Conditioners (3) Other Changes (Review of Disposal Factor)," 2009, https://www.jraia.or.jp/member/seifu/pdf/mite20090317_1-1.pdf
36. Ministry of Economy, Trade and Industry, Ozone Layer Protection Promotion Office, "The situation and future issues surrounding refrigerant HFCs in Japan," p.8, 2009, https://www.jreco.or.jp/data/fm02_hfc.pdf
37. Ministry of the Environment/Working Group on Measures against Fluorocarbons and Related Substances, Chemical Substance Policy Subcommittee, Manufacturing Industry Subcommittee, Industrial Structure Council・Subcommittee on Measures against Fluorocarbons and Related Substances, Global Environment Committee, Central Environmental Council, "Direction of measures to improve the recovery rate of fluorocarbons at disposal," p.10–11, 2019, http://www.env.go.jp/council/06earth/r0615-01/h-1-190212.pdf
38. Ministry of the Environment, "Global warming countermeasures plan" (Cabinet decision on 22 October 2021), Annex 4-2, 2021, https://www.env.go.jp/content/900440195.pdf
39. Ministry of Economy, Trade and Industry, Manufacturing Industries Bureau, Chemical Substance Management Division, Ozone Layer Protection Promotion Office, "Results of 2021 allocation operation based on the Ozone Layer Protection Act," p.2, 2022, http://www.env.go.jp/council/06earth/r0615-01/h-1-190212.pdf

Appendix 2: Performance Evaluation Methodology for Next-Generation Heat Pump Technology

OVERVIEW

As introduced in Section 2.1 of this book, Japan has set a target to achieve carbon neutrality by 2050. Energy saving and decarbonization are required to achieve carbon neutrality. In this context, heat pump technology is an energy saving and decarbonization technology that can generate a wide range of heat from cold to warm with little electricity and can be applied to a variety of uses, so further technological progress and market expansion is expected in the future.

From the perspective of promoting energy saving and decarbonization, performance evaluation, which quantitatively assesses energy saving performance and CO_2 emissions, is important. Heat pump equipment has a long service life and it's important to select a model with high performance. Therefore, accurate performance evaluation is necessary to facilitate appropriate equipment selection.

In this chapter, focusing mainly on air conditioning, the importance and types of equipment performance evaluation is first discussed. Next, the prerequisites for equipment performance evaluation and issues with the measurement environment are listed as issues for equipment performance evaluation. The new performance evaluation model for the future, that is, evaluation based on the actual installation environment and the use of simulation, along with medium- and long-term evaluation are described.

A2.1 WHAT IS EQUIPMENT PERFORMANCE EVALUATION?

A2.1.1 Importance of Equipment Performance Evaluation

In Section 1.2 of this book, energy efficiency, cooling/heating capacity, etc. were listed as important equipment performance values, but only when appropriate equipment performance evaluation is carried out and disclosed on retail displays etc. can it be used as a basis for decision-making when selecting equipment.

The reasons for selecting air conditioning and other heat pump equipment include replacement due to breakdowns, new installation in new buildings, and renewal during refurbishment. At such times, without appropriate performance evaluation, the advantages of high-efficiency equipment cannot be recognized and a high-performance heat pump is unlikely to be selected.

In the Japanese market, the Japanese Industrial Standard (JIS) defines the methodology for equipment performance evaluation, and manufacturers carry out and publish their evaluations in accordance with the JIS. In addition, whether the cooling/heating capacity and energy-saving performance are in line with the manufacturer's published values is verified by spot tests.

Here, measuring the amount of sensible heat (temperature) and latent heat (humidity) in the cooling capacity evaluation and the airflow rate are required and are measured in a large measurement room. In order to avoid measurement errors due to differences in equipment, in Japan, the measuring equipment of the Japan Air Conditioning and Refrigeration Testing Laboratory is designated as the standard, and quasi-originals are defined for each company's testing facilities to ensure that errors are maintained within ±3%.

A2.1.2 Types of Equipment Performance Evaluation Methodology

There are different types of performance measurement methods and evaluation indicators required to assess equipment performance. Focusing on energy consumption efficiency, which represents energy-saving performance, there is the coefficient of performance (COP), which is an instantaneous value such as the rated point, and annual performance factor (APF), which is a performance indicator throughout the year (see Section 1.2).

Calorimetric or air enthalpy measurements are mainly used to measure the performance of air conditioning and heat pump equipment. An overview is given in Figure A2.1.

The air enthalpy method measures only the capacity of the part through which airflows, whereas the calorimeter method also evaluates heat leakage from pipes, etc. as capacity.

However, the calorimeter method requires time for the temperature and other parameters to stabilize, making performance evaluation time-consuming. The calorimeter method is heavily used in small air conditioners, where the effect of heat leakage is relatively large, while the air enthalpy method is often used in medium and large air conditioners.

A2.1.3 Usage Scenarios for Equipment Performance Evaluation

Equipment performance evaluation is used as information measured and published in accordance with standards (that define testing methods and other related matters), and is used as a decision-making tool for selecting equipment, but it is also used as an indicator to judge the achievement of targets in various systems that require energy saving.

Method	Balanced Ambient Room-Type Calorimeter (calorimeter type)	Air Enthalpy Measurement Instrument (Air Enthalpy Method)
Image	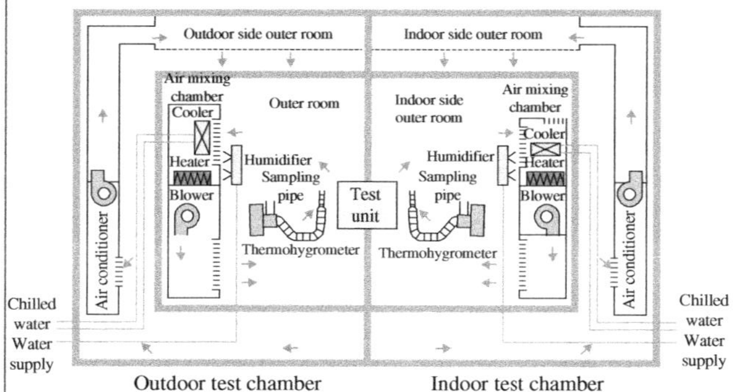	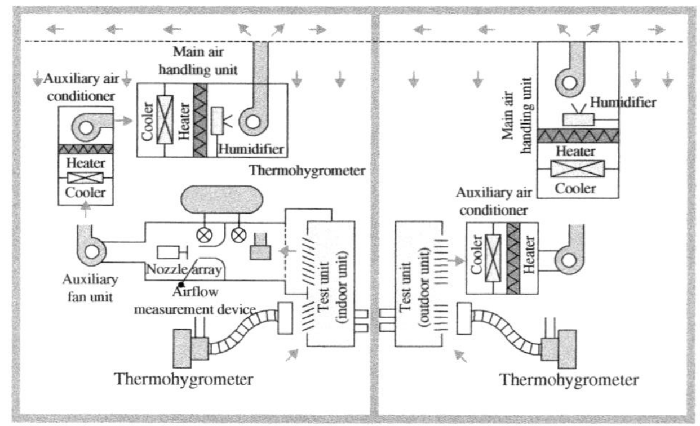
Procedure (Cooling)	The heater output is adjusted so that the added heat and the heat removed by the air conditioner are balanced, and the measured heater output is taken as the cooling capacity of the air conditioner.	Calculates the amount of heat cooled by the air conditioner by measuring the difference in enthalpy (heat value) between the inlet and outlet of the air conditioner and the air flow rate.

FIGURE A2.1 Types of equipment performance measurement methods. (From Japan Air Conditioning and Refrigeration Testing Laboratory, "Introduction to test equipment" [1]).

Under the Top Runner program of the Law Concerning the Rational Use of Energy (Energy Conservation Law), which is administered by the Ministry of Economy, Trade and Industry, residential air conditioners are specified in terms of APF as their energy saving performance.

The latest regulations in the Top Runner program call for achieving an APF of 6.6 as a weighted average of shipped products by 2027, except for cold-region specifications of products of 2.8 kW or lower. The Energy Conservation Law also requires that measured performance (APF, cooling capacity, etc.) be displayed at the point of sale, etc.

An overview of the Top Runner program under the Energy Conservation Law and the retailer labeling system is shown in Figure A2.2. Retail labeling must display the degree to which energy-saving standards are being met and the annual energy consumption under standard use. These allow consumers to ascertain, as a guide, whether the energy-saving performance and electricity consumption of air conditioners and other products are commensurate with their price.

On the other hand, the Ministry of Land, Infrastructure, Transport and Tourism, which has jurisdiction over the energy consumption of the entire building, requires the calculation of building energy consumption and the achievement of targets based on the rated capacity and COP at the rated point of air conditioning. In the net zero energy building (ZEB) assessment, which is the energy standard for the whole building, COP is used as the energy-saving performance, and equipment performance is simulated by using COP and the part-load performance of the equipment as a method to empirically estimate the performance of the equipment from its performance at the rated point.

A2.2 CHALLENGES IN EQUIPMENT PERFORMANCE EVALUATION METHODOLOGY

A2.2.1 Significance of Establishing Equipment Performance Evaluation Methodology

Product-specific performance evaluations in Japan and overseas have been carried out by fixing the frequency of compressors to ensure stable conditions in order to improve measurement accuracy. However, this method has a large deviation from actual operation, which includes transient conditions such as during start-up, and does not reflect the effects of the control of each piece of equipment and other factors. There is also room for improvement, for example, in the load assumption when evaluating the period with APF.

In recent years, in Europe and North America, studies have been conducted on performance evaluation methodologies under non-fixed compressor conditions, in response to suggestions that the operating conditions during actual use may not be reflected. In Japan, Waseda University is conducting research and development on a dynamic performance evaluation methodology using emulators, and publishing the results as appropriate.

These developments call for the establishment of performance evaluation methods under conditions close to those of actual operation[1].

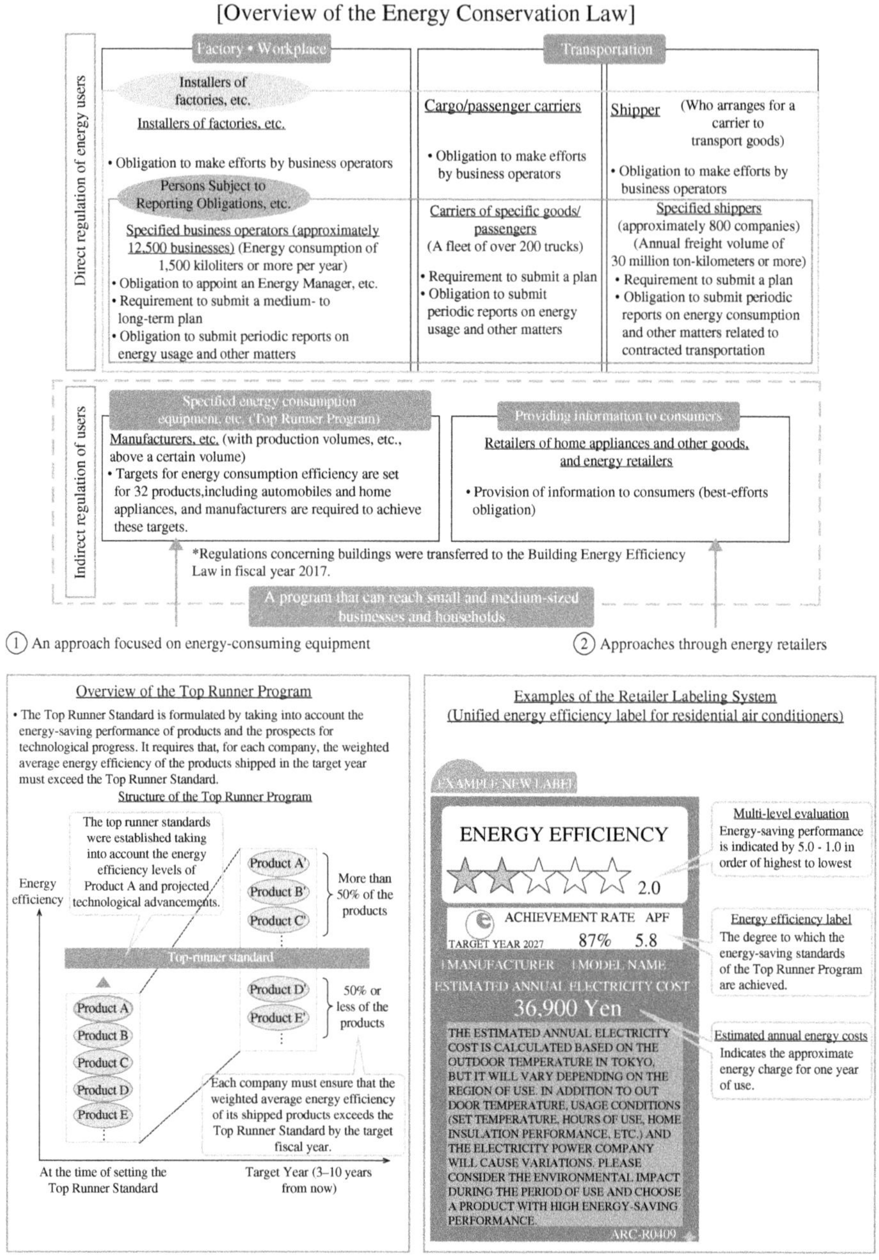

FIGURE A2.2 Overview of the Top Runner program of the Energy Conservation Law and the retailer labeling system. (From Ministry of Economy, Trade and Industry, Agency for Natural Resources and Energy, "Future energy saving and non-fossil fuel conversion policies on the energy demand side" [2, 3]; the Ministry of Economy, Trade and Industry, Agency for Natural Resources and Energy, "The unified energy-saving label has changed" [4]).

A2.2.2 Issues when Assuming Loads

A2.2.2.1 Challenges for Performance Tests Not Associated with Actual Operation

The performance of an air conditioner is difficult to measure accurately unless it is placed in a measurement chamber, and is evaluated by its performance in the laboratory under specific test conditions. For this reason, current standards and regulations tend to develop products in such a way that if the performance of an air conditioner is enhanced in a test room, the indicated performance can be higher than when the air conditioner is actually installed.

More than 40 years have passed since the introduction of microcomputer control, and while advanced control has become possible, the response of equipment to performance testing has also become more sophisticated. Without eliminating special controls such as the so-called test mode, where measured performance according to standards does not correctly reflect performance in actual operation, accurate performance evaluation is difficult. Performance standards and regulations must be considered in a way that promotes energy savings in actual conditions. Rather than developing air conditioners that perform well only in the test room, there is an urgent need to improve energy-saving performance under actual operating conditions, as a measure to combat global warming and save energy.

A2.2.2.2 Challenges in Measuring Rated Cooling Capacity at 35°C

In the current JIS standards and other standards, cooling period efficiency is calculated assuming that a load equal to the rated capacity[2] is generated when the outdoor temperature is set at 35°C. However, actual loads are often considerably lower than this. Historically, the rated cooling capacity has been measured at 35°C, so removing this measurement condition would make it impossible to contrast with older products. It is therefore likely that periodic efficiency is still calculated under these conditions.

Usually, the capacity of the equipment is selected as the sum of the respective maximum values of the internal load[3] and the external load[4], with an additional margin. However, it is seldom the case that the internal load and the external load generate their maximum values at the same time. For this reason, the current performance calculations are generally excessive as an assumption of the load relative to the outdoor temperature. An analysis of air conditioning field data by the Japan Refrigeration and Air Conditioning Industry Association showed that the average load in actual conditions is about half of the rated capacity at an outdoor temperature of 35°C during cooling (Figure A2.3).

In Figure A2.3, the actual average cooling load at 35°C is about half of the assumed load, but on the other hand, pull-down operation[5] is essential, for example, when returning home from an outing. In such cases, the temperature in the room is lowered in the area where there is a margin of capacity relative to the cooling load.

Therefore, if equipment is selected with a capacity that is too close to the load, the pull-down and pull-up times will become longer, which may bring the level of satisfaction down.

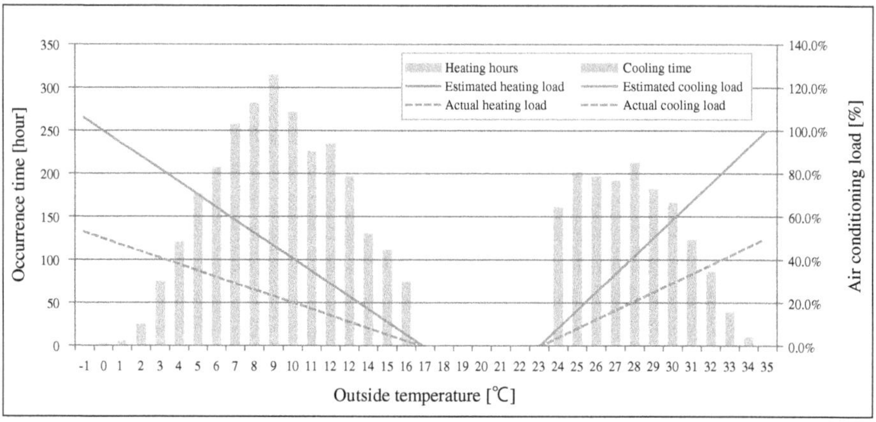

FIGURE A2.3 Assumed and actual loads in ISO 16358. (From Japan Refrigeration and Air Conditioning Industry Association, Survey Bulletin "ISO 16358/JISC 9612" [5]).

If the results of the Japan Refrigeration and Air Conditioning Industry Association's survey are interpreted in reverse, this means that at present, on average, the rated capacity is about twice the load, which simply means that half of the rated capacity is the capacity to perform pull-downs. Even assuming that twice the current pull-down time is acceptable, a margin of about 25% is required.

A2.3 CHALLENGES IN THE MEASUREMENT OF PERFORMANCE EVALUATION

A2.3.1 Challenges in Assuming Operating Conditions

Currently, discrepancies between the performance evaluation of air conditioners according to standards and the actual performance are becoming an issue in many parts of the world. Specifically, this has been pointed out in reports on the preparation for revision by the Energy-related Products (ErP) Directive in Europe.

For this reason, the International Energy Agency (IEA) has started to develop more realistic assessment methods [6]. In addition, as explained in Section A2.2.1, studies into a new performance evaluation, a load fixation test[6], are underway in the USA and Japan [7]. On the other hand, the operating conditions of air conditioner in the actual installation environment are not well known and, in many cases, development is likely to be based on the test conditions in the current standards.

In the actual environment, it is thought that a great variety of operations are carried out, and it is necessary to develop technology and products that minimize energy consumption across all of these operations. Performance evaluation for this purpose must be specified.

Preference will be given to equipment and manufacturers that have made progress in saving energy based on actual conditions, which will lead to further energy savings. Looking toward the future, advances in Internet of Things (IoT) technology are making it possible to secure data on the actual operating conditions of the large

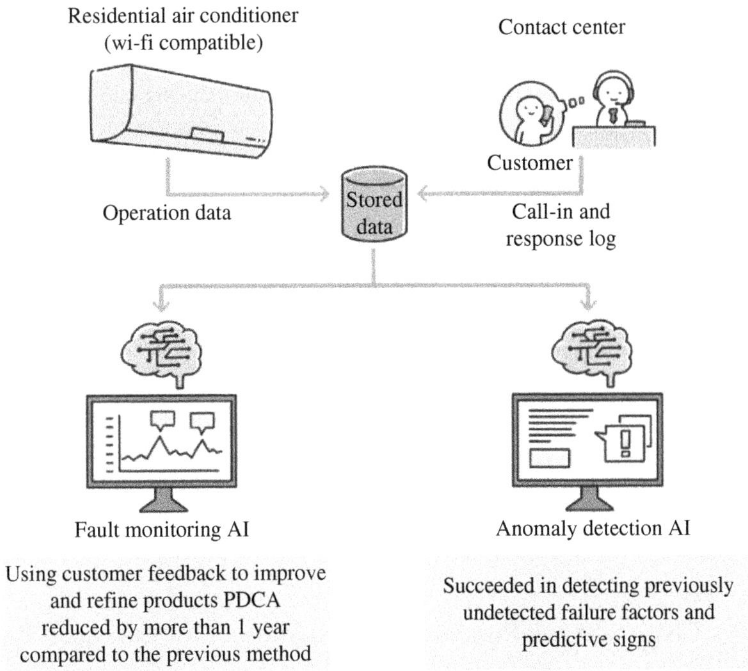

PDCA is a continuous improvement cycle of planning, doing, checking, and improving

FIGURE A2.4 Overview of joint development by Daikin Industries and JDSC. (From Daikin Industries, "Daikin Industries and JDSC jointly develop fault monitoring and operation abnormality prediction detection AI using IoT data of air conditioning equipment" [8]).

number of air conditioners on the market. The analysis of this data is just beginning to make real progress [8] (Figure A2.4).

The performance of air conditioners at present is usually measured by fixing the frequency of the compressor, etc., even for products with variable speed compressors. This has been the measurement method since the days when air conditioners were only available with constant speed compressor. It is also a remnant from the days when the main purpose was to check whether the actual maximum capacity was met. For this reason, the current situation is unavoidable in some respects, but as a standard for evaluating efficiency, it is often criticized for not reflecting the actual conditions of the installation environment.

The adoption of inverter compressors, fans, and electronic expansion valves has given a lot of freedom to air conditioner controls. Performance is measured in a test mode of operation with fixed operating conditions, and period performance is calculated based on these results. There is no problem if the control pattern in this test mode matches the control pattern that occurs in continuous control during actual operation. However, energy savings in this test mode alone are merely an improvement in the performance of the product in the test mode, and there may be a discrepancy between the control in normal operation and the control in the test mode.

A2.3.2 Challenges with Humidity Conditions

Considering the air conditioning load, it is necessary to evaluate not merely the sensible heat but also the latent heat load. In actual operation during cooling, dehumidification is almost inevitably required due to the rise in humidity levels caused by people perspiring and breathing. JIS and ISO standards have no particular restrictions on operations in which dehumidification does not occur, so the properties of such operations are evaluated, which has become problematic in other countries. In order to make dehumidification mandatory, there have been moves to set limits on sensible heat ratios[7] and proposals to set limits on airflow per capacity, and some of these have been standardized.

On the other hand, the humidification required during heating differs from dehumidification during cooling. As cooling is a completely separate function from heating, it is not covered as outside the scope of normal air conditioner performance standards.

When considering the functions required for air conditioning and the energy consumption associated with its operation, humidification should be included in the assessment.

A2.3.3 Challenges for Performance Evaluation in the Low Load Range

It was pointed out that the maximum capacity of heat pumps is often marginally higher than the actual load, but ensuring a margin of maximum capacity also leads to inefficiencies at low loads.

Generally, when the load falls below a certain level, heat pumps operate intermittently, resulting in reduced efficiency. In intermittent operation, losses occur when the compressor stops, as the compressed refrigerant flows from the high pressure side to the low pressure side and part of the compressor's work when it is restarted is used to restore this back-flowing refrigerant and transferred heat.

When the maximum capacity increases in a heat pump, the minimum capacity, which is intermittent operation, increases at the same time. Therefore, if the maximum capacity has an unnecessary margin, the low load range of intermittent operation will also increase, causing a reduction in overall efficiency.

The coefficient that expresses the reduction in efficiency of intermittent operation relative to continuous operation is called the coefficient of degradation (CD) value. For example, Figure A2.5 shows the CD value for cooling operation in the performance evaluation of residential air conditioners, in which the CD value = 0.25, i.e. a 25% reduction in efficiency compared to continuous operation.

Another problem is the difficulty of measuring actual CD values.

Currently, it is difficult to measure the actual CD value of inverter air conditioners due to the small absolute value of the capacity to be measured, and default values are actually used in the majority of cases.

However, CD values vary according to operating conditions and equipment capacity, so accurate evaluation and handling of start-stop operation is necessary. In Japan, the CD value has already been reviewed in JIS B8615, the performance standard for commercial air conditioning, and 0.5 has been adopted, but there are

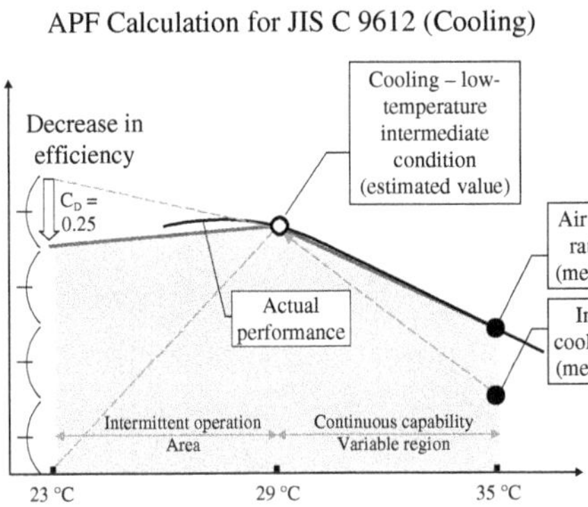

Tests used for APF calculations

	Outside temperature	Indoor temperature	Output	JIS C 9612: 2013
Minimum cooling	35°	27°	Minimum	○
Intermediate cooling	35°	27°	Intermediate	●
Rated cooling	35°	27°	Rated	●
cooling – low-temperature intermediate	29°	27°	Intermediate	○
Minimum heating	7°	20°	Minimum	○
Intermediate heating	7°	20°	Intermediate	●
Rated heating	7°	20°	Rated	●
Low-temperature maximum heating	2°	20°	Maximum	●

● Required test/○ Optional test

FIGURE A2.5 CD value during cooling operation in performance evaluation of residential air conditioners. (From Ministry of Economy, Trade and Industry and Agency for Natural Resources and Energy, "Current Status of Air Conditioners" [9]).

discussions that a larger value should be used. In the USA, on the other hand, the value has been changed from 0.25 to 0.2.

The CD value on inverter air conditioners has not been discussed or examined because of the narrow start-stop timeframe and the fact that it does not have a significant impact on APF. However, if the average load is reduced by reviewing the load distribution, the impact of the CD value may become larger and needs to be re-examined.

However, measurement on individual products is not easy in practice. This is because it is difficult to ensure measurement accuracy as the measurements are carried out at low loads and under start-stop conditions.

For this reason, it is anticipated that the calculation will be based on the control specifications of the expansion valve and the amount of refrigerant filled.

A2.3.4 Challenges in the Measurement Environment

A2.3.4.1 Set Temperature of the Measurement Environment

As mentioned above, the equipment for measuring the performance of air conditioner is quite large and requires an investment of up to 100 million yen for one facility. These facilities blow air indoors at a relatively high airflow rate to eliminate temperature irregularities to ensure stable readings (Figure A2.6).

As a result, the temperature distribution is much smaller than in a normal room, creating conditions conducive to measurement. On the other hand, due to the high airflow, the influence of the room walls on the heat levels tends to be reflected in the air temperature in a short time. The temperature response is very slow compared to a normal room where air conditioner and other equipment is actually installed. For this reason, if the currently proposed load fixation tests are simply carried out with a calorimeter, they often record very different behavior to that of the actual room.

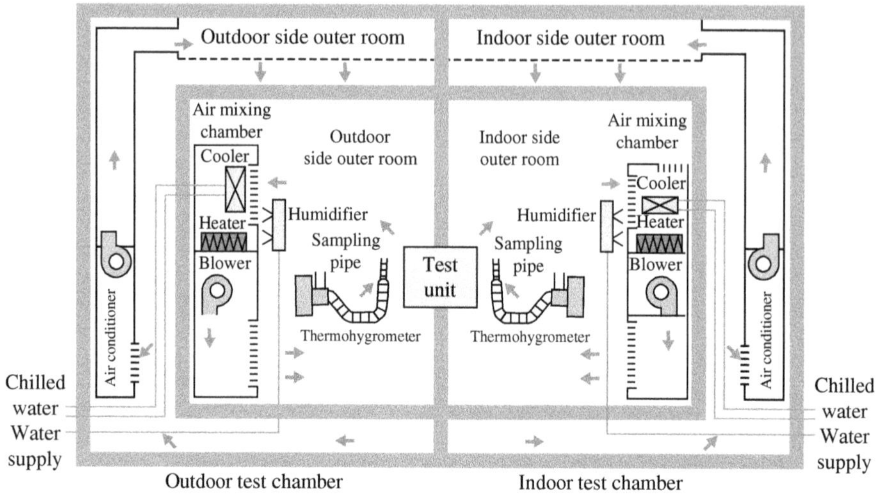

FIGURE A2.6 Example of a measurement environment. (From Japan Air Conditioning and Refrigeration Testing Laboratory, "Introduction to test facilities" [10]).

A2.3.4.1.2 Setting Conditions for Outdoor Unit Suction Temperature

The set temperature of air conditioner is controlled so that the temperature is set near the center of the room.

In wall-mounted indoor units, the actual suction temperature is usually higher than that of the center of the room, and control is carried out in anticipation of this.

On the other hand, the test room is controlled so that there is as little temperature difference as possible. Therefore, the temperature difference between the center of the room and the inlet area is eliminated and the average temperature of the room is higher when left to the control of the equipment. If the load is reduced, it is assumed that the temperature difference between the center of the room and the intake will be smaller, so the air conditioner will be controlled so that the temperature at the intake is lower. If this is done in the test room, the overall room temperature will be lower because there is no temperature difference with the center of the room.

A temperature difference is the normal temperature controllability and is considered to be the condition that should be evaluated, but this temperature difference is likely to vary from one test room to another. This would result in a loss of reproducibility of the suction temperature. This forces the test room to eliminate temperature irregularities, but it must be recognized that this is an operation that differs from the actual situation, and is a difficult task.

A2.3.4.1.3 Setting Outdoor Temperature and Outdoor Unit Suction Temperature

Another error in load assumption is the difference between outdoor temperature and suction temperature. Current efficiency calculations are based on the assumption that outdoor temperature = air conditioner suction temperature, but the average

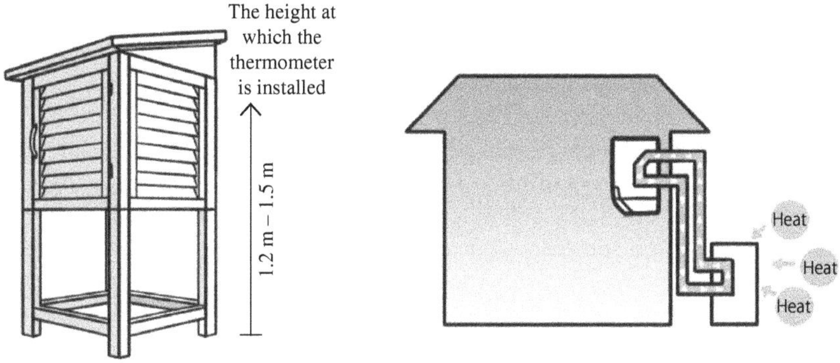

FIGURE A2.7 35°C outside air temperature (in a Stevenson screen/Hyakuyoubako) and air conditioning suction position.

suction temperature of actual air conditioners is usually shifted a few degrees to the more challenging temperature side.

An outdoor temperature of 35°C means that the air temperature inside a Stevenson screen (white painted box or "Hyakuyoubako") in the shade on a lawn at a height of 1.2–1.5 m is 35°C. However, the air conditioner intake is usually much lower and is affected by the ground temperature.

During the summer when used for cooling, air is sucked in through asphalt and concrete surfaces that have been heated by the sun, and during winter, for heating use, cool air in contact with the ground surface, which has cooled below the air temperature through radiative cooling, is also sucked in. For this reason, a simple assessment such as "distribution of air temperatures = distribution of air conditioner intake temperatures" leads to wide margins of error. The actual suction temperature should be considered to operate under conditions several degrees Celsius stricter than the air temperature (Figure A2.7).

A2.4 NON-AIR CONDITIONING PERFORMANCE EVALUATION CHALLENGES

A2.4.1 Issues for Commercial Refrigeration and Freezing Equipment

The energy consumption of commercial refrigeration and freezing equipment is also considerable and of high importance in the fight against energy saving and global warming. However, as the majority of these handle food and pharmaceutical products, hygiene and safety requirements are absolute, and awareness of the need to improve energy-saving performance tends to be relatively low. The diversity of products also hinders performance evaluation.

In freezing and refrigeration applications, there is a wide variety of objects to be cooled, differing from air conditioning, which aims to cool and heat only the living environment. Since the target temperatures and humidity are different, the product forms are also diverse. For this reason, performance standards have been established for built-in showcases, where the configuration of the equipment is specified, and they are regulated as Top Runner target equipment. However, separate type showcases, which have various system configurations combining condensing units and showcases, have not yet been standardized.

On the other hand, in the case of separate type showcases, condensing units and showcases often come from different manufacturers, and in many cases the performance and energy efficiency when actually operated in combination is unknown. Although it may be difficult to require these indications now, if, for example, the registration of operating characteristics was inputted into a common database and promoted as a government initiative, users would be able to evaluate energy efficiency and greenhouse gas reduction effects. If the number of years over which the increased initial investment can be recovered through reduced running costs etc. becomes clear, it can be expected that users' energy-saving awareness and behavior will improve and change even for separate type showcases.

A2.4.2 Challenges for Heat Pump Water Heaters

Heat pump water heaters are heat storage water heaters developed primarily to meet water heating demand by utilizing electricity late at night, when electricity demand is lowest, in order to equalize the electricity load.

On the other hand, the recent expansion of photovoltaic (PV) power generation as a renewable energy source has created a need to increase self-consumption of electricity. For this reason, products such as the Ohisama Eco Cute[8] have been commercialized in response to the future expansion of PV installation.

Performance evaluation of water heaters involves measuring energy consumption when hot water is delivered according to a standardized water heating pattern, which is a very time-consuming test because of the learning function used with heat pump water heaters.

Generally, it is more energy efficient to reduce the amount of hot water stored, but if the hot water is used up, the original water heating function is not satisfied. Learning control and other methods are used to minimize the hot water storage volume, but this complicates performance measurement. The current JIS is a very time-consuming test and should be simplified by supplementing the characteristics through simulation and AI.

A2.4.3 Challenges in Hot Water Heating

In Japan, the use of hot-water heating is limited, with only a few products having a hot-water floor heating function added. In Europe, on the other hand, hot-water heating has historically been used with hot water from boilers, and there is a large market for hot-water heating. The market is also expanding in China, where measures are being taken to promote hot water heat pump heating from the perspective of preventing air pollution. If the heat pump heating market is to develop in earnest in cold regions where air conditioning is usually used to heat the air, the hot water sector will likely need promoting.

On the other hand, hot water systems have a large heat capacity, which provides an adequate heat source for defrosting, avoiding the problem of the cold air feeling of defrosting operations in air conditioning. For full-scale heating in cold climates, air conditioning systems may be more likely to be more widespread than air conditioning systems that heat the air directly.

A2.4.4 Challenges Related to Aging of Performance in Actual Operation

Currently, performance evaluation indicators for heat pump equipment are evaluated during constant operation or over a certain period of time after operating performance has stabilized during product manufacture. When categorized by evaluation timing, the main focus is on advance information at the time of production, but actual performance evaluation after installation and actual operational evaluation over a long period of time, such as the degree of degradation and failure rate, are not often carried out.

Essentially, equipment performance changes with the passage of operating time, heat exchangers degrade, and compressor performance reduces over time, as do the maintenance and installation status. The evaluation indicators currently used are evaluations at the time of manufacture—they do not look at actual operation. An overall life cycle assessment is required and will need to be considered in the future.

A2.5 EVALUATION OF TEST CONDITIONS CLOSER TO THE ACTUAL OPERATING ENVIRONMENT

The current status and issues of equipment performance have been discussed. In order to carry out more appropriate performance evaluation, the following three points, classified according to evaluation purpose and timing, are described in terms of evaluation methodology for the future

1. Evaluation based on test conditions closer to the actual operating environment: Section A2.5
2. Evaluations based on the actual installation environment: Section A2.6
3. Assessments based on medium- and long-term aging: Section A2.7

A2.5.1 Assessment under Conditions Appropriate to Actual Conditions

A2.5.1.1 Top Runner Program for Energy Saving

The Top Runner program has been adopted in Japan since 1998 and is one of the evaluation criteria for determining the energy consumption efficiency of equipment. It is an energy-saving measure that requires the top performing product at a certain point in time to exceed that standard in a few years' time, based on the weighted average of units shipped by each manufacturer (Figure A2.8).

The third round of energy-saving obligations under this system was set for FY2022, and the system is becoming saturated.

In residential air conditioners, performance evaluation is based solely on performance at the midpoint and rated point, and performance tends to be optimized

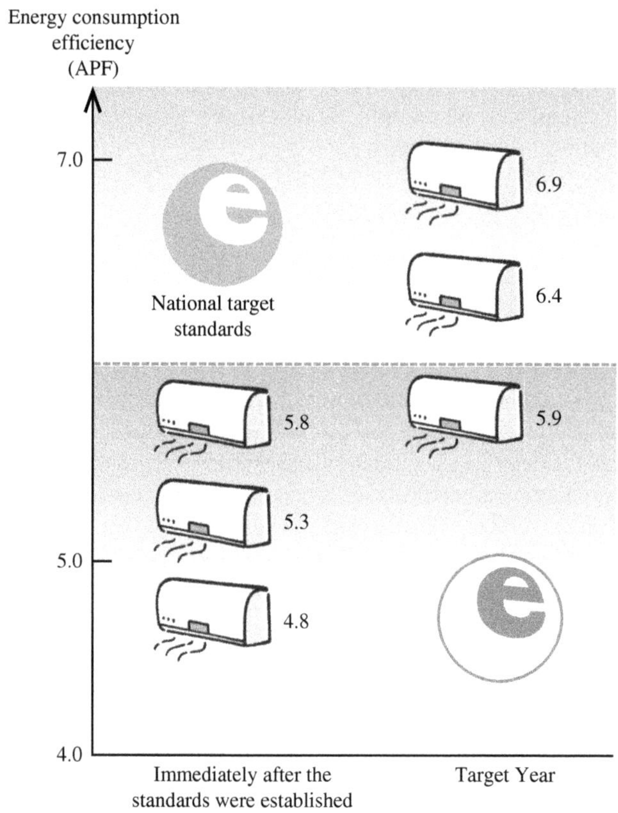

FIGURE A2.8 Top Runner program. (From Ministry of Economy, Trade and Industry, Agency for Natural Resources and Energy, "Current status of energy-saving equipment—Systems to promote the spread of energy-saving equipment" [11]).

between these two positions. However, in actual operation, the majority of the time, the air conditioner is operated at loads below this intermediate capacity, and it is more important to operate under conditions that are outside the measured points, which are supposed to be optimized according to current standards.

It is therefore desirable to bring the performance evaluation points used for these periodic performance evaluations closer to the conditions that are more likely to occur in actual operation. By doing so, it will be possible to evaluate points that have not been paid much attention so far, and it is thought that energy savings can be promoted.

Evaluation of performance at a load of less than 50% has not been carried out for residential air conditioners, but it was confirmed that the operating time is relatively important, so there is considerable room for improvement, including operating at the minimum frequency. In particular, the evaluation during start-stop operation is rarely evaluated in practice, as the impact has conventionally been small and the default value, the CD value, is calculated at 0.25, which does not have much of an adverse effect. However, the reality is that, depending on the product, the ratio of operation in the start-stop area is often high, even for inverter air conditioners.

A2.5.1.2 Performance Evaluation Considering Energy Efficiency and Product Cost

As mentioned above, domestic and international standards will be revised to promote energy efficiency and conservation from a new perspective by reviewing the load distribution in line with the actual situation and by making it mandatory to measure performance in the start-stop area and at low loads.

In addition, there are many conditions where not only sensible heat load but also latent heat load needs to be treated, and dehumidification capacity also needs to be evaluated. Limits on sensible heat ratios have already been standardized in the USA, and there are proposals in Europe to limit sensible heat ratios by limiting the airflow per capacity. If the load is to be closer to the actual load, a certain latent heat load would be added to the test room, and the humidity achieved during air conditioner operation would be used to evaluate the load.

However, the display of humidity alone may be difficult for general consumers to understand, and it is necessary to display a discomfort index or other information to enable them to understand the comfort level. Performance evaluation including latent heat capacity in some way seems essential.

In particular, in the case of desiccant air conditioning with latent–sensible separation, it is difficult to make comparisons unless the latent heat load is accurately given and measured. However, as new technology development is required on both the measurement and product technology sides, it should take some time to revise standards and change regulations.

Specific product changes include lowering the excluded volume of the compressor and rating the higher load of the operating frequency, which is expected to improve the performance to a lower load. The motor can also be downsized as the required operating torque can be reduced. For larger products, design changes are foreseen, such as the installation of multiple compressors to allow continuous operation up to low loads. These specification changes will increase product costs, but they will also reduce energy consumption, which will naturally reduce their cost.

Ultimately, reducing CO_2 emissions as a measure against global warming involves investing costs in equipment to increase performance and reduce power consumption. The conventional balance between initial and running costs would be toward the input of product costs and the reduction of energy costs.

A2.5.1.3 Performance Evaluation of Air Conditioner in Terms of Heating and Cooling Load Criteria

Globally, the heating and cooling load of air conditioning is estimated much higher, as is ISO 16358. For example, the European EN 14825 has similar estimates to ISO 16358.

The Air-Conditioning, Heating, and Refrigeration Institute (AHRI) standard allows a certain amount of leeway in capacity selection, but perhaps because it is based on the American idea that air conditioners should be operated all the time, the leeway in selection is not large, and similarly in Europe and the USA, the actual load is probably much lower than the assumed load in the period of performance calculation.

Conventional energy saving has required high performance in medium to large load operations. The standard was such that a large amount of refrigerant had to be used in large heat exchangers.

By reviewing the loads and bringing them closer to reality, the load distribution will shift to the low load side. This will lead to an increase in start-stop operation, and it will become clear that increasing the amount of refrigerant with an emphasis on large-load performance, as is now the case, is in fact worsening performance under start-stop conditions.

In any case, it is hoped that standards and regulations will be developed for the future to promote energy savings in actual conditions.

A2.5.2 Performance Evaluation Taking Operating Performance into Account

A2.5.2.1 Issues in Test Conditions

In relation to the optimization of test conditions, it is desirable to update the current performance evaluation to be more in line with the actual conditions of use, even for issues that are evaluated during constant operation after the operating performance has stabilized or for a certain period of time.

As mentioned in Section 2.3.1, the performance evaluation in constant operation ignores the control. The effects of frequency fluctuations caused by discrepancies between actual loads and equipment capacity, for example, are not evaluated at all, which is recognized as an issue both in Japan and overseas.

In response to this issue, research and discussions have been conducted both in Japan and elsewhere to reflect this issue in the performance indicators of the load fixation test and dynamic test (dynamic performance evaluation), which are close to the actual conditions of use. In Japan, Waseda University is continuing the technical development of the load fixation test [12]. Outside Japan, load fixation tests are also being studied in the USA [13] and Germany [14]. In addition, building-related institutes in the country are continuing to try to fix loads, and measure capacity and efficiency [15].

If load fixation tests could be carried out easily and accurately, the issues would be solved at once, but in practice it is not so easy. It is necessary to rectify the differences in dynamic characteristics between the air conditioner performance test room and the actual installation environment. Despite these issues, it seems essential to bring the air conditioning load closer to the actual situation and to evaluate the energy-saving performance of the air conditioners, including the control of the equipment.

A2.5.2.2 Performance Evaluation in Automatic Operation Mode

To measure performance in conditions that reflect actual operation, it may be necessary to utilize automatic operation mode and evaluate operation.

As mentioned earlier, the operating condition of air conditioner varies with the load, outdoor temperature conditions, indoor temperature, set temperature, and airflow rate settings. Evaluating capacity and efficiency by varying these parameters is not realistic as it would require a very large number of measurements.

On the other hand, current performance evaluations of air conditioners use maximum or considerably high airflow settings. In some respects, this is unavoidable since there were no inverter air conditioners and only constant speed air conditioners were available in the past. However, if energy savings are to be promoted based on actual facts and figures, performance evaluation should be carried out using operational settings that are actually used and in large numbers. In fact, IoT data confirms that a significant number of users operate in automatic mode or at low airflow.

In addition, as air conditioners have many operating parameters, such as air direction, air volume, and set temperature, it may be necessary to evaluate air volume and other parameters in automatic mode, as well as encouraging users to operate in automatic mode as much as possible during actual use.

A2.6 EVALUATION BASED ON THE ACTUAL INSTALLATION ENVIRONMENT

A2.6.1 Performance Evaluation Using Digital

Assuming that the digital environment, including communication and information processing, will develop further in the future, actual performance evaluation using digital technology after installation of the equipment, which is not often done at present, is expected to be carried out.

One of the expected methods of utilizing actual performance assessments is the establishment of a method based on the real-time operational status of the equipment. In the current performance evaluation, results measured under acceptable conditions are used as-is, as performance data. However, by utilizing operating data after installation, performance evaluation can be carried out in a more realistic manner.

As for the use of operating data, in addition to performance evaluation by directly utilizing the data, performance evaluation that is fused with simulation, known as digital twin, can also be considered (Figure A2.9).

By calculating the coefficients required for simulation from the actual measurement results of performance measurements, it is possible to predict the performance

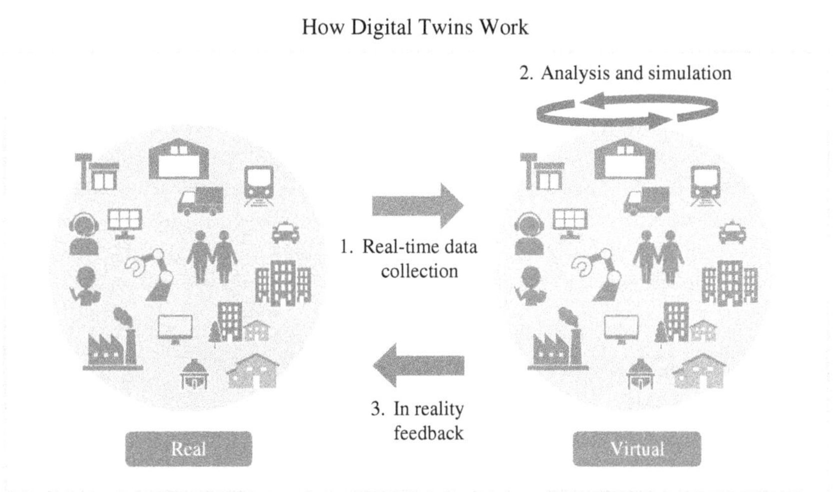

FIGURE A2.9 How digital twins work. (From NTT DOCOMO BUSINESS Inc., ICT Business Online,"Surprisingly, what is the IT trend term digital twin?" [16]).

of a wide variety of operating ranges based on the actual measurement results of several points. Errors in operating conditions within acceptable limits are corrected, which is expected to improve the accuracy of the performance of each measurement point.

Applying this calculation method and experimental results to building simulations would also solve the problem of data mapping etc. In addition, if real-time evaluations can be carried out, energy consumption and environmental impact, which until now could only be estimated, can be understood in more detail.

This will make it possible to update operational control to match the energy-saving, comfort, and health orientation on the part of the user, based on an understanding of power consumption and environmental impact.

A2.6.2 Improving Efficiency Based on Actual Energy Consumption

When considering improving performance under specific load patterns and improving average performance over a wide range of loads, the latter is naturally more difficult to achieve. For example, when considering efficiency in motor operation with an inverter, COP varies considerably depending on what kind of voltage and compressor frequency relationship is used.

Under specific conditions, it is possible to improve performance by tuning, but if efficiency is to be increased over a wide range of operating conditions, the only way is to increase base performance by improving the input quantity and quality of materials.

If the relationship between temperature and load can be assessed, including the actual energy consumption of the delivery site, it will be possible to design accordingly, making it relatively easy to achieve a certain degree of efficiency.

A2.6.3 Simulating Operational Performance

One type of equipment other than air conditioner for which there is expected to be a need for evaluation based on the actual installation environment is the separate type showcase.

Currently, separate type showcases are often made by different manufacturers for condensing units and showcases, making it difficult to understand their performance and energy efficiency in advance. Separately installed showcases are obliged to register their performance characteristics in a public database. Simulating operational performance to help visualize the actual combinations could raise user awareness and promote the spread of energy-saving products.

A2.7 EVALUATION OVER TIME IN THE MEDIUM AND LONG TERM

A2.7.1 Improvement in Performance by Addressing Filter Blockages

It is anticipated that high performance will only be produced for a short period of time, but it is also expected that controls will be required to ensure reliability even when the operating mode is limited to 50 or 100 hours from the start of operation.

This may require responses such as monitoring changes in equipment control on the market. Otherwise, if performance is applied for in a control mode that honestly takes long-term reliability into account, it may be disadvantageous. In addition, from the perspective of medium- and long-term performance evaluation, monitoring changes in long-term capacity and COP using IoT data could lead to new energy-saving possibilities.

Currently, the filter cleaning function is not evaluated at all in terms of performance evaluation. However, in actual operation, it should contribute to performance improvement in the medium to long term, as airflow is maintained by suppressing filter clogging.

Such a medium-term performance evaluation is not easy but should be considered in future performance evaluations closer to the actual situation.

In the draft ErP Lot 10 regulation, the EU energy efficiency regulation, there is an attempt to incorporate the response to filter blockages into the evaluation as part of the performance evaluation indicators [17].

A2.7.2 Performance Evaluation in Commissioning

As an example of medium- to long-term performance evaluation, commissioning[9] may be carried out periodically after handover (i.e. after operation) in large commercial buildings (Figure A2.10).

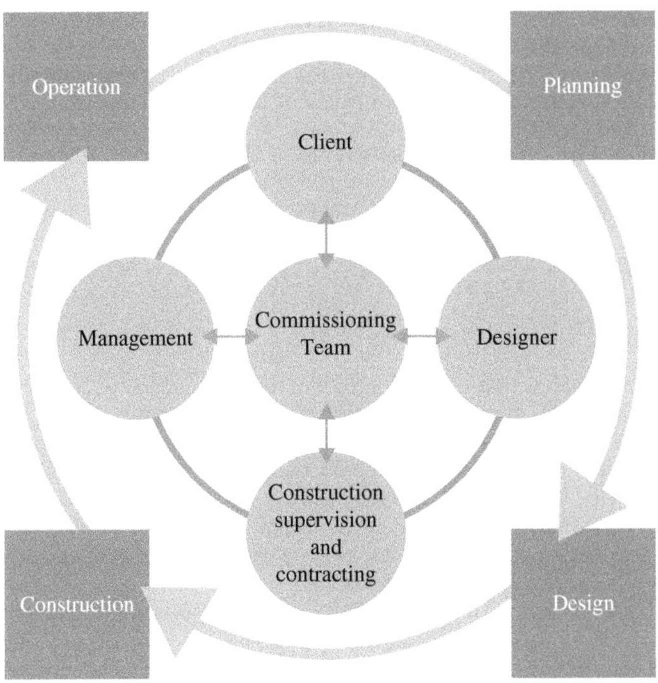

FIGURE A2.10 How commissioning works. (From NPO Building Services Commissioning Association, "What is commissioning?" [18]).

Commissioning is not limited to heat source equipment such as heat pumps, but covers the entire building. The aim is to diagnose and verify performance as required and, if necessary, to provide performance improvement methods so that the building is kept in optimum condition for the user.

The need for such post-operational performance verification is expected to increase in the future.

NOTES

1. Conditions close to actual operation: Includes not only environmental assessment conditions but also operating conditions that reproduce the dynamic movement of the equipment.
2. Rated capacity: The capacity that the equipment can stably produce when it is operated continuously under conditions based on the JIS conditions.
3. Internal loads: Loads generated indoors, such as people and lighting.
4. External loads: Loads, mainly heat loads, that enter from outside the room through walls, etc.
5. Pull-down operation: Cooling from a room temperature equal to or higher than the outdoor temperature to a comfortable temperature.
6. Load fixation test: Operation of air conditioner and other equipment without fixing the operating state of the equipment, leaving it to the control of the equipment and keeping the load constant. By keeping the load constant and leaving the operation to the equipment control, it is an attempt to prevent high-performance operation only in special test modes.
7. Sensible heat ratio: The ratio of the sensible heat portion of the heat to the sum of the sensible and latent heat portions of the heat to be pumped out.
8. Ohisama Eco Cute: Heat pump water heater that uses PV surplus electricity to meet water heating demand.
9. Commissioning (Cx): Commissioning, also known as performance evaluation, is the compilation of the required performance of the target system from environmental, energy, and user-friendliness points of view, as required by the user. Throughout the process of design, construction, and delivery, it advises, reviews, and confirms the decisions and actions of those involved in performance verification in order to realize the performance and document the necessary requirements. It is also defined as carrying out functional performance tests to verify that the system to be delivered is in a state of proper operation and maintenance.

REFERENCES

1. Japan Air Conditioning and Refrigeration Testing Laboratory, "Introduction to test equipment," 2023, https://www.jatl.or.jp/testing-facilities.html
2. Ministry of Economy, Trade and Industry, Agency for Natural Resources and Energy, "40th meeting of the subcommittee on energy saving under the subcommittee on energy saving and new energy, comprehensive resource and energy study committee, secretariat document "future energy saving and non-fossil fuel conversion policies on the energy demand side," p.4, 2023, https://www.meti.go.jp/shingikai/enecho/shoene_shinene/sho_energy/pdf/040_00_04.pdf
3. Ministry of Economy, Trade and Industry, Agency for Natural Resources and Energy, "41st meeting of the subcommittee on energy saving under the subcommittee on energy saving and new energy, comprehensive resource and energy study committee, secretariat document "future energy saving and non-fossil fuel conversion policies on the energy demand side," p.6, 2023, https://www.meti.go.jp/shingikai/enecho/shoene_shinene/sho_energy/pdf/041_00_04.pdf

4. Ministry of Economy, Trade and Industry, Agency for Natural Resources and Energy, "The unified energy conservation label has changed," 2023, https://www.enecho.meti.go.jp/category/saving_and_new/saving/enterprise/retail/touitsu_shoenelabel/
5. Japan Refrigeration and Air Conditioning Industries Association, Preliminary Report "ISO16358/JIS C 9612," 2023.
6. European Commission, DG ENER C.3, "Space and combination heaters," Ecodesign and Energy Labelling, Review Study, Task 4 Technologies, Final Report, 2019, https://www.eceee.org/static/media/uploads/site-2/ecodesign/products/Space%20and%20combination%20heaters/boilers_task_4_final_report_july_2019.pdf
7. International Energy Agency (IEA) Technical Cooperation Programme Cadeo Group, "Domestic air conditioner test standards and harmonization, final report," 2020, https://www.iea-4e.org/wp-content/uploads/2020/12/AC_Test_Methods_Report_Final_V2_incl_JP_KO.pdf
8. Daikin Industries Ltd., "Daikin industries and JDSC jointly develop fault monitoring and operation abnormality prediction AI using IoT data of air conditioning equipment," 2023, https://www.daikin.co.jp/press/2022/20220301_2
9. Ministry of Economy, Trade and Industry, Agency for Natural Resources and Energy, "2nd meeting of the subcommittee on energy saving under the subcommittee on energy saving and new energy, comprehensive resource and energy study committee, working group on judgment criteria for air conditioners and electric water heaters," Appendix 4 Current Status of Air Conditioners, p.27, 2019, https://www.meti.go.jp/shingikai/enecho/shoene_shinene/sho_energy/air_denki/pdf/002_04_00.pdf
10. Japan Air Conditioning and Refrigeration Testing Laboratory, "Introduction of test facilities," 2023, https://www.jatl.or.jp/testing-facilities.html
11. Ministry of Economy, Trade and Industry, Agency for Natural Resources and Energy, Energy Conservation Portal Site, "Current status of energy-saving equipment—Systems to promote the spread of energy-saving equipment," Top Runner program, 2024, https://www.enecho.meti.go.jp/category/saving_and_new/saving/general/choice/
12. Niccolo Giannetti, Shun Matsui, Ryohei Mori, Jongsoo Jeong, Hifni Mukhtar Ariyadi, Yoichi Miyaoka, Eisuke Togashi, Kiyoshi Saito, Energy and Buildings Vol. 273, No. 1, pp. 3–24. Buildings Vol. 273 'Emulator-type load-based tests for dynamic performance characterisation of air conditioners', 2022, https://www.sciencedirect.com/science/article/abs/pii/S0378778822005825
13. Parveen Dhillon, Purdue University, 18th International Conference on Refrigeration and Air Conditioning, "Validation of a load-based testing method for characterizing residential air-conditioner performance," 2021, https://docs.lib.purdue.edu/cgi/viewcontent.cgi?article=3256&context=iracc
14. BAM, "Test guideline for dynamic performance testing and calculation of the seasonal coefficient of performance, Air conditioners with electrically driven compressors for space heating and cooling," 2019, https://netzwerke.bam.de/Netzwerke/Content/DE/Downloads/Evpg/Heizen-Kuehlen-Lueften/bam%20test%20guideline%20-%20dynamic%20testing%20of%20air%20conditioners.pdf.pdf?—blob=publicationFile
15. Masato Miyata (National Institute for Land and Infrastructure Management), KAKEN (Grant-in-Aid for Scientific Research, Ministry of Education, Culture, Sports, Science and Technology), "Grant-in-Aid for Scientific Research (C) Development of a Method for Evaluating the Energy Consumption Performance of VRF Air Conditioners Considering Dynamic Response to Load Fluctuations," 2021–2023, https://kaken.nii.ac.jp/ja/grant/KAKENHI-PROJECT-21K04387/
16. NTT DOCOMO BUSINESS Inc., ICT Business Online, "Surprisingly, what is the IT trend term digital twin?," 2023, https://www.ntt.com/bizon/glossary/j-t/digital-twin.html

17. EU Commission, EUR-Lex, Commission Regulation (EU) No. 206/2012, "Implementing directive 2009/125/EC of the European Parliament and of the Council with regard to ecodesign requirements for air conditioners and comfort fans (text with EEA relevance)," 2012, https://eur-lex.europa.eu/legal-content/EN/TXT/?uri=CELEX%3A32012R0206
18. Building Services Commissioning Association, "What is commissioning?," 2023, http://www.bsca.or.jp/outline/commissioning02.html

Appendix 3: Resource Circulation in Next-Generation Heat Pump Technology

OVERVIEW

In response to the social demand to create a circular economy, it has become necessary to review the supply chain and other aspects of resource circulation with a heat pump technology. Given the possibility of future price increases for resources such as copper, and concerns that the uneven distribution of fluorite, the raw material for CFs refrigerants, will make it impossible to ensure a stable supply, it is important to circulate and recycle resources as much as possible, without relying on virgin resources.

With this in mind, it is worth noting that residential air conditioners already have a high recommodification rate of 92%, so there is little room for solutions through product design, etc. However, increasing the recovery rate, which is only 38%, is an urgent issue. This is difficult to achieve if the existing way of selling equipment is still based on the current commercial distribution channel, and a new circular supply chain using AI and Internet of Things (IoT) needs to be established.

In this chapter, three scenarios are examined as directions for achieving this, namely the introduction of leasing and subscription services, stricter market equipment management through the introduction of an individual management system, and the promotion of resource circulation by strengthening the management of used equipment. A vision for the future is also proposed.

A3.1 NECESSITY OF SUPPLY CHAINS SUPPORTING RESOURCE CIRCULATION

A3.1.1 Relationship between Heat Pump Equipment and the Circular Economy

When considering the future supply chain of heat pump appliances, it is necessary to take into account the social trends toward the realization of a circular economy in addition to climate change measures.

Heat pump appliances can contribute to maintaining a comfortable lifestyle while reducing CO_2 emissions due to the high heat efficiency of their operating principle.

On the other hand, heat pump appliances use finite resources such as resin, metal, and refrigerant, and it is important to create a supply chain with a circular economy in mind.

Domestic and international trends toward the transition to a circular economy are described in Section 2.5, but it can be said that it is still taking some time for full-scale diffusion. The following challenges can be attributed to this, as pointed out by Khaled Soufani and others [1]

1. Consumers and policymakers are only now becoming interested in the concept of circular supply chains but are not yet at the point of taking concrete action
2. Most consumers are not willing to give up their function (e.g. convenience. etc.) for environmental sustainability at the moment

Overcoming this consumer mindset (orientation) is a major challenge for heat pump equipment manufacturers to address in their circular supply chains. In addition, as the circular economy progresses, production and consumption efficiency will increase. The rebound effect [2], whereby the increase in production and consumption is offset by the effect of reducing environmental impact, needs to be kept in mind when implementing it in society.

The issues concerning economic rationality and social aspects are unavoidable in developing a circular economy, but opinions differ greatly depending on the position taken, so this section will be explained from a technical perspective.

A3.1.2 Circular Supply Chains to Realize the Circular Economy

In considering resource circulation of heat pump appliances, this book proposes a supply chain, referring to the image of the circular economy presented in the Circular Economy Vision 2020 published by the Ministry of Economy, Trade and Industry (METI) in May 2020 [3], and describing both the current supply chain and the ideal one for heat pump appliances (Figure A3.1).

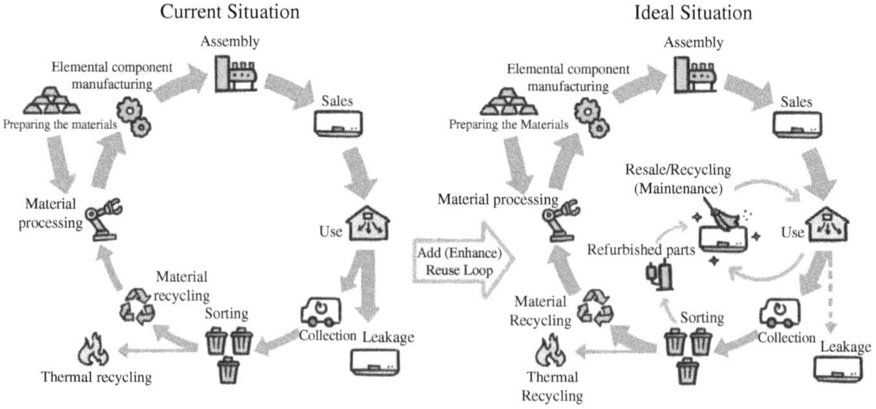

FIGURE A3.1 The current situation and the image of an ideal circular supply chain.

Our ideal circular supply chain is to incorporate a mechanism to promote recycling and re-use of the equipment itself in the sequence of procurement, manufacturing, sales, and consumption. On top of this, the supply chain loop is to have diversity and multiplicity, and a mechanism to promote product and resource circulation is to be implemented in society.

From the perspective of a circular economy and carbon neutrality, life cycle assessment (LCA) is a concept to quantitatively assess the degree of impact of a product on the global environment and is used for product environmental impact assessments.

In heat pump technology, studies on refrigerants, which are already widely known to have a significant impact on global warming, have been conducted in advance. During the development of heat pump equipment, especially when selecting and developing refrigerants, the product life cycle climatic load (life cycle climate performance (LCCP)) [4] has been used as an evaluation indicator. However, it is necessary to conduct an environmental impact assessment using LCA, not only focusing on LCCP for refrigerants alone, but also including other materials used in the equipment, and to consider the ideal circular supply chain with a low environmental impact from a comprehensive perspective.

The next section details the current supply chain for heat pump equipment and the challenges of creating a circular supply chain.

A3.2 PROCUREMENT AND PRODUCTION OF RAW MATERIALS

A3.2.1 Importance of Copper as a Resource

According to a report on the risk of metal resource depletion compiled by the National Institute for Materials Science in 2007 [5], many metals needed in the current market are facing resource depletion challenges. Furthermore, metal resources such as copper and iron, which are often used in heat pump equipment, were said to be in demand by 2050, exceeding or equaling the reserves that can be mined with the current economic viability.

Copper, in particular, will exceed economically viable mining by 2035, and the problem of resource depletion is imminent. According to the United States Geological Survey (USGS) estimate for 2022 [6], global economically minable copper reserves are estimated at 880 million tonnes. Furthermore, according to the USGS 2015 estimate shown in Table A3.1, there are 2.1 billion tonnes as currently proven copper resources, 3.5 billion tonnes as undiscovered deposits, and total copper resources[1] of 5.6 billion tonnes. In contrast, cumulative mine production is estimated at 600 million tonnes, so resource depletion is not an urgent issue.

While improvements in mining and smelting technology, in addition to improved exploration technology for copper resources, are thought to have contributed to the diminishing concern over resource depletion, the rise in copper prices is also considered to be a major factor. The price of copper has shown a steady upward trend over the past 20 years, with minor rises and falls, while maintaining stable market stocks, as shown in Figure A3.2.

TABLE A3.1
Global copper assessment results (Comparison with other measurements of global and US copper reserves, production, and consumption)

Parameter	Copper (Mt)	Comment	Source
World refined copper consumption	20.4 21.4 22.9	For 2012 For 2013 For 2014	International Copper Study Group (2015a)
World mine production	17.0 17.9 18.7	For 2012 For 2013 For 2014	Edelstein (2013) Brininstool (2014) Brininstool (2015)
Cumulative world mine production	600	Cumulative world copper production from 1879 to 2012	D.L. Edelstein, US Geological Survey, written commun., 2013
	567	Cumulative world copper production from 1892 to 2011	Mudd and others (2013)
World reserve	680 690 700	For 2012 For 2013 For 2014	Edelstein (2013) Brininstool (2014) Brininstool (2015)
Identified world resources	2,100	Past production and measured, indicated, and inferred resources at the lowest reported cutoff grade	Singer and Menzie (2010)
	1,860	Resources identified in 730 copper projects as of 2010, includes operating mines as well as deposits in development and exploration; all types; includes 80.4 Mt Chinese national resources	Mudd and others (2013)
Undiscovered world resources	3,100	Porphyry copper deposits (upper 1 km)	This study
	420	Sediment-hosted stratabound copper deposits (upper 2.5 km)	
Theoretical world resources	170,000	Porphyry copper deposits	Kesler and Wilkinson (2008)
	89,000	All deposit types (upper 3.3 km) recoverable	
	300,000	All types (entire crust)	
US copper resources in all deposit types	290	Mean undiscovered US deposits minable with existing technology as of 1998	US Geological Survey National Mineral Resource Assessment Team (2000, 2002)
	350	Identified US resources (260 Mt) and past production (91 Mt) as of 1998	

Note: Mt: million metric tonnes.
Source: Based on Jane M Hammarstrom et al., "Assessment of Undiscovered Copper Resources of the World, 2015" [7].

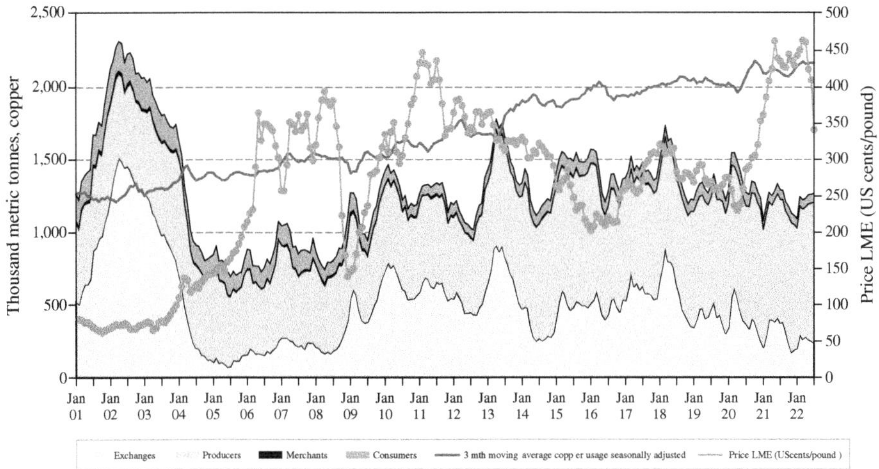

FIGURE A3.2 Copper stocks, prices, and uses (Jan. 2001–Jul. 2022). *Note:* Copper weight unit: thousand metric tonnes/price unit: London Metal Exchange (LME) price (US cents/lb). (From ICSG, The World Copper Factbook 2022 [8]).

Potential future mining targets include deposits that cannot be accommodated by conventional mining costs, such as submarine hydrothermal deposits that form near underwater volcanoes.

Furthermore, the rising price of copper has significantly influenced the gradual expansion of recoverable deposits, as it allows for the absorption of the costs associated with mining copper ore. This is considered a major factor in the increase in recoverable reserves.[2]

Copper is used in heat pump technology as a key material in heat exchangers and piping, and rising copper prices have a significant impact on production costs, which is an issue that requires attention. Various studies have been conducted due to depletion concerns [9], but as mentioned above, copper prices are on the rise, which should make it easier to absorb recycling costs. As active use of recycled materials is considered an effective means of stabilizing prices, further use of recycled materials is expected in the future.

Furthermore, as copper is a material suitable for recycling, the establishment of a system to recycle materials has been promoted for some time, including the analysis of resource circulation using material flow analysis (MFA) (Figure A3.3).

Analysis such as in Figure A3.3 shows that copper is being actively recovered and recycled, taking advantage of its excellent recycling properties. The proportion of copper recovered from heat pump products that is recycled again as parts of heat pump equipment is unknown, as material recycling is mainly based on shredding.

On the other hand, if the electrification of vehicle is expected as a major social change in the future, and high-grade copper resources are directed to the vehicle, where the unit price of products is higher, the supply and demand for copper will become more constrained.

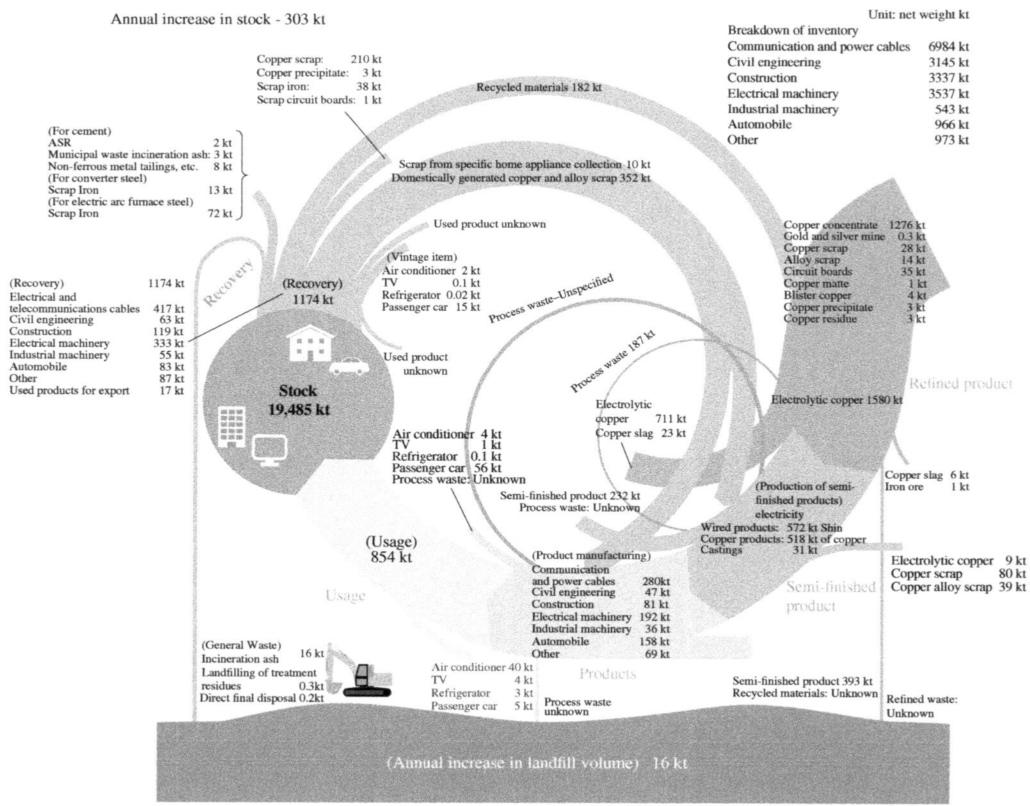

FIGURE A3.3 Estimation of copper stocks and material flows in Japan. (From Mitsubishi UFJ Research & Consulting Co., Ltd, "Summary of the Survey Results on Mineral Resources Recycling Flows and Stocks in FY2021" [10]).

In view of the above situation, the proportion of horizontal recycling, where waste material from heat pump equipment is reused as material for heat pump components, needs to be increased in order to actively secure recycled material. This will ensure freedom to choose materials in the design and manufacture of heat pump equipment, as well as a desirable cycle of stable prices and product quality.

A3.2.2 Importance of Resin as a Resource

Petroleum, the raw material of resin, is well known as a resource with a limited, minable amount[2], but in order to achieve carbon neutrality, attention is being paid to reducing the use of resins derived from fossil resources and to resource circulation of resins.

The material flow of resin materials is shown in Figure A3.4. While material recycling accounts for 22%, the amount used as thermal energy (thermal recycling) is approximately 63%. However, the petroleum used as a resin feedstock is not directly used as fuel, but the resin product is used once for another purpose and disposed of, and is also used for heat utilization as fuel.

One of the challenges of material recycling is that more costly processes may be required to meet the same quality requirements as virgin materials. On the other hand, fluctuations in the price of virgin materials derived from fossil fuel resources make it difficult to create stable cost benefits from recycled materials.

Focusing on PET drinks bottles, the recycling rate reached 86.9% in 2022 in Japan [12], the highest recycling rate among resin products, with bottle-to-bottle[3] also reaching about 44.6% in 2022 [13].

The bottle-to-bottle ratio, which was around 13% in 2012, improved significantly in the following 5 to 6 years. This is a clear indication of the technological development efforts of the relevant industries, and the selective recovery and recycling of homogeneous resins of the same quality for the same purpose of use can also contribute significantly to the maintenance and stability of quality after recycling and utilization.

Resin materials used in heat pumps are selected with a primary focus on the functionality of the components, cost, and stable quality.

In many cases, devices are manufactured by combining parts made of different materials. It is currently very difficult to convince consumers that increased use of recycled parts and materials will result in a decrease in product integrity, reduced functionality, and worsened ease of use. Therefore, recycled materials in every component is not easy.

However, when considering resin materials to be used in heat pump equipment in the future, the possibility of reducing the environmental impact will also be an indicator for material selection. Reviewing the choice of materials, such as biomass plastics, recycled materials, and materials other than resin, will be an important design issue.

A3.2.3 Environmental Impact Assessment of Materials Used in Heat Pump Equipment

Various studies are being carried out to reduce the environmental impact of material use in heat pump equipment, with resource circulation taking place.

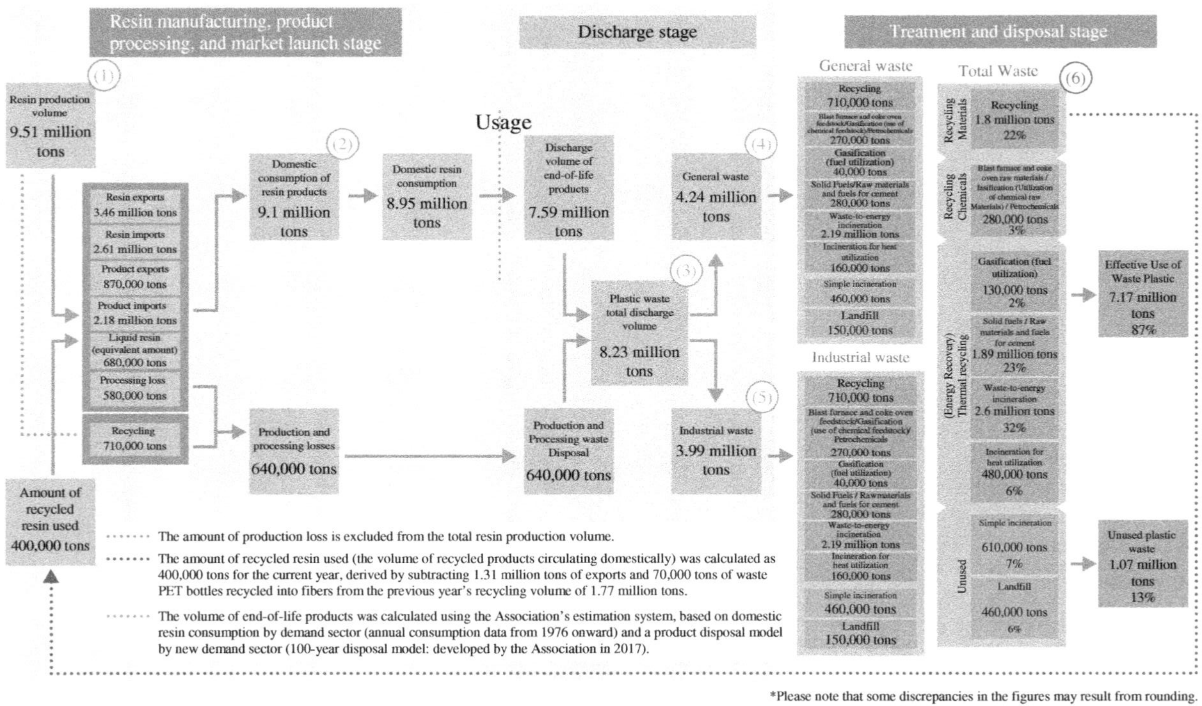

FIGURE A3.4 Material flow for plastic products in 2022. *Notes:* (3) The total amount of waste plastic disposed is classified into (4) "general waste" and (5) "industrial waste." (4) "General waste" includes, in addition to general waste, PET bottles and white trays from the business (voluntary collection) route, treatment residues from the refuse collection route, and waste plastic mixed with business general waste. (5) "Industrial waste" includes unused production and processing losses, and waste plastic that is traded for value. (6) Gasification is divided into chemical raw material usage and fuel usage according to the use of the recycled product, with chemical raw material usage included in chemical recycling and fuel usage in thermal recycling. (From Plastic Waste Management Institute, "2022 Status of Production, Disposal, Recycling, Treatment and Disposal of Plastic Products Material Flow Chart" [11]).

According to a report [14] that examined the relationship between the performance transition and resource utilization of residential air conditioners using the total amount of substances involved[4] (or total material requirement (TMR)) as an indicator, copper alloys, which are often used in heat exchangers and piping, account for about 85% and are a major environmental impact factor. This is a clear indication of the high environmental impact of copper mining.

However, it is possible that resource efficiency is deteriorating due to larger heat exchangers as a result of improved performance, as opposed to lower energy consumption during the usage phase due to yearly performance improvements [15].

Figures A3.5 and A3.6 show a case study of one representative commercial air conditioning (multi-air conditioner for buildings) system (one outdoor unit and four indoor units), where the environmental impact during manufacturing was analyzed from the perspective of LCA.

LCA estimation methods are still being actively researched and various calculation models for environmental impacts are expected to be proposed in the future. While the latest models are considered to be the best models, it is considered necessary to use fair and impartial LCA methods with appropriate calculation conditions set.

It is important to establish a mechanism to ensure fairness, such as standardizing a method with enhanced suitability for heat pump equipment, while assuming dthat the LCA method is in accordance with ISO 14040 and ISO 14044[5].

The assessment in Figure A3.6 is an example of an assessment made using the Japanese version of the damage-assumption type impact assessment method

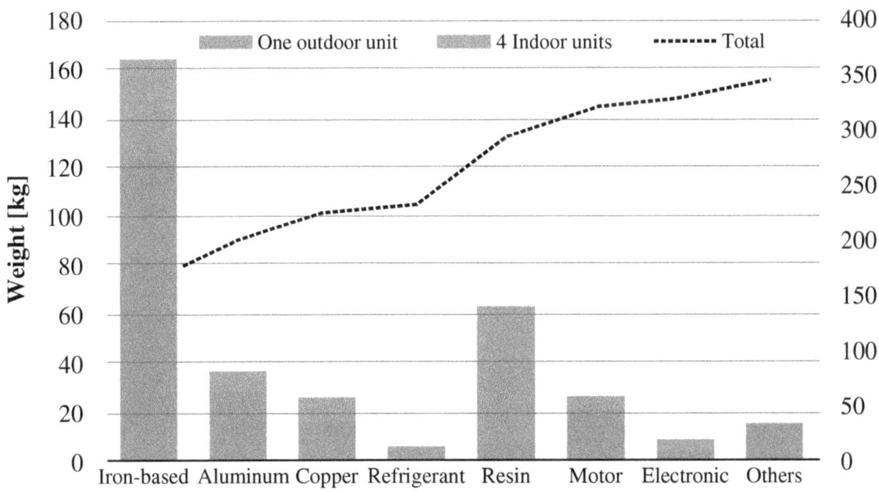

FIGURE A3.5 Weight composition by material for one representative commercial air conditioning system (one outdoor unit and four indoor units). Weight composition at factory shipment. (From Takayuki Nishizawa (Daikin Industries Ltd) et al., "Hotspot analysis on the environmental impact of the manufacturing stage of commercial air conditioners" (2022) [16]).

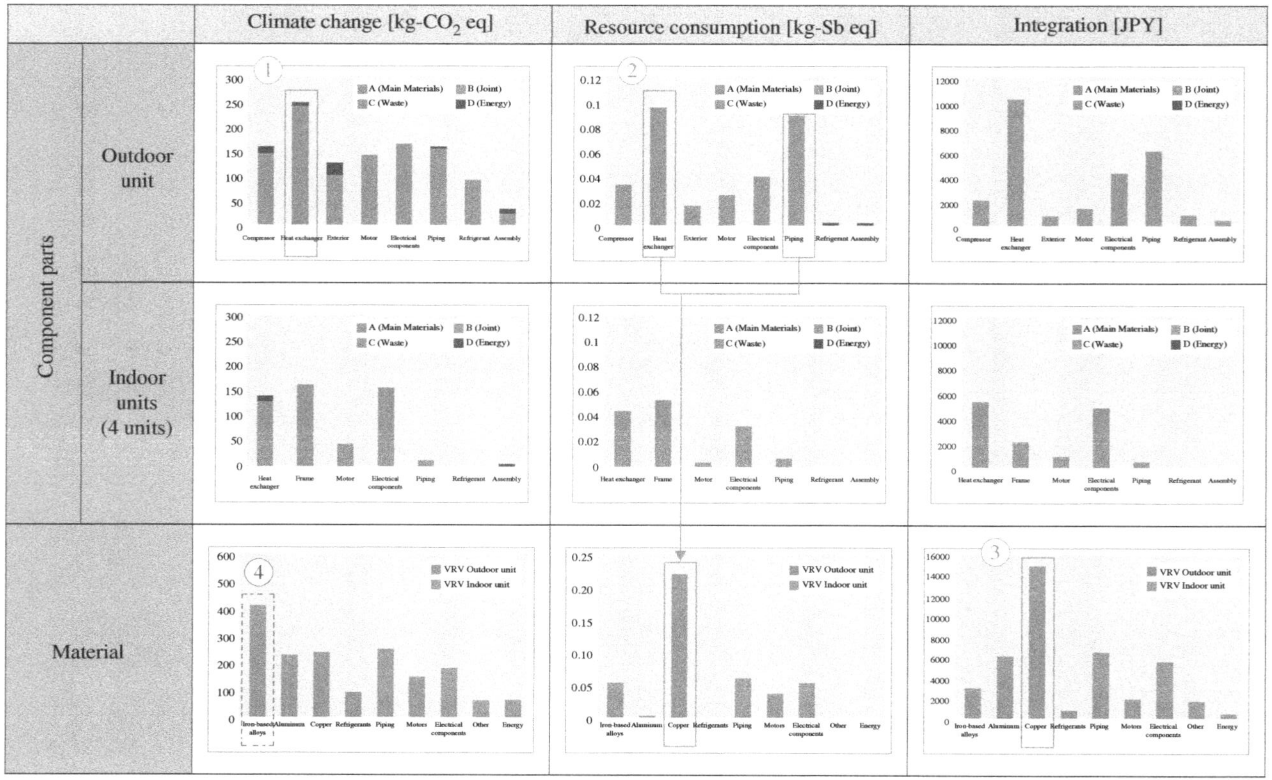

FIGURE A3.6 LCA analysis of environmental impact during manufacture of one representative commercial air-conditioning system (one outdoor unit and four indoor units). (From Takayuki Nishizawa (Daikin Industries Ltd) et al., "Hotspot analysis on the environmental impact of the manufacturing stage of commercial air conditioners" [17]).

(Life cycle Impact assessment Method based on Endpoint modeling (LIME 3)[6]. Figure A3.6 can be summarized as follows

- Climate change impact: Assessment of greenhouse gas emissions as a footprint, with a significant impact on ferrous materials
- Resource consumption: The impact of depletion is assessed based on antimony (Sb), a material found in the earth's crust, and the impact of copper on the environment is significant
- Integration: The social impact is converted into a single indicator, a monetary value, and the overall social cost is calculated, with copper having a significant impact.

Based on these results, the environmental impact of the use of copper in heat pump equipment is significant. In other words, the environmental impact of copper recycling could be significant. It is important to ensure that copper used in heat pump equipment is properly recovered and placed in the recycling loop.

In addition, it is also necessary to consider replacing copper with less environmentally hazardous materials such as aluminum and resin as a further measure, and to continue to consider the sustainability of products used.

A3.3 PROCUREMENT AND PRODUCTION OF RAW MATERIALS FOR REFRIGERANTS

When considering the issue of refrigerant circulation, the issue of refrigerant recovery cannot be avoided.

This section is explained from the perspective that refrigerants are one of the resources that should be in circulating use.

A3.3.1 Importance of Refrigerants as a Resource

A3.3.1.1 Resources of Refrigerants Mined in the World

Fluorocarbons (FCs), commonly used as refrigerants for heat pump technology, are compounds of carbon and fluorine, such as hydrochlorofluorocarbons (HCFCs), hydrofluorocarbons (HFCs), and hydrofluoroolefins (HFOs). Due to their chemical structure and composition of Fluorocarbon (FC) on the market, many types are used. Fluorocarbons are mainly produced from acid grade fluorite[7] or hydrogen fluoride.

Fluorite is mined as a mineral resource but is known to be a highly unevenly distributed resource. As shown in Figure A3.7, Mexico, China, and South Africa hold about half of the reserves, with over 60% produced in China, creating an oligopoly in certain countries.

Considering the amount of production relative to reserves, the situation does not seem alarming for resource depletion, but the extreme uneven distribution of resources is also worrying for supply disruptions and price increases due to geopolitical reasons, such as the ongoing invasion of Ukraine.

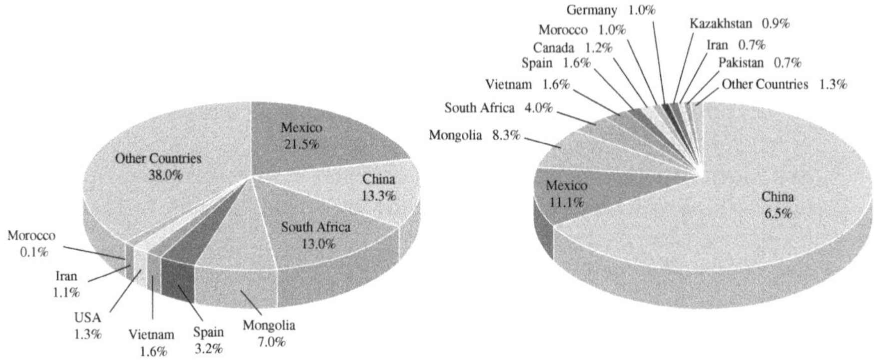

FIGURE A3.7 (Left) Fluorite reserves by country (total 320 million metric tons) and (right) fluorite production by country (total 8.24 million metric tons (2020)). (Based on US Department of the Interior/US Geological Survey, "Mineral Commodity Summaries 2022" [18].)

A3.3.1.2 Potential of Fluorine as a Material

Fluorine is imported in various forms, including hydrogen fluoride and fluorite (Figure A3.8). Acid grade fluorite is also primarily processed into hydrogen fluoride once imported and then reprocessed into fluorocarbons, resins, and rubber.

About 40% of the demand for fluorocarbons is as refrigerants, foaming agents, and cleaning agents, while the rest is used as secondary products in fluorocarbon resins and fluororubber, mainly for automotive parts.

Fluorine is also used in semiconductor cleaning agents (wet etching, wet cleaning, etc.) and lithium-ion batteries, making fluorine a material with diverse needs (Figure A3.9).

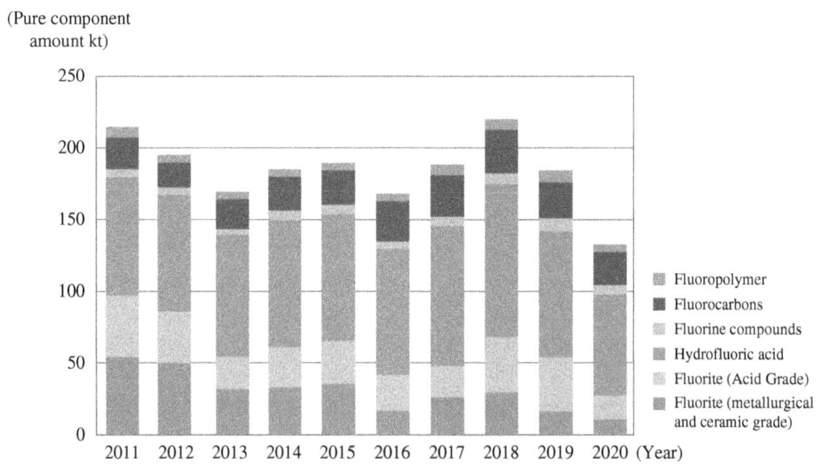

FIGURE A3.8 Fluorine import quantity. (From Japan Energy, Metals and Minerals Corporation (JOGMEC), "Mineral Resources Material Flow 2021 25. Fluorine (F)" (2023) [19]).

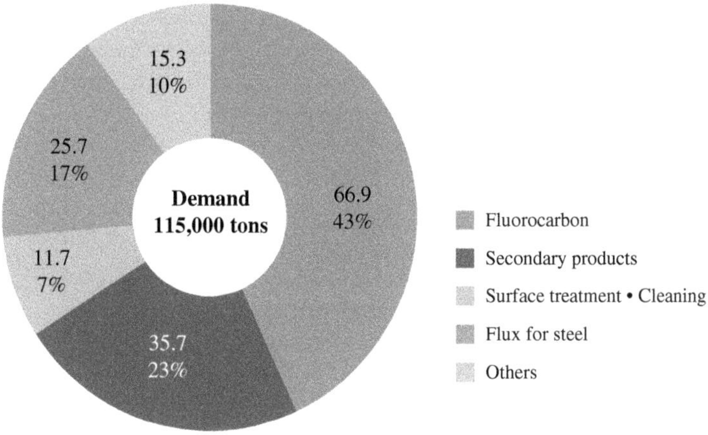

FIGURE A3.9 Domestic demand for fluorite (CaF2). (From JOGMEC, "Mineral Resources Material Flow 2017" [20]).

As shown in Figure A3.10, Japan relies on imports from China for much of its fluorine as a refrigerant feedstock, accounting for about 88% of its total imports, and over 98% if hydrogen fluoride alone is taken. Vietnam, the second largest importer, has seen a significant increase in fluorite imports in recent years, but is still far from importing as much as it does from China.

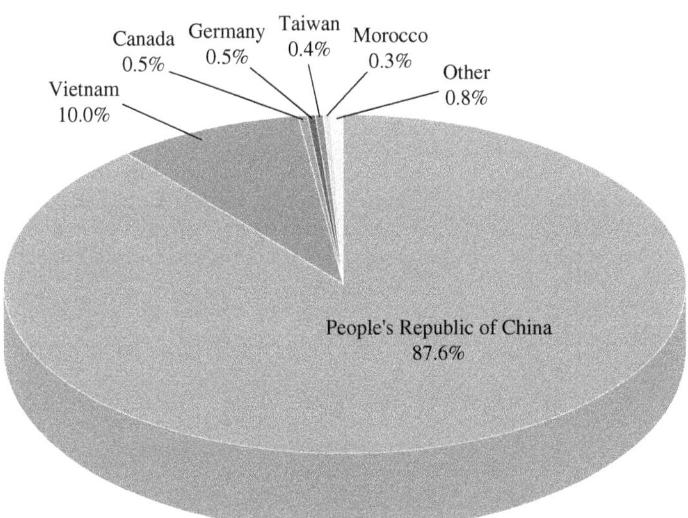

FIGURE A3.10 Fluorine import ratio by partner country (total net weight of fluorine in fluorite (metallurgy and ceramic grade), fluorite (acid grade), and hydrogen fluoride) *Note:* Of the fluorite, 47.4% is acid grade and 95% is hydrogen fluoride. These are the values used in the JOGMEC "Mineral Resources Material Flow 2021 25. Fluorine (F)." (From Ministry of Finance Trade Statistics, "Table A1: By Commodity and Country" [21]).

Thus, the resources of fluorocarbons refrigerants are excessively unevenly distributed in China, and it is technically and cost-prohibitive to change this situation. It is also positioned as a valuable imported resource, not only for refrigerants, but also for semiconductor and automotive applications.

Depending on developments in China, there is a significant risk of supply disruptions, and stable procurement is one of the important issues.

A3.3.2 Initiatives to Promote Refrigerant Recycling

A3.3.2.1 Establishment of Digital Platforms

In addition to reducing refrigerant leakage during equipment use and emissions at disposal, as mentioned in Chapter 6, and ensuring that refrigerants are recovered, it is also important to improve user awareness, for example by promoting recycling to prevent the recovered refrigerants from being destroyed and disposed of.

Daikin Industries Ltd has begun to create value by developing a digital system to share awareness of the importance of refrigerants and not waste them, and to propose and operate a value chain (Figure A3.11).

A digital platform to manage refrigerant circulation will be developed that will connect information on the entire circulation cycle of refrigerants, from manufacturing to recovery, regeneration and destruction, through the use of blockchain technology, and to visualize information management and the recovery and regeneration process at the time of refrigerant filling and during equipment use. In this way, a

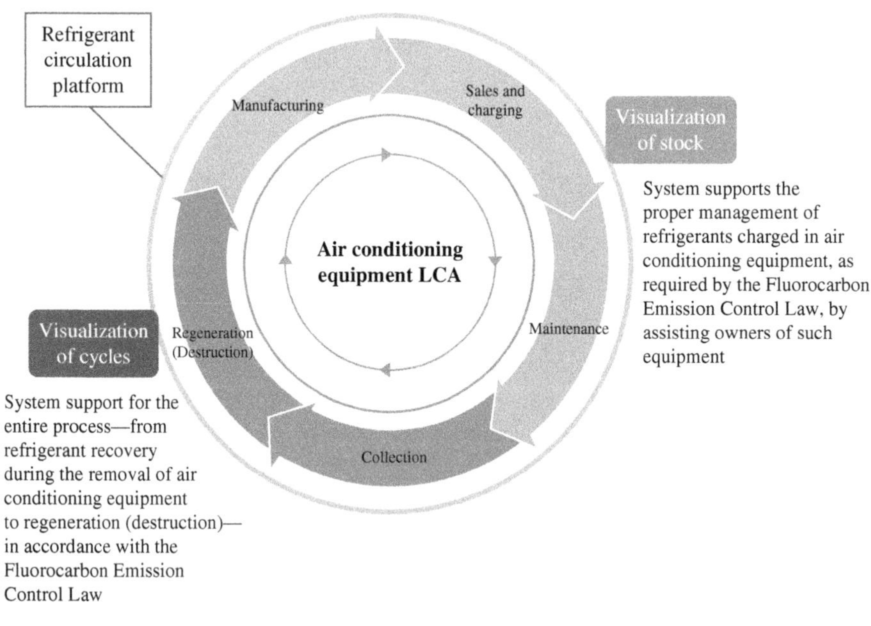

FIGURE A3.11 Overview of the refrigerant circulation digital platform. (From Daikin Industries Ltd, "Start of demonstration of digital platform for refrigerant circulation" [22]).

mechanism is proposed to ensure transparency of the quantity of refrigerant used, its history and quality.

The realization of quantity management in the circulation cycle will make the amount of refrigerant leaks and emissions visible, which will be an effective means of raising user awareness and encouraging behavioral change to prevent leaks and emissions.

A3.3.2.2 Fluorite Recycling Efforts

In addition to reusing recovered refrigerants as refrigerants, technologies are being developed to break them down with steam and air, react the resultant gases with inexpensive calcium carbonate and recycle them as fluorite (CaF2) in a dry process [23]. These techniques open up the possibility of using refrigerants that are not suitable for reuse as materials for other fluorocarbons products.

However, ensuring price competitiveness against current fluorocarbon production methods using natural resources is a major challenge. Although there are examples of developments to improve the economics of fluorine recycling technology [24], its use is limited to purposes requiring expensive raw materials, such as fluorite lenses. Moreover, there are indications that they are struggling to secure cost advantages [25].

A3.4 COLLECTION OF USED EQUIPMENT FROM RESIDENTIAL AIR CONDITIONERS FOR AND COMMERCIAL USE

A3.4.1 Collection Status of Residential Air Conditioners

A3.4.1.1 Legal Collection Issues

Residential air conditioners are subject to the Specified Home Appliance Recycling Law (Home Appliance Recycling Law) [26] and recycling efforts are underway. For commercial air conditioners (e.g. packaged air conditioners and multi-air conditioners for buildings) and showcases for shops, efforts are underway to hand over used equipment and recover refrigerants under the Fluorocarbons Emissions Control Law [27].

Residential air conditioners are responsibility of retailers and others when consumers dispose of them under the Home Appliance Recycling Law, with the consumer paying recycling, collection, and transportation fees. Waste home appliances must be recycled or otherwise treated at home appliance recycling plants.

With the implementation of the law, the status of recycling is reported every year, and changes over time in the recovery rate based on the number of units shipped are shown in Figure A3.12 for the four products subject to the Home Appliance Recycling Law.

The recovery rate for residential air conditioners was 38.1% in 2020; this is less than half that of the other three home appliance categories, and the low recovery rate is noticeable, although the volume of all four categories is increasing year by year.

A3.4.1.2 Difficulties in Recycling from Disposal

The analysis of waste product flows for residential air conditioner shown in Figure A3.13 shows that only 38.2% of the amount of material generated by consumers

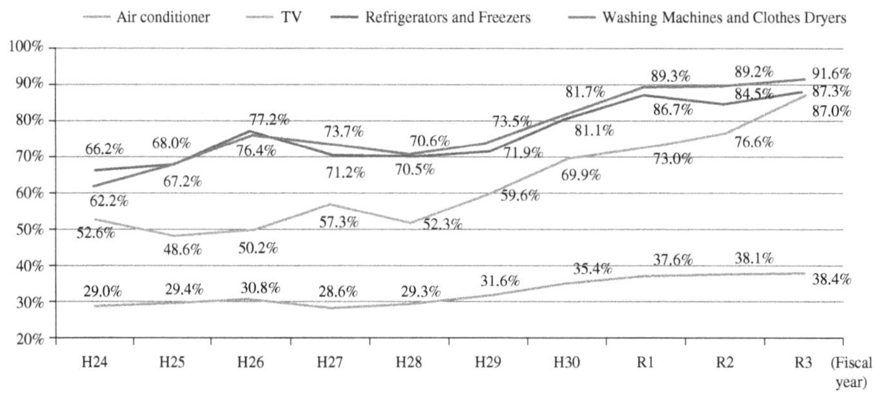

FIGURE A3.12 Year-to-year comparison of recovery rates (per number of units shipped) by four product categories. (From METI/Ministry of the Environment, "Status of recycling based on the home appliance recycling law in FY2021" [28]).

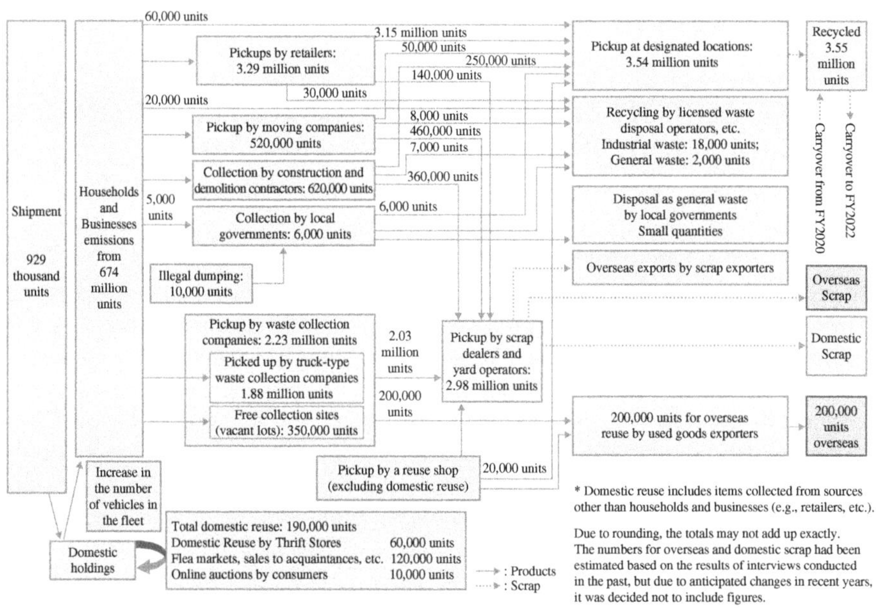

FIGURE A3.13 Flow estimation results (air conditioners: FY 2021). (From METI/Ministry of the Environment, "Status of recycling based on the home appliance recycling law in FY2021" [29]).

is recommodified by collecting the equipment at designated pick-up locations or via waste disposal licensed contractors. It can be seen that a large amount of the material volume flows to scrap via scrap dealers and yard operators, and to overseas reuse.

This may be due to the fact that, as a result of the rise in the purchase price of copper due to the increase in the price of metallic resources, people are aware that a large amount of copper is used in residential air conditioners and may choose to sell them as scrap metal to scrap dealers, rather than paying to recycle them.

A3.4.1.3 Challenges in Recommodification

The recommodification envisaged by the Home Appliance Recycling Law refers to the recovery of parts and materials after dismantling, shredding, and other processes and their re-use as scrap materials, i.e. material recycling. Therefore, the act of reusing equipment itself or components of equipment as second-hand goods is not included in recommodification [30]. If the equipment is reused as used equipment, it is excluded from the Home Appliance Recycling Law, both in Japan and abroad.

The flow in Figure A3.13 shows that there is also a lot of re-use, with 200,000 units exported overseas and 190,000 units reused domestically. However, it is thought that it is not possible to ascertain whether appropriate disposal processes are subsequently carried out.

The means by which individuals can sell used goods, such as on online auctions and flea market websites, have diversified, and a large number of items can be found for sale. The possibility cannot be ruled out that the recovery rate is stagnating due to an environment in which an increasing number of consumers are not opting for recycling, which involves monetary payments.

On the other hand, there are also examples of consumer electronics retailers pioneering the re-use movement [31], mainly through retail outlets, for the purpose of resource circulation. The key is how to promote the construction of a system that uses products and components to the end of their usable life in a cost-effective manner.

In addition, the Home Appliance Recycling Law, which was created on the premise of the previous system, is a system designed to be operated on paper. This means that, at present, the procedures are undeniably complicated and it is difficult to say that the environment is conducive to users choosing to recycle at the time of disposal (Figure A3.14).

A3.4.2 Challenges to Improving the Recovery Rate of Residential Air Conditioners

Efforts to improve the recovery rate of used residential air conditioners have been discussed in a joint meeting between the Ministry of the Environment and the METI [34].

The structure of ownership of residential air conditioners by businesses shown in Figure A3.15 indicates that there are about 20–30% of residential air conditioners owned by businesses, and that there is a wide range of operational modes.

On the other hand, the disposal method of residential air conditioners by businesses shown in Figure A3.16 indicates that only about 20% of cases are disposed of using recycling tickets in line with the Home Appliance Recycling Law.

The businesses surveyed that disposed of residential air conditioners, shown in Figure A3.16, are rental management companies, inns, and hotels, etc. Approximately 2,500 companies were surveyed and 224 companies (those with experience of disposing of residential air conditioners) responded to the survey. Of these 224 companies,

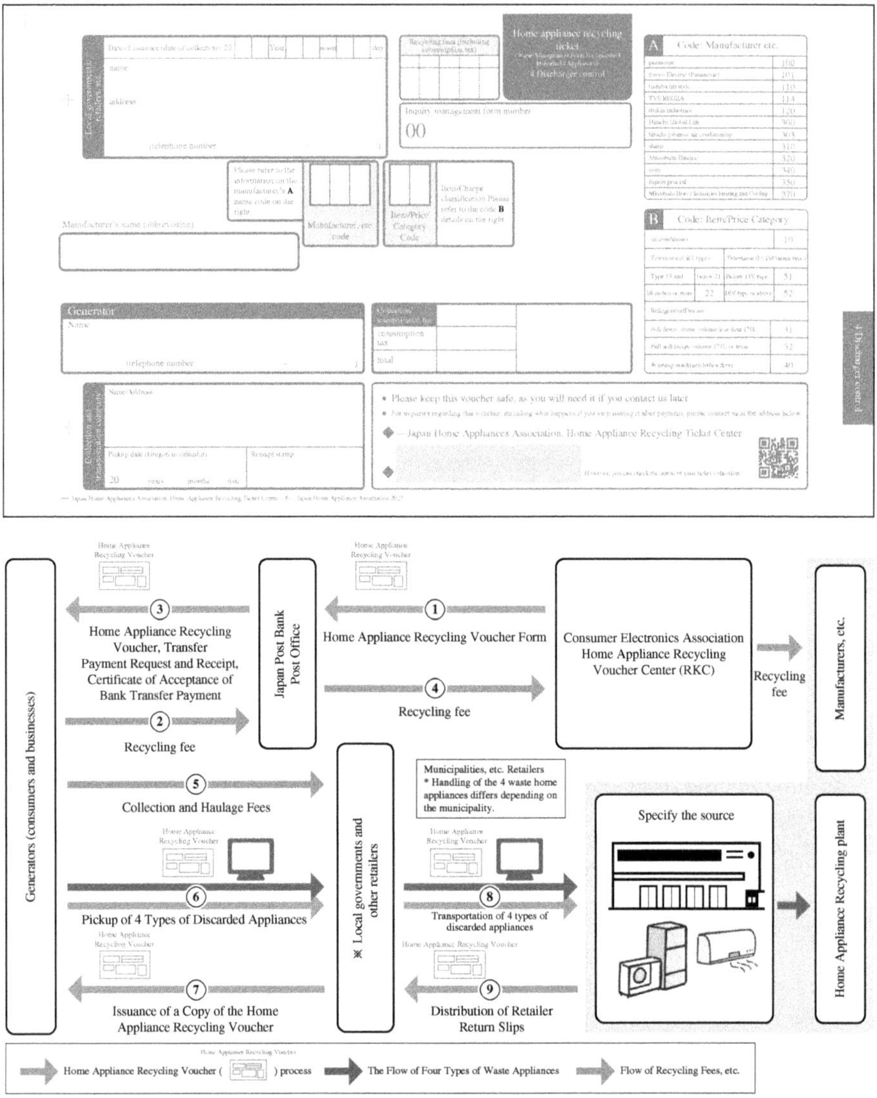

FIGURE A3.14 Home appliance recycling ticket and procedural flow for the post office fee transfer method. (From Association for Electrical Home Appliance of Japan, Home Appliance Recycling Annual Report FY2021 (21st period) [32, 33]).

71%, or 159 companies, entrusted the disposal to or outsourced the process to licensed industrial waste disposal companies.

In addition, special treatment standards (Ministry of the Environment Notification No. 56) have been established for the treatment of industrial waste based on the Waste Disposal and Public Cleansing Law for home appliances, including the separation and recovery of resources and the recovery and destruction of fluorocarbons. The Ministry of the Environment conducted a survey of 38 licensed industrial

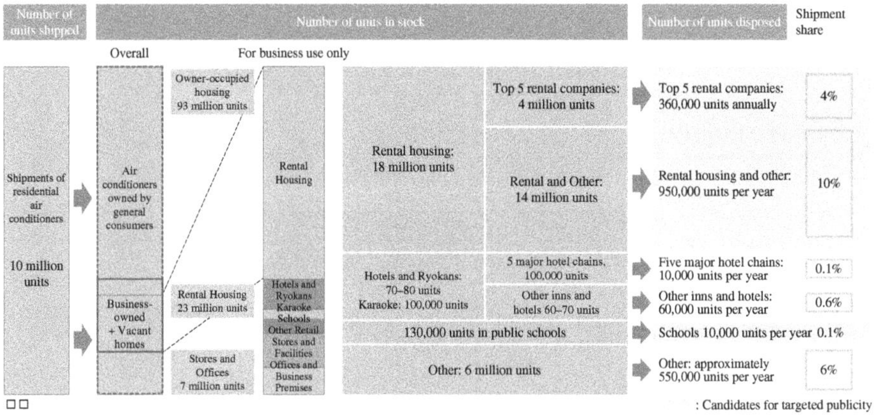

FIGURE A3.15 Ownership structure of residential air conditioners by businesses. (From METI, "Efforts to Raise Public Awareness and Conduct Surveys to Improve the Recovery Rate of Used Air Conditioners" [35]).

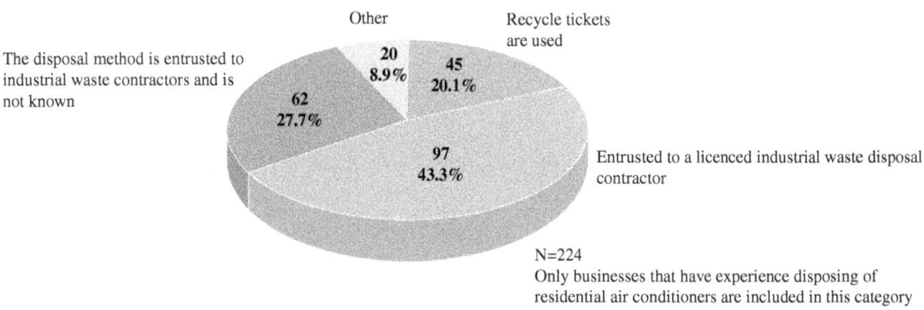

FIGURE A3.16 Results of the survey on disposal methods for residential air conditioners at businesses. *Note:* ※ 1. The businesses surveyed included rental management companies, inns, and hotels, etc. Approximately 2,500 companies were surveyed, and the results of the 224 companies that responded (those with experience of disposing of residential air conditioners) are summarized above. (From METI, "Efforts to Raise Public Awareness and Conduct Surveys to Improve the Recovery Rate of Used Air Conditioners" [36]).

waste disposal companies nationwide, and of the 19 companies that responded, 26% (5 companies) confirmed they are carrying out treatment that complies with the standards for home appliances, and it is possible that many of them are carrying out treatment that does not meet the necessary standards. On the other hand, it could not be confirmed whether the industrial waste disposal licensees that did not respond to the survey are carrying out treatment that complies with the treatment standards for home appliances.

Factors contributing to this situation have also been analyzed, and there are cases where the disposer is unaware of the appliance recycling ticket itself, and where the business is also unaware of the availability of the recycling ticket. Furthermore, various factors have been pointed out, such as the existence of cases where separate contractors are involved in the purchase and disposal of the product, or where the

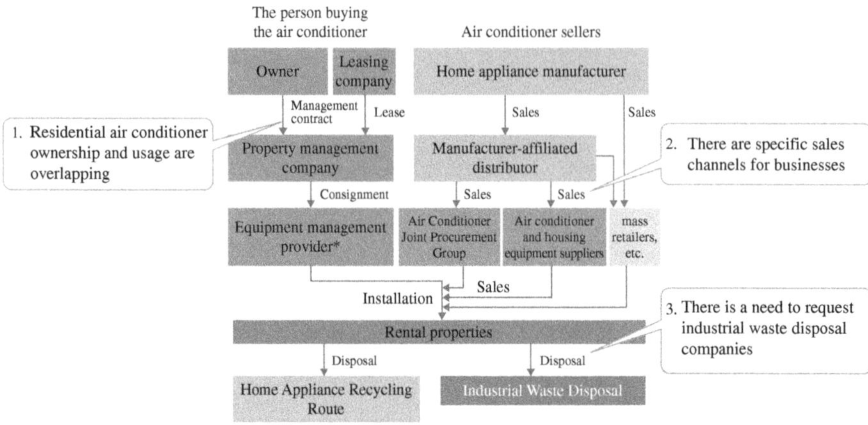

FIGURE A3.17 Flow from product sale to disposal: An example of a rental management company. (Based on the METI, "Efforts to Raise Public Awareness and Conduct Surveys to Improve the Recovery Rate of Used Air Conditioners" [38].)

product is outsourced to a contractor together with the disposal of waste materials during renovation or demolition work.

The most likely factor is that the flow of products from sale to disposal is very complex and multi-layered with sellers, owners, and users all involved (Figure A3.17). There is ambiguity as to who is a retailer under the Home Appliance Recycling Law. As a result, identifying those responsible for disposal is also ambiguous, resulting in cases where the correct disposal route is not taken.

In addition, the increase in the number of vacant houses has been pointed out in recent years and has been regarded as a problem [37], and the existence of old residential air conditioners abandoned along with vacant houses is also considered to be an example of ambiguity in ownership.

A3.4.3 Status of Treatment after Equipment Collection

The status report for recycling implementation compiled by the METI shows that the recommodification rate of residential air conditioners after equipment recovery is 92%, which is much higher than the statutory standard of 80% and higher than that of other items (Figure A3.18).

Furthermore, there are cases where appliance manufacturers have their own recycling plants and are actively engaged in resource recycling [39, 40]. The recommodification rate of residential air conditioner at these recycling plants is as high as approximately 93–95% [41, 42].

Therefore, it is important to first focus on increasing the recovery rate of residential air conditioners. In addition, it is necessary to arrange various means of re-use, including re-use of product units and components after collection. In this case, it is also important to take measures at the development and design stages, such as integrating easy disassembly design[8], as one of the methods to facilitate re-use.

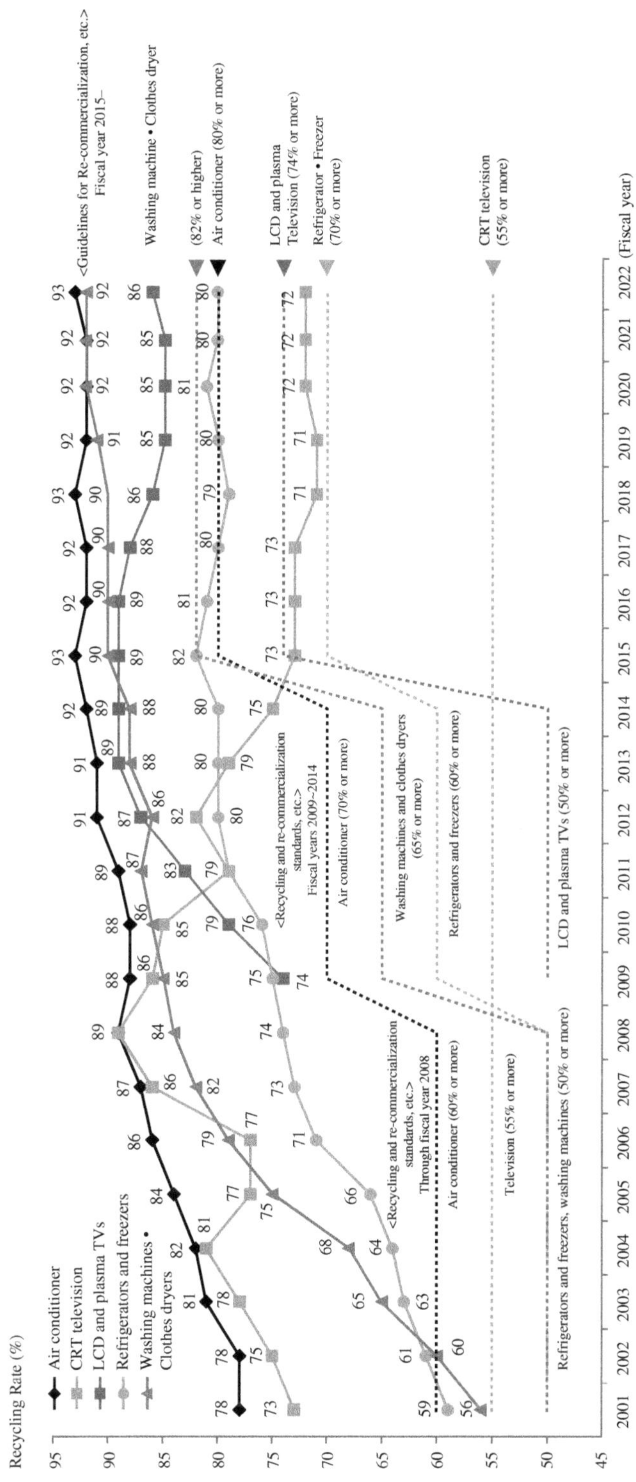

FIGURE A3.18 Trends in recommodification rates at manufacturers, etc. (Based on Association for Electrical Home Appliance of Japan, "Annual Report on Home Appliance Recycling 2022" [43].)

A3.4.4 Collection Status of Commercial Heat Pump Equipment

Other heat pump equipment not bound by the Home Appliance Recycling Law is subject to refrigerant management obligations at the time of disposal under the Fluorocarbons Emissions Control Law, and the refrigerants from this equipment must either be recycled or disposed of safely. There is an obligation to report on the recovery of refrigerant, its quantity and calculated leakage when the equipment is serviced (repaired), or at the time of disposal.

As mentioned in Aection A1.4, the actual recovery of refrigerants has not always achieved a good recovery rate, and efforts are being made to improve this (Figure A3.19).

It is assumed that equipment that has been processed for refrigerant extraction is taken to a scrap dealer or yard operator for recycling of materials as scrap, but the actual status of processing and recycling rates are not known. Issues have been raised regarding the recycling of used electrical and electronic equipment via this route [45]. Appropriate treatment methods for used equipment require further study.

In the future, refrigerants used in heat pump equipment will be required to have even lower global warming potential (GWP). In this context, products using refrigerants such as propane and ammonia, which are natural refrigerants and are strongly assumed to be hazardous in the event of leakage, are likely to be introduced to the market. In such cases, it is necessary to consider methods in which the manufacturer/vendor of the equipment takes responsibility for management of the equipment throughout its operating life and up to the point of disposal, rather than leaving it to the user. With this in mind, it is also necessary to consider a system to clarify management responsibility for all aspects of the equipment life cycle.

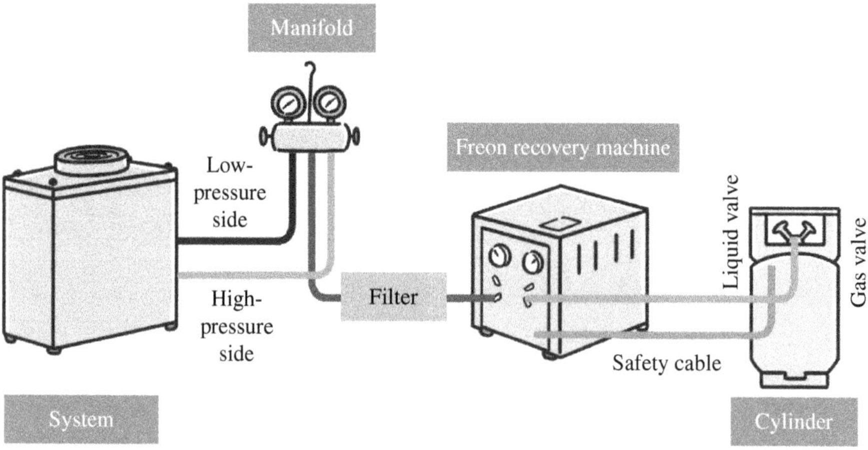

FIGURE A3.19 Refrigerant recovery methods. (Based on Japan Refrigeration and Air Conditioning Industry Association, "Connection method of fluorocarbons recovery machine (residential air conditioning)" [44].)

A3.5 FUTURE DEVELOPMENTS IN THE SUPPLY CHAIN

A3.5.1 Possibilities for Supply Chains to Expand with AI and IoT

A3.5.1.1 Using Communication Devices to Understand Operating Conditions

When considering the operational challenges of the Home Appliance Recycling Law and Fluorocarbons Emissions Control Law, it is impossible to talk about the significant changes in the communication environment in recent years without considering how to incorporate them.

Residential air conditioners with wi-fi connectivity were beginning to be widely used from around 2020, but network connectivity is restricted to environments with wi-fi connections in place. Steady progress is being made in diversifying communication methods, improving convenience, and increasing communication speeds to meet the needs of the times. Continuous improvements in communication speeds are also expected in post-5G network technologies, which began to be applied commercially from 2020 [46].

Increasing improvements in communication speed and capacity will also lead to lower communication costs.

Furthermore, if the objective is to monitor the operational status of heat pump equipment, only a relatively small amount of information needs to be communicated. Therefore, by connecting individual heat pump equipment directly to the public network without wi-fi, a low-cost and stable connection can be guaranteed at all times (Figure A3.20).

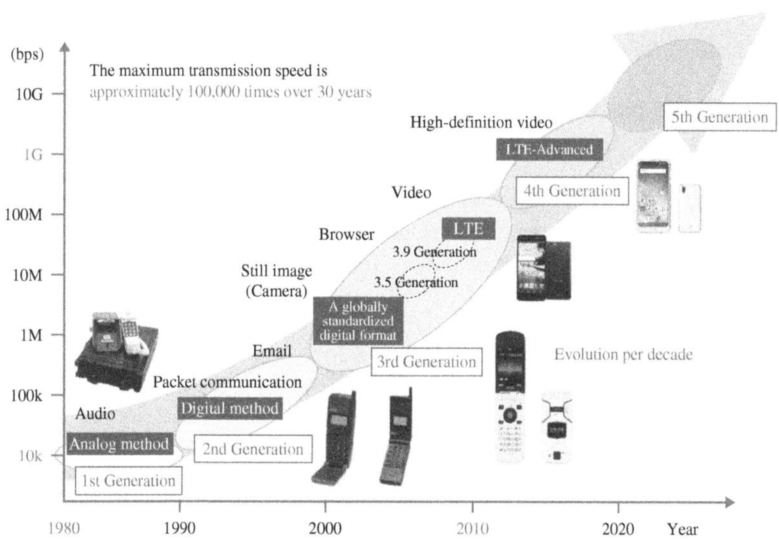

FIGURE A3.20 Evolution of mobile communication systems (1st–5th generation). (Based on Naohiko Ogiwara (Ministry of Internal Affairs and Communications, Telecommunications Infrastructure Bureau), "5th generation mobile communication system (5G) now and future prospects" [46].)

A3.5.1.2 Product Management Using Data

The use of 2D barcodes (QR codes), for example, enables large amounts of information, including traceability information per production lot, to be held in an efficient manner.

Combining information on usage and defects obtained from connected devices and analyzing big data may make it possible to estimate the degree of wear and tear of products and components.

This could help to dispel concerns that users have about the re-use of used parts when promoting the re-use of products and components. It also makes it easier to judge whether a product cannot be reused as a product, but can be reused as individual components. Therefore, not only does this greatly help to expand the re-use possibilities, it could also be used as a criterion for determining the price of recycled parts.

In addition, AI will be used to understand and manage the operating status of big data that accumulates information on the operation of equipment. In doing so, the use of blockchain technology to ensure confidentiality, including personal data such as administrator and user information, will enable both secure individual management of equipment and the protection of confidential information about the user.

Such data utilization strongly supports the promotion of re-use and reliable recycling. Establishing a mechanism that provides a basis for data usage that encompasses the whole supply chain is an extremely important initiative, and it is expected that the design of an efficient mechanism will be promoted through collaboration between industry, government, and academia.

Among these, refrigerant management service/RaMS[9] and the digital platform for refrigerant circulation initiated by Daikin Industries Ltd can be regarded as concrete initiatives to create new value through the simplification of management and procedural work by utilizing the advantages of networks and new technologies such as blockchain technology.

A3.6 ESTABLISHMENT OF A CIRCULAR SUPPLY CHAIN AIMING FOR OVERALL OPTIMIZATION

Heat pump appliances are expected to continue to expand their use by utilizing their high energy efficiency. Under such circumstances, we will consider the ideal form of a circular supply chain that is necessary for heat pump equipment to contribute to maintaining a comfortable lifestyle, while promoting energy savings and reducing environmental impact.

Three scenarios are presented in this section, but on the assumption that they are means to achieve total optimization. The issues to be considered in common with these three scenarios are summarized below.

1. Ensuring reliable product recovery: Improving recovery rates
2. Building a system that contributes to the end-of-life use of equipment as either a product or component: Appropriate material recycling and promotion of re-use.
3. Ensuring the idea is economically rational: Guarantee of permanence.
4. Developing measures that enable users to realize benefits: Ensure user benefits.

A3.6.1 Scenario 1: Introduction of leasing/subscription Services

To enable secure product recovery, all equipment needs to be strictly managed. In the traditional equipment sale business, it is difficult for the seller to manage the equipment after the sale, as ownership is transferred to the user.

In addition, for commercial heat pumps and residential air conditioners that are not used in ordinary households, the delivery route is complex and multi-layered, and there are many cases where the user and owner are different. It is clear that the existence of such cases hinders the improvement of equipment and refrigerant recovery rates.

One possible solution to the above issues is to clarify the equipment manager (the person responsible for the appropriate disposal at the end-of-use stage) by introducing a lease format or subscription service.

There are already some leading examples, such as subscription services for residential air conditioners [47, 48] and rental services for industrial heat pumps [49]. Each equipment will become more connected in the future, which is expected to improve the degree of freedom of individual management of equipment and enable a more accurate understanding of their operating status.

Furthermore, if it becomes possible to estimate the remaining life of products and components from the information collected on a large scale, the value of reused equipment and components (including information such as usage history) can be properly explained to users. This would enable the appropriate valuing of reused products in subscription services, which would both dispel concerns about the quality and reliability of reused equipment and parts among users and promote the benefits of their use, ensuring business benefits through appropriate valuing.

On the other hand, for residential air conditioners and other products, this is a measure that will significantly change the current commercial distribution system, and there are many business challenges that need to be resolved. It is also clear that customer retention is an extremely difficult challenge for companies entering the subscription business [50]. Research analysis shows that only about 20% of companies succeed in improving customer retention [51], so it can be said that introducing subscriptions is not expected to easily solve the issues, and careful consideration must be given to the system design and other aspects.

A3.6.2 Scenario 2: Stricter Market Equipment Management through the Introduction of an Individual Management System

A possible measure to find answers to the issues while retaining the good parts of the current commercial distribution system is to introduce an individual management system based on the premise that all heat pump equipment is connected.

Such a system not only manages the attributes of the equipment, as in a digital product passport, but also records the management status of the equipment, including location, owner, refrigerant management, etc., so that it can be accessed elsewhere.

In this case, equipment management is the responsibility of the user and, for equipment with a multi-layered ownership and utilization structure, it is particularly important to identify the person responsible. Ensuring the reliable recovery of used equipment by making it possible to trace the person responsible, for example through legislation, will be achieved.

In doing so, consideration should be given to the protection of users' personal data. An environment in which both convenience and strict information management can be achieved is already in place, for example through the use of blockchain technology, which was also used in the refrigerant circulation digital platform described in Section A3.3.2.1 of Appendix 3.

In addition, according to the Fluorocarbons Emissions Control Law, users of equipment with Class I specific fluorocarbons emissions and freezing equipment (commercial refrigeration and air conditioning equipment) are required to create a logbook for each refrigerant system and keep a history of refrigerant filling and recovery for the equipment.

From the perspective of ensuring effective use of resources and safety, the applicable products for this logbook will be expanded to include not only equipment using fluorocarbons but also flammable refrigerants such as R290, toxic refrigerants such as NH_3, and high-pressure CO_2 refrigerants, which are expected to be used in the future. At the same time, it is likely to be Starting with commercial equipment and expanding not only to consumer equipment but also to household equipment. In addition, it would be useful to record a wider range of information, including not only refrigerant recovery and filling, but also the history of equipment failures. To this end, it is one idea to consider the development of current legislation to expand the scope of application.

A3.6.3 Scenario 3: Promoting Resource Circulation by Strengthening Management of End-Of-Life Equipment

While Scenarios 1 and 2 are measures to change the existing commercial distribution system and equipment management, scenario 3 comes from the perspective of resource circulation with a possible approach focusing on strengthening the management of end-of-life equipment. Collecting equipment once it has been used, and ensuring that equipment and parts are reused and materials recycled is the last point for promoting the wider use of heat pumps.

As mentioned in Section A3.4 [31], there are examples of the promotion of re-use, with distributors playing a key role in promoting it. If a mechanism to promote re-use can be established by entities other than manufacturers and retailers, it would be a measure that would have little impact on the current commercial flow of new equipment sales, if not only residential air conditioners but also commercial heat pump equipment is to be addressed.

In order to encourage equipment users to discard of used equipment into recognized disposal routes that facilitate re-use and recycling, it is also necessary to consider creating benefits for the user themselves. In particular, in the recycling of residential air conditioners, the current system whereby users pay a recycling fee at the time of disposal does not prevent collectors who claim to take back the equipment free of charge or users who sell the equipment to scrap dealers as valuable metal.

For example, a possible incentive program would be to have users purchase a recycling coupon at the same time as the appliance purchase, in a way that the coupon cost is refunded when the appliance is transferred for re-use (the next user bears the cost). However, this may require a review of the current Home Appliance Recycling Law.

Conversely, for commercial heat pump equipment, not only refrigerant management obligations under the Fluorocarbons Emissions Control Law, but also a legal system for equipment handover and recycling, such as the Home Appliance Recycling Law, could be developed.

A system whereby a recycling fee is paid at the time of purchase has already been introduced, e.g. under the Automobile Recycling Law. The Automobile Recycling Law contains some crossover with home appliance recycling legislation, such as the collection of fluorocarbons gas, and so there are many points of reference in the creation of an operational mechanism.

In any case, to promote resource circulation, the individual management system described in Scenario 2 will clarify the whereabouts of the equipment and who is responsible for its management. Sufficient recovery of used equipment is a prerequisite for maximizing circulating use, including re-use and recycling.

In combination with these measures, it is important to create a system that encourages behavioral changes of equipment users as described so far. In doing so, it is necessary to make a comprehensive effort, not only through the efforts of those directly involved in the development, manufacture, installation, and sales of equipment, but also through the overall optimization of laws, standards and regulations, and a concerted effort by all stakeholders and relevant parties.

NOTES

1. Total copper resources: The sum of identified world resources plus undiscovered world resources.
2. Recoverable amount: The amount of ore, crude oil, etc. that can be mined from the current technological and economic point of view; it is treated as the basis for overall company planning, as the minable amount is used as the basis for trading in the mountains from which the raw materials are mined.
3. Bottle to bottle: The proportion of waste PET bottles that are used again as PET bottle material.
4. Total amount of substances involved: Also referred to as total material requirement (TMR). The total amount of substances involved is the resources that include hidden flows emitted by resources extracted or mined for other purposes or waste due to resource extraction, for example. It is one of the guidelines for quantifying the sustainability of resource use and the environmental impact it has on the global environment.
5. ISO 14040/ISO 14044: Provisions related to LCA set out by the International Organisation for Standardisation; ISO 14040 is "Environmental management—Life cycle assessment—Principles and framework" and ISO 14044 is "Environmental management—Life cycle assessment—Requirements and guidelines." In the Japanese Industrial Standards, they are referred to as JIS-Q-14040 and JIS-Q-14044 respectively.
6. LIME 3: Abbreviation for Life cycle Impact assessment Method based on Endpoint modeling.

7. Acid grade fluorite: Fluorite whose CaF2 content has been upgraded to 97% or more; fluorite with CaF2 content of 97% or less is called metallurgy and ceramic grade and is used for other applications.
8. Easy disassembly design: The disassembly of products should be considered from the design stage. By making products easy to dismantle, dismantling work is facilitated and recycling efficiency is improved. Design for easy dismantling and sorting is also necessary to promote resource circulation of rare metals such as copper.
9. RaMS: A refrigerant management system provided as a software as a service (SaaS) that complies with the Fluorocarbons Emissions Control Law.

REFERENCES

1. Khaled Soufani and Christophe Loch, Diamond Inc., Diamond Harbour Business Review, "Why Circular Supply Chains Aren't Popular: Two Barriers Companies Face," 2021, https://www.dhbr.net/articles/-/7863 (Original: https://hbr.org/2021/06/circular-supply-chains-are-more-sustainable-why-are-they-so-rare)
2. Trevor Zink and Roland Geyer, Journal of Industrial Ecology, Vol. 21, "Circular Economy Rebound," 2017, https://www.researchgate.net/publication/313371834_Circular_Economy_Rebound
3. Ministry of Economy, Trade and Industry, "Circular Economy Vision 2020," 2020, https://www.meti.go.jp/press/2020/05/20200522004/20200522004.html
4. Purdue University, International Refrigeration and Air Conditioning Conference, "Harmonization of Life Cycle Climate Performance (LCCP) Methodology," 2016, https://docs.lib.purdue.edu/iracc/1724/
5. National Institute for Materials Science, "Global Resource Constraints by 2050," 2007, https://www.nims.go.jp/news/press/2007/02/p200702150.html
6. US Department of the Interior/US Geological Survey, "Mineral Commodity Summaries 2022," Copper, p.54–5, 2022, https://pubs.usgs.gov/periodicals/mcs2022/mcs2022.pdf
7. Jane M. Hammarstrom/Michael L. Zientek/Heather L. Parks/Connie L. Dicken/U.S. Geological Survey Global Copper Mineral Resource Assessment Team,. USGS "Assessment of Undiscovered Copper Resources of the World, 2015," Scientific Investigations Report 2018-5160, p.20 (Table A), 2018, https://pubs.usgs.gov/publication/sir20185160
8. ICSG (International Copper Study Group), "The World Copper Factbook 2022," p.34, 2022, https://icsg.org/copper-factbook/
9. K Kenji Koshizaki/Hideki Kobayashi/Toshiba Corporation, Japan Science and Technology Agency (JST), J-Stage (The 6th Annual Meeting of the Japan Society for Life Cycle (Life Cycle Studies, Session ID: A2-18), "Consideration of the Direction of Measures against the Risk of Copper Depletion," p.148–9, 2011, https://www.jstage.jst.go.jp/article/ilcaj/2010/0/2010_0_75/_pdf
10. Mitsubishi UFJ Research & Consulting Co., Ltd., "Summary of the Survey Results on Mineral Resources Recycling Flows and Stocks in FY2021," p.8, 2022/2023, https://mric.jogmec.go.jp/wp-content/uploads/2022/04/survey_mineral_recycling_flow_stock.pdf
11. Plastic Waste Management Institute of Japan, "2022 Status of Production, Disposal, Recycling, Treatment and Disposal of Plastic Products Material Flow Chart," p.2, 2023, https://www.pwmi.or.jp/pdf/panf2.pdf
12. The Council for PET Bottle Recycling, "Applications of Recycled PET Resin (FY2022)," 2024, "PET Bottle Recycling Annual Report 2023," "6 Recycle—(1) The recycling rate of FY2022 was 86.9%," "Figure 3: Overview of recovery and recycling," 2023, https://www.petbottle-rec.gr.jp/data/use.html
13. The Council for PET Bottle Recycling "Applications of recycled PET resin (FY2022)," 2024. https://www.petbottle-rec.gr.jp/data/use.html

14. Keita Hamasuna, Kiminori Sato, Shoki Kosai, Shunsuke Kashiwakura, and Eiji Yamasue (Ritsumeikan University), The 16th Annual Meeting of the Japan Society for Life Cycle Analysis, 1-C2-04 Presentation document "Resource Efficiency from the Perspective of Air Conditioner Mining Activities," 2021.
15. Ministry of Economy, Trade and Industry (METI), Agency for Natural Resources and Energy, Energy-Saving Product Information Site "Energy Saving Performance Catalog 2021 Edition: How to Use and Choose Energy-Saving Home Appliances "Air Conditioners," Figure "Changes in Period Power Consumption, (kWh/year)," p.18, 2023, https://seihinjyoho.go.jp/.frontguide/pdf/guide_aircon.pdf
16. Takayuki Nishizawa, Yoichi Handa, Satoshi Yoneda, and Q. Fang (Daikin Industries, Ltd)/Satoshi Ohara, Yuko Oshita, Yoshiman Fujii, Yasunori Kikuchi (The University of Tokyo), Proceedings of the 17th Annual Meeting of the Japan Society for Life Cycle Analysis, 1-A1-01 Presentation document "Hotspot analysis on the environmental impact of the manufacturing stage of commercial air conditioners," p.4, 2022.
17. Takayuki Nishizawa, Yoichi Handa, Satoshi Yoneda, and Akira Katakata (Daikin Industries, Ltd)/Satoshi Ohara, Yuko Oshita, Yoshiman Fujii, Yasunori Kikuchi (The University of Tokyo), Proceedings of the 17th Annual Meeting of the Japan Society for Life Cycle Analysis, 1-A1-01 Presentation "Hotspot analysis on environmental impact of manufacturing stage of commercial air conditioners," p.6, 2022.
18. U.S. Department of the Interior/U.S. Geological Survey, "Mineral Commodity Summaries 2022," p.63, (2022), https://pubs.usgs.gov/periodicals/mcs2022/mcs2022.pdf
19. Japan Organization for Metals and Energy Security (JOGMEC), "Mineral Resources Material Flow 2021," "Fluorine (F)," p.6 (Figure 3-1-1), 2023, https://mric.jogmec.go.jp/wp-content/uploads/2023/03/material_flow2021_F.pdf
20. Japan Organization for Metals and Energy Security, "Mineral Resources Material Flow 2017," "Fluorine (F)," "Table 1-2 Domestic Supply and Demand of Fluorine (Chemical Applications)," p.3, 2017, https://www.meti.go.jp/meti_lib/report/2019FY/000287.pdf
21. Ministry of Finance, Trade Statistics, "A-1 Appended table of countries by category," Designation of imports and exports: Imports, Designation of statistical year: Year-to-date, 2021, year (Jan–Dec), Designation of items: Reference designation, 2529.22-000, 2811.11-000, https://www.customs.go.jp/toukei/srch/index.htm?M=01&P=0
22. Daikin Industries Ltd, "Start of demonstration of digital platform for refrigerant circulation," 2022, https://www.daikin.co.jp/press/2022/20220623_2
23. Shinji Yasui, "Decomposition of Fluorinated Greenhouse Gases for Carbon Neutrality and Recycling of Fluorine for SDGs," p.22–24, 2021, http://yasui-lab.web.nitech.ac.jp/research/pdf/フッ素リサイクル.pdf
24. Iwatani Corporation, "World's first! Establishment of the world's first calcium fluoride (fluorite) synthesis technology: Realisation of a stable supply as a raw material for lenses that does not depend on natural resources," 2014, https://www.iwatani.co.jp/jpn/news/2014/jpn_newsrelease_detail_1183.html
25. Hidehisa Kume, Nikkei Business Publications, Inc./Nikkei Cross Tech, "NE Report: Artificial Synthesis Technology of Fluorite Liberated from China Risk," p.3, 2014, https://xtech.nikkei.com/dm/article/MAG/20141203/392110/
26. Ministry of Economy, Trade and Industry, "Home appliance recycling law (Act on Recycling of Specified Kinds of Home Appliances)," 2023, https://www.meti.go.jp/policy/it_policy/kaden_recycle/index.html
27. Ministry of Economy, Trade and Industry, "Outline of the METI Fluorocarbons Emissions Control Law," 2023, https://www.meti.go.jp/policy/chemical_management/ozone/law_furon_outline.html
28. Ministry of Economy, Trade and Industry/Ministry of the Environment, "Status of recycling based on the home appliance recycling law in FY2021," p.17, 2023, https://www.env.go.jp/content/000128191.pdf

29. Ministry of Economy, Trade and Industry/Ministry of the Environment, "Status of recycling based on the home appliance recycling law in FY2021," "Flow estimation results (total of 4 items: FY2021)," p.23, (2023), https://www.env.go.jp/content/000128191.pdf
30. Katsumi Nakai (Fukushima University), "Current Status and Issues of the Home Appliance Recycling Law," Ritsumeikan University Policy Science Vol. 13, No. 3, 119–32, p.125, (2006), https://ritsumei.repo.nii.ac.jp/record/4727/files/ps13_3nakai.pdf
31. Yamada Environmental Resource Development Holdings Co., Ltd., "Environmental Development Project: Three Resource Circulations Centered on Retail," (2023), https://www.yamadaerd.jp/projects/
32. Association for Electric Home Appliances, "2021 Home Appliance Recycling Annual Report," p.12, (2022), https://www.aeha-kadenrecycle.com/pdf/report/kadennenji_2021.pdf
33. Association for Electric Home Appliances, "2021 Home Appliance Recycling Annual Report," p.11, (2022), https://www.aeha-kadenrecycle.com/pdf/report/kadennenji_2021.pdf
34. Ministry of Economy, Trade and Industry, Working Group on Electric and Electronic Equipment Recycling, Subcommittee on Waste and Recycling, Industrial Science and Technology Policy and Environment Bureau, Industrial Structure Council, Subcommittee on Evaluation of Home Appliance Recycling System, Recycling-based Society Committee, Central Environmental Council 42nd Joint Meeting, (2021), https://www.meti.go.jp/shingikai/sankoshin/sangyo_gijutsu/haikibutsu_recycle/denki_wg/042.html, https://www.env.go.jp/council/03recycle/post_223.html
35. Ministry of Economy, Trade and Industry, Working Group on Electric and Electronic Equipment Recycling, Subcommittee on Waste and Recycling, Industrial Science and Technology Policy and Environment Bureau, Industrial Structure Council, Subcommittee on Evaluation of Home Appliance Recycling System, Recycling-based Society Committee, Central Environmental Council 42nd Joint Meeting, Document 2 "Efforts to Raise Public Awareness and Conduct Surveys to Improve the Recovery Rate of Used Air Conditioners," p.5, (2021), https://www.env.go.jp/council/03recycle/y032-42/900418822.pdf
36. Ministry of Economy, Trade and Industry, Working Group on Electric and Electronic Equipment Recycling, Subcommittee on Waste and Recycling, Industrial Science and Technology Policy and Environment Bureau, Industrial Structure Council, Subcommittee on Evaluation of Home Appliance Recycling System, Recycling-based Society Committee, Central Environmental Council 42nd Joint Meeting, Document 2 "Efforts to Raise Public Awareness and Conduct Surveys to Improve the Recovery Rate of Used Air Conditioners," p.6, (2021), https://www.env.go.jp/council/03recycle/y032-42/900418822.pdf
37. Ministry of Land, Infrastructure, Transport and Tourism (MLIT), Policy Research Institute for Land, Infrastructure, Transport and Tourism, "Survey and Research on the Unknown Ownership of Land and Buildings in the Problem of Vacant Houses," Journal of Land, Infrastructure, Transport and Tourism Policy, No. 154, (2020), https://www.mlit.go.jp/pri/houkoku/gaiyou/pdf/kkk154.pdf
38. Ministry of Economy, Trade and Industry, Working Group on Electric and Electronic Equipment Recycling, Subcommittee on Waste and Recycling, Industrial Science and Technology Policy and Environment Bureau, Industrial Structure Council, Subcommittee on Evaluation of Home Appliance Recycling System, Recycling-based Society Committee, Central Environmental Council 42nd Joint Meeting, Document 2, "Efforts to Raise Public Awareness and Conduct Surveys to Improve the Recovery Rate of Used Air Conditioners," p.10, (2021), https://www.env.go.jp/council/03recycle/y032-42/900418822.pdf
39. Panasonic Eco Technology Center Co. Ltd, "Panasonic Eco Technology Center," (2023), https://panasonic.co.jp/eco/petec/process/#choice3

40. Hyper Cycle Systems Corporation, "Recycling Guide Home Appliance Recycling—We Are Doing Our Utmost to Achieve Zero Emissions in a Recycling-Oriented Society," 2023, https://www.h-rc.co.jp/
41. Panasonic Eco Technology Center Co. Ltd. "Recycling process," 2023, https://panasonic.co.jp/eco/petec/process/#choice3
42. Hyper Cycle Systems Corporation, "Recycling Guide: Home Appliance Recycling—Making Every Effort to Achieve a Recycling-Oriented Society with Zero Emissions," 2023, https://www.hrc.co.jp/recycle/recycle.html
43. Association for Electric Home Appliance, "Annual Report on Home Appliance Recycling 2021 Edition," p. 16 (Chart II-3), 2022, https://www.aeha-kadenrecycle.com/pdf/report/kadennenji_2021.pdf
44. Japan Refrigeration and Air Conditioning Industry Association, "Connection method of fluorocarbons recovery machine (residential air conditioning)."
45. Masao Takeuchi, Research Centre for Science and Technology Trends, National Institute of Science and Technology Policy/Science and Technology Trends, "Current Status of Recycling by Material Industry and Its Constraining Factors," February, (2009), 2023, https://nistep.repo.nii.ac.jp/records/5758
46. Naohiko Ogiwara (Ministry of Internal Affairs and Communications, Telecommunications Infrastructure Bureau), "5th generation mobile communication system (5G): Now and future prospects," 2019, https://www.soumu.go.jp/main_content/000633132.pdf
47. Baridi Baridi Corporation, "Baridi Baridi," (2023), https://baridibaridi.com
48. Daikin Industries, Ltd., "Commercial air conditioning subscription ZEAS Connect - Air conditioning subscription service," 2023, https://www.daikinaircon.com/shopoffice/zeas_connect
49. Aggreko Japan Ltd., "Rental of industrial cooling systems," 2023, https://www.aggreko.com/ja-jp/products/cooling-global
50. Neil Iyer, "Three Reasons Subscription Services Fail," Diamond Harvard Business Review, 2023, Diamond Inc. (https://dhbr.diamond.jp/articles/-/9047) (https://hbr.org/2022/10/3-reasons-subscription-services-fail)
51. Gartner, Inc., "Top 10 Trends in Digital Commerce," 2023, https://www.gartner.com/smarterwithgartner/top-10-trends-in-digital-commerce

Conclusion

Heat pump technology has become the most important thermal energy technology in response to global demands such as carbon neutrality, and we have stated that the realization of next-generation heat pump technology will greatly contribute to the construction of future societies.

Unlike electrical and information technologies, those related to heat and fluids, such as heat pumps, are still in the early stages of development and require a great deal of know-how. Despite being an extremely advanced thermal energy technology, heat pumps face many operational challenges, such as preventing refrigerant leaks, reducing CO_2 emissions from energy sources, and realizing resource circulation, including equipment and refrigerants. In the future, it will be impossible to avoid the introduction of flammable refrigerants, which will make their handling even more difficult. Precisely because this technology is so difficult, it is essential to optimize the entire value chain, including related technologies and social systems, and this will be the key to establishing next-generation heat pump technology. It is an important point for Japan to continue to lead the world in this field.

It is difficult to be optimistic about Japan's future in these terms, where technological development and environmental measures are still being pursued in a vertically divided manner. This is what led me to compile this book. Just as with mobile phones in the past, products cannot be accepted by users based on hardware performance alone, but it seems that product development in many technological fields continues to lack a user perspective.

With the development of AI and the rapid pace of technological development internationally, even in fields where Japan is currently the most technologically advanced, continuing to develop technology along the same lines as before could quickly lead to technological decline. It is no exaggeration to say that Japanese science and technology are at a major crossroads.

On the other hand, I believe that Japan's cultural assets, such as the diligence of the Japanese people and the advanced ideas, philosophy, and aesthetics that Japan has traditionally possessed, will surely provide an opportunity for revival. As many of you know, Steve Jobs incorporated the Japanese concept of Zen into Apple's products, which led to great success. Japan still has value that can captivate the world.

In compiling this book, I am extremely grateful to the many people, including heat pump manufacturers, who agreed with my ideas and worked with me to produce this book. I hope that by pointing out the future direction of heat pump technology, I can contribute to the future of Japanese science and technology.

I would like to express my sincere gratitude to everyone who supported and cooperated with us in the publication of this book. It is the combination of everyone's expertise and enthusiasm that has made this book possible. I would also like to express my gratitude to the many people outside the team who provided advice and cooperation.

Furthermore, we would like to express our respect and gratitude to the many organizations and individuals who provided illustrations, diagrams, and other materials that were quoted or referenced to enrich the content of this book.

Once again, we would like to express our heartfelt gratitude to all those involved.

Saito Kiyoshi, Waseda University

2025

Index

Note: Page numbers in **bold** denote tables, page numbers in *italics* denote figures. Page numbers of the form XnY denote footnote Y on page X.

For Product Safety Concerns and Information please contact our EU
representative GPSR@taylorandfrancis.com
Taylor & Francis Verlag GmbH, Kaufingerstraße 24, 80331 München, Germany

www.ingramcontent.com/pod-product-compliance
Ingram Content Group UK Ltd.
Pitfield, Milton Keynes, MK11 3LW, UK
UKHW021027031026
476244UK00003B/33
* 9 7 8 1 0 4 1 1 2 8 5 2 6 *